COMPUTATIONAL MODELLING OF REINFORCED CONCRETE STRUCTURES

Edited by:

Ernest Hinton and Roger Owen
*Department of Civil Engineering,
University College of Swansea, U.K.*

PINERIDGE PRESS

Swansea, U.K.

COMPUTATIONAL MODELLING OF REINFORCED CONCRETE STRUCTURES

First Published in 1986 by
Pineridge Press Limited
54 Newton Road, Mumbles, Swansea SA3 4BQ, U.K.

Library of Congress Cataloguing in Publication Data

Computational modelling of reinforced concrete structures:
1. Reinforced concrete construction
2. Finite element method
I. Hinton, E. II. Owen, D. R. J.
624 .1'8341 TA683.2

ISBN 0–906674–46–8

Printed and bound in Great Britain by
Dotesios (Printers) Ltd., Bradford-on-Avon, Wiltshire

CONTRIBUTING AUTHORS

M. AUFAURE	Electricite De France, 6 quai Watier, 78400 Chaton, France.
K.J. BATHE	Department of Mechanical Engineering, Massachusetts Institute of Technology, Cambridge, MA 02139, U.S.A.
N. BICANIC	Department of Civil Engineering, University College of Swansea, Singleton Park, Swansea SA2 8PP, United Kingdom.
R. deBORST	Institute TNO for Building Materials and Building Structures, Software Engineeing Department, PO Box 49, 2600 AA Delft, The Netherlands.
M. CERVERA	Department of Civil Engineering, University College of Swanse, Singleton Park, Swansea SA2 8PP, United Kingdom.
D. CHAUVEL	Electricite De France, 12-14 avenue Dutrievoz, 69628 Villeurbanne Cedex, France.
R.J.COPE	Department of Civil Engineering, Plymouth Polytechnic, Plymouth, United Kingdom,
M.M. ELBADRY	Department of Civil Engineering, The University of Calgary, 2500 University Drive N.W., Calgary, Alberta, Canada T2N 1N4.
J.A. FIGUEIRAS	Department of Civil Engineering, University of Porto, Porto, Portugal.
A. GHALI	Department of Civil Engineering, The University of Calgary, 2500 University Drive N.W., Calgary, Alberta, Canada T2N 1N4.

R. GLEMBERG
Department of Structural Mechanics,
Chalmers University of Technology,
S-41296 Gotheborg,
Sweden.

K.R. GOPAL
Department of Civil Engineering,
Carleton University,
Ottawa, Ontario, Canada.

E. HINTON
Department of Civil Engineering,
University College of Swansea,
Singleton Park, Swansea SA2 8PP,
United Kingdom.

Y.L'HUBY
Electricite De France,
12-14 avenue Dutrievoz,
69628 Villeurbanne Cedex,
France.

D.G. KLEIN
Wolfel,
Otto-Hahn-Strasse 2a,
8706 Hochberg bei Wurzburg,
West Germany.

S.S.J. MOY
Department of Civil Engineering,
University of Southampton,
Southampton SO9 5NH,
United Kingdom.

L. NILSSON
Department of Computer Aided
Analysis and Design,
Lulea University of Technology,
Lulea,
Sweden.

M. OLDENBURG
Department of Computer Aided
Analysis and Design,
Lulea University of Technology,
Lulea,
Sweden.

D.V. REDDY
Department of Ocean Engineering,
Florida Atlantic University,
PO Box 3091,
Boca Raton,
Florida 33431-0991,
U.S.A.

A. SAMUELSSON
Department of Structural Mechanics,
Chalmers University of Technology,
S-41296 Gotheborg,
Sweden.

S. STURE — Department of Civil Engineering,
University of Colorado,
Campus Box 428,
Boulder,
Colorado 80302,
U.S.A.

J.A. SUNBERG — ADINA Engineering AB,
Munkgatan 20D,
S-72212 Vasteras,
Sweden.

K. WILLAM — Department of Civil Engineering,
University of Colorado,
Campus Box 428,
Boulder,
Colorado 80302,
U.S.A.

CONTENTS

PREFACE

This book evolved from the conference 'Computer-aided analysis and design of reinforced concrete structures' which was held in Split, Yugoslavia in September 1984. The conference attracted 260 participants and it became apparent during discussions in Split that there was a real need for a more considered view of recent developments in this general field of research. bearing this in mind, we invited a number of contributions for a state-of-the-art text on the theme of 'Computational modelling of reinforced concrete structures'. The resulting text contains chapters from authors present at the conference and others who are also actively involved in research aimed at the development of efficinet, accurate and reliable computational models for reinforced concrete structures.

The book has two main themes: the first six chapters deal with constitutive modelling of concrete whereas the last six concentrate on the development of numerical models and their application in the analysis and design of reinforced concrete structures. Practitioneers and research workers have long recognised the importance of all aspects of computational modelling. To obtain reliable and realistic solutions from analytical and computational procedures, the constitutive and numerical models and also nonlinear solution techniques must be carefully checked for accuracy and consistency. The contributions in this text reflect the considerable activity in this research area in recent years.

We wish to express our sincere thanks to all of the authors for their cooperation in producing a set of stimulating contributions.

Ernest Hinton and Roger Owen
Swansea, June 1986

PART I

CONSTITUTIVE MODELLING

CHAPTER I

NON-LINEAR ANALYSIS OF REINFORCED CONCRETE SLABS

R. J. Cope
Department of Civil Engineering
Plymouth Polytechnic

SUMMARY

Constitutive models for use in non-linear analytical procedures to predict the post-cracking response of reinforced concrete slabs are reviewed. Many of the proposed models of concrete behaviour are based on the results of tests of plain concrete specimens. In view of the inherent scatter in such results, the variation in material properties throughout a slab, and the influences of reinforcing bars, it is argued that the pursuit of great accuracy is not warranted. The main influences of reinforced concrete behaviour on slab responses are discussed and the current trends and difficulties in the formulation of constitutive models for the cracked composite material are described.

1. INTRODUCTION

To date, procedures for the analysis and design of reinforced concrete slabs have been based mainly on empirical rules, or on simplified treatments of material properties, such as those assumed for linear and rigid-plastic methods [1]. Non-linear methods of analysis could be used to predict, more realistically, the behaviour of non-standard slabs with known patterns of loading. They could also be used to assess the performance of existing slabs under particular overloads and when damaged. Because of the inherent uncertainties concerning concrete properties, analytical predictions are unlikely to be very reliable when the structural response is dependent on those properties. However, for most slabs, structural response is determined mainly by structural cracking and by yielding of reinforcement at high load levels. In such circumstances, non-linear methods of analysis have much to offer.

In this chapter, information on concrete properties in slabs and material models for use in non-linear analytical procedures are reviewed. The text attempts to explain the current state of the art and the current lines of development.

Concrete in a slab is subjected to a variety of stress states. In the immediate vicinity of heavy concentrated loading and near supports, it may be necessary to consider triaxial conditions. However, reliable test data on triaxial reinforced concrete properties are scarce and triaxial stress-strain relationships are not considered in detail here. Away from concentrated loading, it is generally assumed that consideration of biaxial effects is sufficient. However, most biaxial test data are from plain concrete specimens, the behaviour of which may not accurately represent that of a concrete element in a slab which is restrained by reinforcement and by surrounding, less highly stressed concrete. Also, it is well established that the tensile strength of a plain concrete beam or slab is strongly influenced by the strain gradient over its depth. Close to cracks and slab edges, and in zones of slabs subjected to predominantly one way bending, uni-axial conditions may prevail.

2. UNI-AXIAL, COMPRESSIVE STRESS-STRAIN RESPONSE

Consideration of uni-axial stress-strain equations for plain concrete under short term loading is a useful starting point for a discussion of more complex conditions, and a considerable quantity of test data is available. Popovics [2] has reviewed the work of many researchers, and he noted that the average strain is the result of many, small, varying, discrete deformations, that occur in the various constituents of a concrete. Deduced material properties based solely on the cylinder (or cube) strength are, therefore, likely to be in error. This finding has been confirmed by Mirza et al [3], who presented data, obtained from hundreds of laboratory tests, on the scatter of measured tensile strengths and initial elastic moduli about predicted values. When target or characteristic compressive strengths of standard specimens are all that is available to a designer, precision and great sophistication are not warranted in a model of material behaviour. Popovics also noted that the inelastic part of the average strain is greater for the first application of load to a certain level, than for a subsequent application to the same level. This observation implies that constitutive equations based on test data for progressively increasing loading are, strictly, only applicable to the first loading of the concrete in a structure.

In a subsequent paper, Popovics [4] noted that stress-strain curves had different characteristics according to whether the load was applied in stress or strain increments, see Fig. 1a. Using triaxial test data, Gerstle et al [5] have argued that the falling part of the stress-strain curve is mainly dependent on the nature of the testing machine, and, in particular, on the stiffness of its interface with the specimen. If there were no restraint, the concrete would fail completely at its maximum stress. It follows that test data can only be valid for strains below the strain level corresponding to the maximum stress. In the general case, this probably corresponds to all strain states up to a strain state corresponding to the minimum volume of a specimen.

In a concrete structure, a failing element is restrained by surrounding concrete, which may be less highly stressed, and by steel reinforcement. To date, there is no test data known to the author that relates the falling branch of the stress-strain curve to the stiffness of the surrounding material. There is ample test data, however, to show that concrete in the compression zone of a beam, or slab, does not fail completely when the strain corresponding to the maximum uni-axial stress is reached, and that considerably greater strains can be accommodated.

From the test data he studied, Popovics proposed the following uni-axial stress-strain formula

$$\sigma = \sigma_0 \left(\frac{\varepsilon}{\varepsilon_0}\right) \frac{n}{(n - 1 + (\varepsilon/\varepsilon_0)^n)} \qquad (1)$$

where $n = 0.58\,\sigma_0 + 1.0$ for normal weight concretes; σ_0 is the maximum stress (N/mm^2) and ε_0 is the corresponding strain. When ε_0 is not known, which is the usual case facing a designer, Popovics suggested that, for normal weight concretes, it could be evaluated using:

$$\varepsilon_0 = 9.368 \times 10^{-4}\ \sqrt[4]{\sigma_0} \qquad (2)$$

Although Eq. (2) was fitted to data exhibiting considerable scatter, Popovics demonstrated that when combined with Eq. (1), it led to good predictions of uni-axial stress-strain curves for normal weight concretes of varying strength, including their descending branch (sic).

The tangent modulus is given by differentiating Eq. (1). Thus,

$$E_t = \frac{d\sigma}{d\varepsilon} = \frac{\sigma_0}{\varepsilon_0} \frac{(1 - (\varepsilon/\varepsilon_0)^n\ n(n - 1)}{(n - 1 + (\varepsilon/\varepsilon_0)^n)^2} \qquad (3)$$

Tests on plain concrete specimens under eccentric loading [6] have indicated that the stress-strain curves for concrete in a flexural member are similar to those derived from axially

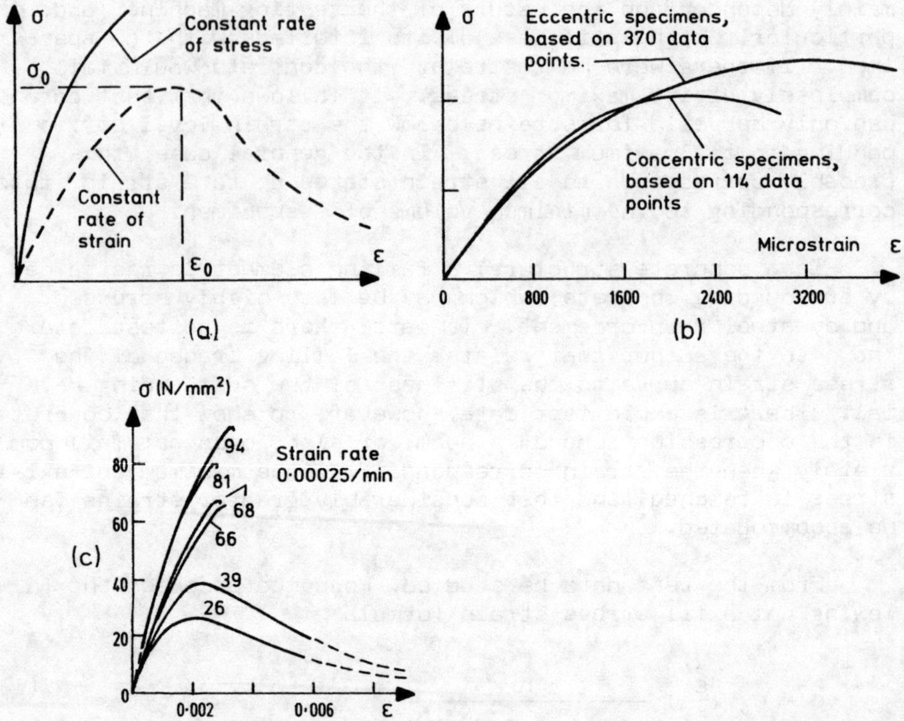

Fig. 1 Uni-axial stress-strain for plain concrete
(a) based on [4]
(b) based on [7]
(c) based on [9,26]

loaded cylinders. The results also showed that, up to a maximum compressive strain of about 0.003, there was no visible cracking or spalling. Other tests, see, for example, [7] have indicated even greater ductility and strength in the presence of strain gradient see Fig. 1b. These improvements in behaviour are probably due to the confining actions of the less highly stressed concrete, which delays the appearance of transverse micro-cracking. Blume et al [8] recommend the use of a maximum strain of 0.004 for ultimate curvature calculations involving concrete not confined by reinforcement.

Caution should be exercised when extrapolating stress-strain equations to concretes outside the range for which they have been validated. High strength concretes have recently been introduced for bridges in the USA and Japan and, when unconstrained, these may be relatively brittle [9]. Figure 1c compares typical stress-strain curves, from 100 x 200 mm cylinder tests, for different strength concretes made with the same aggregates. Although the strength range is large, it can be seen that there are relatively small changes in the initial modulus and in the strain corresponding to the peak failure stress. Although the higher strength concretes appear to be relatively brittle, the lack of a descending branch may be more a function of the testing machine than of the material [26].

The author has obtained reasonable predictions of slab behaviour using Eq. (1) for concrete, and by assuming either complete loss of strength for $\varepsilon \geq \varepsilon_0$ [10], or by assuming a constant compressive stress for $\varepsilon_0 < \varepsilon \leq 0.0035$, followed by complete loss of strength for $\varepsilon > 0.0035$ [11]. However, it should be noted that the examples studied were not very sensitive to the compressive strength or stiffness of concrete.

3. UNI-AXIAL, TENSILE STRESS-STRAIN RESPONSE

The uni-axial tensile stress-strain response of plain concrete is approximately linear. When tested under strain control, however, some softening becomes apparent, and specimens can resist a falling tensile force with strains that are several times the strain at maximum stress.

The flexural-tensile strength determined from a modulus of rupture test tends to be higher than the tensile strength of a split cylinder, which in turn tends to be higher than the tensile strength obtained from a direct tension test. The range of values obtained from specimens cast from the same concrete can be large, as can the scatter about values predicted from the compressive strength [3].

There is a number of contributory reasons for these variations. In a direct tension test, concrete on the weakest plane fails and there is a large number of potential failure planes. In a split cylinder test, the stress is not uniform and the test is effectively performed on a single plane. Concrete in a modulus of rupture test is subjected to a linearly distributed strain over the depth. The theory assumes a corresponding linear stress distribution, which is not strictly true. Also, it is known that a strain gradient enhances the strength of a specimen. There are many potential failure planes, but concrete fibres subjected to the maximum strain are limited to the soffit zone of a test beam.

The tensile behaviour of concrete in a structure may be different to that of a standard specimen. When tensile concrete in a structure is restrained by adjacent portions of concrete which are less highly stressed (or are in compression, as may be the case with composite construction) the ultimate tensile strain may be increased significantly. However, when concrete is reinforced, the bars may act as stress raises and adversely effect compaction locally. Also, restraint to early thermal contraction, while the concrete is weak, may lead to the presence of additional internal tensile stresses and micro-cracking. Indirect evidence that the direction of weakening of the effective tensile strength is biased by the reinforcement directions in a slab has been reported [12].

The predicted, non-linear response of a reinforced member to the first application of loading is quite sensitive to the specified tensile strength of the concrete [12]. In the author's experience of beam and slab analyses, reasonable predictions are usually obtained when the split cylinder strength of a standard specimen is used for the tensile strength of the concrete. However, a lower value may be needed when there is transverse reinforcement positioned close to the soffit (as in a model of a slab), and running in a direction close to that of the expected cracking.

From the above discussion, it can be appreciated that the effective tensile strength of the concrete in a structure may be difficult to determine. Even its average value and coefficient of variation cannot be determined with any degree of precision.

3.1 Tensile Toughness

Concrete is not a perfectly brittle material and possesses a property that Hillerborg has called, 'tensile toughness' [13]. This toughness plays a substantial role in controlling the shear capacity of a member and the spacing and width of bending and shrinkage cracks.

For a detailed analysis, a single stress-strain curve cannot be used for tension, as the difference in behaviour between concrete in a fracture zone and that outside the zone is too great. Hillerborg has proposed the use of a stress-strain curve for concrete outside the fracture zone and a stress "crack width" relationship for the cracking zone. The concept is illustrated for a tensile member loaded under displacement control in Fig. 2a-b. The deformation on a gauge length ℓ outside the fracture zone is $\ell\varepsilon$, whereas that on the same gauge length including a fracture zone is $\ell\varepsilon + w$. After the peak tensile stress, concrete between fracture zones is predicted to experience reducing strains, while the cracks form.

Hillerborg has argued that it is reasonable to assume the $\sigma - w$ curve to be a material property for concrete. A typical stress-deformation curve over a gauge length of 100 mm is shown in Fig. 2c. The area below the curve gives the energy absorbed per unit area over a 100 mm length and is a measure of the tensile toughness. From the shape of the curve, it can be apppreciated that the toughness is mainly dependent on the $\sigma - w$ curve.

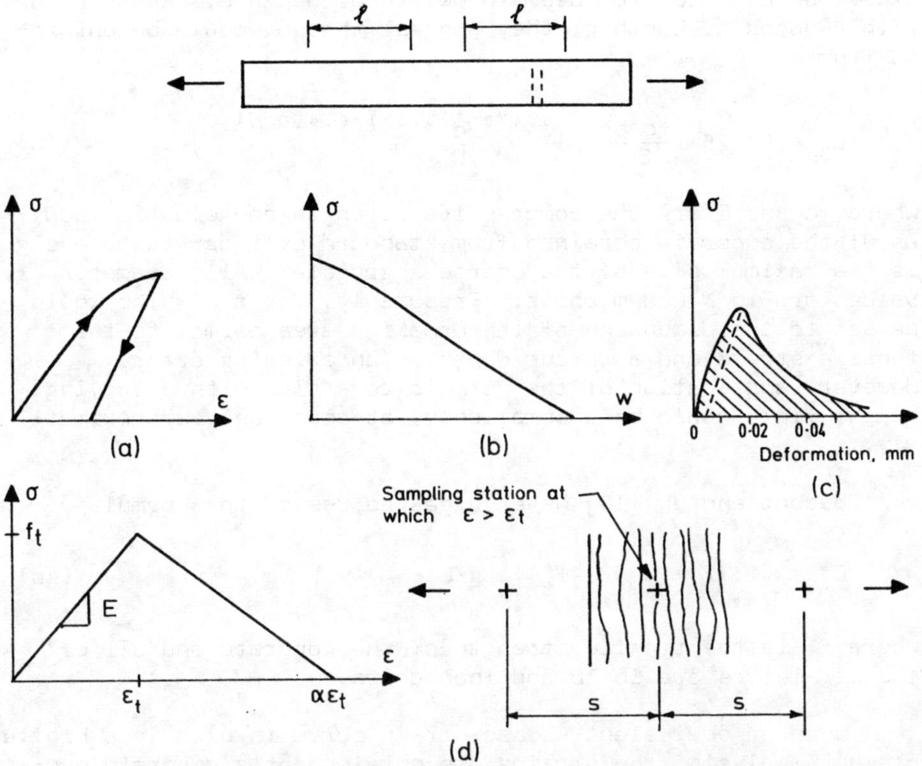

Fig. 2 (a) Stress-strain response outside fracture zone; (b) Stress-deformation response of fracture zone; (c) Typical stress-deformation curve for concrete on a gauge length of 100 mm (based on [13]) (d) Equivalent stress-strain curve.

Experiments indicate that the average length of a fracture zone in the stress direction is of the same order as the maximum aggregate size [14]. For analytical purposes, it can therefore be considered as a fictitious crack that precedes the formation of a discrete crack in a tensile member and as a fictitious extension of a flexural crack in a bending member. In the smeared crack analytical approach, the σ - w curve has to be converted to an equivalent stress-strain curve.

The material parameters used to do this are the modulus of elasticity, E, the tensile strength, f_t and the fracture energy G_f. The latter quantity is given by the area under the σ - w curve and can be obtained from test data. However, unfortunately, these exhibit considerable scatter. For normal weight concretes, typical values for G_f lie in the range 200 f_t^2/E to 400 f_t^2/E N/mm.

Nallathambi et al [15] have studied the effects of member and crack sizes, water-cement ratio and coarse aggregate texture on the fracture toughness of concrete. Based on their experimental data from beams of width b, depth d, length L and with a notch of depth g, they suggest that G_f could be obtained from

$$G_f = 0.125 \left(\frac{fc^2 b}{E}\right)\left(\frac{g}{b}\right)^{0.173}\left(\frac{d}{L}\right)^{0.631+0.4(a/d)} \tag{4a}$$

where fc and E are the compressive strength and elastic modulus of the concrete obtained from standard cylinder tests and g is the maximum size of the coarse aggregate. All parameter values are in N and m units. Presumably, for a slab, b would be set to the dimension of the cracking area normal to the tensile stress and a to the depth of an existing crack. However, application of the formula to a slab with a varying moment field could be accomplished, at best, only approximately.

Bazant and Oh [16] have suggested use of the formula

$$G_f = (2.72 + 0.0214 f_t) f_t^2 \, g/E \tag{4b}$$

where f_t is the tensile strength of the concrete and all of the parameters are in lb and inch units.

When an equivalent stress-strain curve is used in a Finite Element analysis, the strains are obtained at a discrete number of sampling stations. Constitutive equations in the form of stress-strain curves are imposed at these stations and the spacings of the stations give the gauge lengths over which average deformations are considered. To obtain mesh independent solutions, it is necessary to take the sampling station spacing into account when specifying the equivalent post peak, tensile stress-strain curve [16].

If the spacing of the sampling stations is s, and a bi-linear equivalent stress-strain curve is adopted, see Fig. 2d, the area under the stress-deformation curve is given by $A = \int \sigma(s d\varepsilon) = \alpha s f_t \varepsilon_t/2$. Making the assumption that this area is a constant for the material, G_f, gives

$$\alpha = 2G_f/(s f_t \varepsilon_t) \tag{4c}$$

Bazant [17] has suggested that when α makes the descending branch vertical, a reduced value of f_t should be used to retain the essential shape of the bi-linear curve.

4. BI-AXIAL, STRESS-STRAIN RESPONSE

In general, for concrete in compression, the presence of an orthogonal compressive stress leads to a stiffer response and to greater strength. This is due to the effect of Poisson's ratio and, at high stress levels, to the delay of transverse micro-cracking. In tension, the presence of an orthogonal compressive stress reduces the tensile stiffness and strength, as cracking is encouraged. Bi-axial tensile strengths are similar to the uni-axial strength.

The constitutive equations for concrete under plane-stress conditions have the general form:

$$\begin{bmatrix} \varepsilon_1 \\ \varepsilon_2 \\ \varepsilon_3 \\ \gamma \end{bmatrix} = \begin{bmatrix} b_{11} & b_{12} & b_{13} & b_{14} \\ b_{21} & b_{22} & b_{23} & b_{24} \\ b_{31} & b_{32} & b_{33} & b_{34} \\ b_{41} & b_{42} & b_{43} & b_{44} \end{bmatrix} \begin{bmatrix} \sigma_1 \\ \sigma_2 \\ \sigma_3 \\ \tau \end{bmatrix} \quad (5)$$

where σ_1 and σ_2 are the orthogonal in-plane stresses; ε_1 and ε_2 are the corresponding direct strains; σ_3 and ε_3 are the normal stress and strain, respectively; and τ and γ are the in-plane shear stress and strain, respectively.

There are too many coefficients to handle economically and simplifying assumptions are usually made. For plane stress conditions, $\sigma_3 = 0$. Therefore, the coefficients in the third column of the matrix are not of interest. If it is assumed that energy is conserved (no frictional losses, for example), the matrix of coefficients must be symmetric about the leading diagonal. If it is further assumed that the concrete is orthotropic, with the directions of material orthotropy coinciding with the principal stress directions, and that no shear strains are induced by the application of increments of the principal stresses, then

$$b_{41} = b_{42} = b_{43} = b_{34} = 0$$

With these assumptions, the constitutive equations reduce to:

$$\begin{bmatrix} \varepsilon_1 \\ \varepsilon_2 \end{bmatrix} = \begin{bmatrix} b_{11} & b_{12} \\ b_{12} & b_{22} \end{bmatrix} \begin{bmatrix} \sigma_1 \\ \sigma_2 \end{bmatrix} \quad (5a)$$

$$\varepsilon_3 = b_{31}\,\sigma_1 + b_{32}\,\sigma_2 \quad \text{and} \quad \gamma = b_{44}\,\tau \quad (5b,5c)$$

The coefficients of these equations can be expressed in terms of material properties by considering uni-axial loadings. (i) Apply σ_1, with $\sigma_2 = \tau = 0$, then $b_{11} = \varepsilon_1/\sigma_1 = 1/E_1$, $b_{21} = \varepsilon_2/\sigma_1 = -\nu_{21}/E_1$, and $b_{31} = \varepsilon_3/\sigma_1 = -\nu_{31}/E_1$, where E_1 is the elastic modulus of the concrete and ν_{21}, ν_{31} are the values of Poisson's ratio in the orthogonal in-plane and normal directions, respectively. When total stresses and strains are used, the properties take secant values. When incremental quantities are used, tangential property values are taken. (ii) Apply σ_2, with $\sigma_1 = \tau = 0$, then $b_{22} = 1/E_2$, $b_{12} = -\nu_{12}/E_2$, $b_{32} = -\nu_{32}/E_2$.

Thus
$$\begin{bmatrix} \varepsilon_1 \\ \varepsilon_2 \end{bmatrix} = \begin{bmatrix} 1/E_1 & -\nu_{12}/E_2 \\ -\nu_{21}/E_1 & 1/E_2 \end{bmatrix} \begin{bmatrix} \sigma_1 \\ \sigma_2 \end{bmatrix} \tag{5d}$$

and $\varepsilon_3 = -(\nu_{31}/E_1)\,\sigma_1 - (\nu_{32}/E_2)\sigma_2$

If it is assumed that σ_1 and σ_2 have the same effect on ε_3, then $\nu_{31}/E_1 = \nu_{32}/E_2$. If it is further assumed that σ_1 has the same effect on ε_2 and ε_3, then $\nu_{21}/E_1 = \nu_{31}/E_1$. Similarly, considering the effects of σ_2 on ε_1 and ε_3 to be the same gives $\nu_{12}/E_2 = \nu_{32}/E_2$. Thus, $\nu_{21} = \nu_{31} = \nu_1$ and $\nu_{32} = \nu_{12} = \nu_2$, and the constitutive equations can be written in terms of three independent material parameters.

$$\begin{bmatrix} \varepsilon_1 \\ \varepsilon_2 \end{bmatrix} = \frac{1}{E_1} \begin{bmatrix} 1 & -\nu_1 \\ -\nu_1 & \nu_1/\nu_2 \end{bmatrix} \begin{bmatrix} \sigma_1 \\ \sigma_2 \end{bmatrix} \tag{5e}$$

and $\varepsilon_3 = -\nu_1/E_1\,(\sigma_1 + \sigma_2)$

To obtain an expression for b_{44} in terms of ν_1, ν_2 and E_1, the effect of loading case $\sigma_1 = -\sigma_2$ on an element with sides at 45° to the principal directions is considered. The shear stress-strain relationship is then assumed to be the same for all directions [27]. With this assumption it can be shown that

$$b_{44} = \frac{1}{E_1} + \frac{2\nu_1}{E_1} + \frac{\nu_1}{E_1\nu_2} \tag{5f}$$

Equations 5e and 5f can be written in the alternative form

$$\begin{bmatrix} \sigma_1 \\ \sigma_2 \end{bmatrix} = \frac{1}{1 - \nu_1^{\,2}(E_2/E_1)} \begin{bmatrix} E_1 & E_2\nu_1 \\ E_2\nu_1 & E_2 \end{bmatrix} \begin{bmatrix} \varepsilon_1 \\ \varepsilon_2 \end{bmatrix} \tag{6a}$$

and $$\tau = \frac{E_1\,E_2}{E_1 + E_2 + 2E_2\nu_1}\,\gamma. \tag{6b}$$

If isotropic properties are assumed, $E_1 = E_2 = E$ and $\nu_1 = \nu_2 = \nu$, then

$$\begin{bmatrix} \sigma_1 \\ \sigma_2 \end{bmatrix} = \frac{E}{1 - \nu^2} \begin{bmatrix} 1 & \nu \\ \nu & 1 \end{bmatrix} \begin{bmatrix} \varepsilon_1 \\ \varepsilon_2 \end{bmatrix} \qquad (7a)$$

and $\tau = \dfrac{E}{2(1 + \nu)} \gamma$ (7b)

4.1 Poisson's Ratio Effect

The effect of Poisson's ratio can be most easily appreciated by considering elastic isotropic conditions. Equation (7a) can be manipulated to give effective uni-axial constitutive equations. Thus,

$$\sigma_1 = \frac{E\varepsilon_1}{1 - \nu(\sigma_2/\sigma_1)} \qquad \sigma_2 = \frac{E\varepsilon_1}{1 - \nu(\sigma_1/\sigma_2)} \qquad (7c)$$

When the orthogonal stress and the stress under consideration are of the same type (compressive or tensile) the effect of Poisson's ratio is to increase the effective stiffness. When of the opposite type, the effective stiffness is reduced.

Although concrete is not elastic, and a description of material properties should ideally reflect the stress history, reasonable degrees of success have been achieved by using total stress-strain relationships for the analysis of reinforced concrete slabs. A simple model of orthotropic behaviour can be produced by assuming that the principal strain directions define the axes of material orthotropy and that the value of Poisson's ratio is a constant. Equations (7a) then become

$$\begin{bmatrix} \sigma_1 \\ \sigma_2 \end{bmatrix} = \begin{bmatrix} E_1(\varepsilon_1{*}) & 0 \\ 0 & E_2(\varepsilon_2{*}) \end{bmatrix} \begin{bmatrix} \varepsilon_1{*} \\ \varepsilon_2{*} \end{bmatrix} \qquad (7d)$$

where

$$\begin{bmatrix} \varepsilon_1{*} \\ \varepsilon_2{*} \end{bmatrix} = \frac{1}{1 - \nu^2} \begin{bmatrix} 1 & \nu \\ \nu & 1 \end{bmatrix} \begin{bmatrix} \varepsilon_1 \\ \varepsilon_2 \end{bmatrix}$$

Equation (7d) has been used for the non-linear analysis of reinforced concrete slabs failing in flexure [10]. The equivalent principal strains ($\varepsilon_1{*}$, $\varepsilon_2{*}$) were evaluated and then used to obtain the corresponding stresses from the uni-axial relationship given by Eq. (1). As the analyses were based on the initial stress method, the unsymmetrical nature of the constitutive equations after stressing did not create undue difficulties.

4.2 Confinement of Micro-cracking Effect

Although the above approach recommends itself by its simplicity and ease of implementation, it ignores the stiffening at high stress levels caused by the orthogonal compressive stress restricting the growth of micro-cracks. To account for both micro-crack confinement and Poisson's ratio effect, Liu et al [18] recommended the use of a more complicated stress-strain formula, of the type given by Eq. (7c), that gives acceptable fits to bi-axial test data for plain concrete.

$$\sigma_i = \frac{E\varepsilon_i}{\left(1 - \frac{\nu\sigma_j}{\sigma_i}\right)\left[1 + \left(\frac{1}{1 - \frac{\nu\sigma_j}{\sigma_i}}\frac{E}{E_s} - 2\right)\left(\frac{\varepsilon_i}{\varepsilon_0}\right) + \left(\frac{\varepsilon_i}{\varepsilon_0}\right)^2\right]} \qquad (7e)$$

where i = 1, 2 for j = 2, 1; E is the initial, uni-axial modulus; and E_s is the secant modulus at maximum stress ($E_s = \sigma_0/\varepsilon_0$). Because σ_i and σ_j are not known when the right-hand side of Eq. (7e) is evaluated, an iterative solution is required. However, as the influence of the stress ratio is relatively small, the author has used the equation with values evaluated for the previous strain estimate [19]. As was the case for Eq. (7d), the formula is, strictly, only applicable when there is no change in the direction of the principal axes.

Both equations assume that knowledge of the previous strain history is not important. After the first loading and for stresses in the working range, this assumption is reasonable, but, if a previous stress state has approached the failure envelope, considerable errors would be introduced. The formula has been assessed against test data for loading applied monotonically, and with the principal stress ratio fixed [18].

To use Eq. (7e), values of the maximum stress and corresponding strain under bi-axial conditions are required. For plain concrete, some test data are available. Kupfer et al [20] have summarised efforts prior to the 1970s and have published results of carefully conducted tests that have been very influential. They provided strength envelopes for concretes with a range of uni-axial strengths. Loading was, however, limited to fixed ratios of monotonically increasing principal stresses. The results indicated increases in compressive strength of up to 27% in the presence of an orthogonal compressive stress. However, when similar specimens were tested with untreated bearing platens, that provided additional restraint to micro-cracking, the strength enhancement was up to 50%. Presumably, failing concrete in a structure, which is restrained by reinforcement crossing the micro-cracks and by surrounding less highly stressed material, would have a strength greater than that of an unrestrained specimen.

When using Eq. (7e) for slab analyses, the author has used the lower bound to the unrestrained failure stress envelope and the corresponding maximum strains recommended by Liu, et al [18] and by Tasuji et al [21]. Although the strain data show considerable scatter and are incomplete, the stress-strain curves produced are in good agreement with the test data of both Kupfer and Liu. However, it should be emphasised that concrete in a structure, which is restrained by reinforcement and surrounding less highly stressed material, is likely to have higher failure strains.

Equations (7d) - (7e) are basically extensions of uni-axial formulations (Eq. (7e) reduces to the uni-axial equation suggested by Saenz [22], when $\sigma_j = 0$) and they provide orthotropic constitutive equations in principal strain directions. It is emphasised, that when used in a slab analysis, it is implicitly assumed that the stress history of the uncracked concrete does not have a significant influence on the response, and that the orthotropic material axes can rotate to follow the principal strain directions.

Test data based on monotonically increasing loading with fixed principal stress ratios cannot provide sufficient information to determine values for all of the possible material constants for bi-axial conditions. However, Kupfer et al [20] considered their data supported the use of a constant value of Poisson's ratio for stresses up to at least 80% of the ultimate stress. At higher stress levels, micro-cracking parallel to the failure stress direction increases the apparent value of Poisson's ratio.

Kupfer, and Gerstle [23] have suggested that the available test data permit the use of isotropic constitutive equations for bi-axial compression conditions. They studied relationships between the test values of the hydrostatic and deviatoric stress and strain components and proposed formulae to evaluate the secant and tangent shear and bulk moduli (G and K respectively). The formulae require the stress state to be known, and thus pose the same difficulties for implementation as Eq. (7e). They also depend on constants that have only been evaluated for three concrete strengths. The constitutive equations take the form:

$$\begin{bmatrix} \sigma_x \\ \sigma_y \\ \tau_{xy} \end{bmatrix} = 4G\left(\frac{3K+G}{3K+4G}\right) \begin{bmatrix} 1 & \frac{3K-2G}{2(3K+G)} & 0 \\ \frac{3K-2G}{2(3K+G)} & 1 & 0 \\ 0 & 0 & \frac{3K+4G}{4(3K+G)} \end{bmatrix} \begin{bmatrix} \varepsilon_x \\ \varepsilon_y \\ \gamma_{xy} \end{bmatrix} \qquad (8a)$$

Using their own test data, Kotsovos, and Newman [24] have proposed alternative formulae for the shear and bulk moduli. Although their equations also require the stress state to be known, they are non-dimensionalised with respect to the concrete strength. In their tests, loading was applied at a constant rate of stress (in contrast to a constant rate of principal strain in Kupfers' tests), and several different load histories were considered.

To overcome the difficulty in implementing equations expressed in terms of the stress ratio, Cedolin and Mulas [25] have suggested a secant model up to the peak stress, based on the test results of Kupfer et al [20] and expressed in terms of the in-plane principal strains. Their equations are, strictly, only applicable to monotonically increasing loading and relate principal stresses and strains. Thus,

$$\begin{bmatrix} \sigma_1 \\ \sigma_2 \end{bmatrix} = \frac{2G}{3K + 4G} \begin{bmatrix} 6K + 2G & 3K - 2G \\ 3K - 2G & 6K + 2G \end{bmatrix} \begin{bmatrix} \varepsilon_1 \\ \varepsilon_2 \end{bmatrix} \qquad (8b)$$

where K and G are specified functions of the compressive strength, initial value of Young's modulus and Poisson's ratio for the concrete, and of ε_1 and ε_2. A tangential form of Eq. (8b) is also given. This, However, is non-symmetric, and requires an iterative solution when the principal strain direction is not constant.

In the late 1970s, an international study was undertaken to identify the influences of the variations of concrete properties and of test methods on the large scatter of results for plain, laboratory controlled concrete specimens tested under multi-axial stress states [5]. All of the laboratories involved used specimens from the same mix, cast and cured in one laboratory. The researchers concluded that all of the test methods produced results with considerable scatter and that the material properties obtained could vary over a considerable range.

As there was considerable scatter from such carefully controlled specimens, and as the influence of restraints due to testing conditions and surrounding material in actual structures has not been quantified, it can be argued that the material models for use in non-linear analysis should be as simple, as is consistent with the trends indicated by test results.

The test data from monotonically-loaded biaxial specimens appear to support the use of either isotropic or orthotropic material models for the compression-compression and tension-tension states. This suggests that the degree of orthotropy is not large, at least until the failure envelope is approached. Support for this supposition is found in the experimental results of Kupfer, as reported by Link, J. [27]. From comparisons of values of ν_1 and ν_2, and of E_1 and E_2, Link concluded that there was no noteworthy effect of stress related anisotropy in the compression-compression states, for strains up to about 80% of the breaking strain (i.e. until at least one stress was close to the uni-axial strength). However, with uni-axial loading and in compression-tension stress states, he concluded that stress related anisotropy is significant.

4.3 Loss of Poisson's Ratio Effect

When concrete cracks in tension, it is customary to set Poisson's ratio to zero. From Eqs. (7a-b) it can be seen that this creates a sudden change in the stresses corresponding to a particular strain state. As discontinuities can lead to numerical problems, Cope and Rao [10, 11] assumed elastic properties during iterations, and only introduced new cracks with a branched unloading line, and with the loss of Poisson's ratio effect, at the start of the next load increment. This procedure leads to minor violations of the constitutive equations during the iterative process and to out of balance forces at the end of the iterations for a load increment, but is relatively stable.

4.4 Compression-Tension States

Maekawa and Okamura [28] have reported results of a carefully conducted and detailed study of bi-axially loaded concrete in the compression-tension zone. By adopting two loading paths, they were able to evaluate all four of the constants in Eq. (5a).

In one series of tests, a compressive stress was applied in one direction and then held constant while the orthogonal tensile stress was gradually increased. Tensile cracking failures resulted when the orthogonal compressive stress was less than about 90% of the uni-axial compressive strength. It was considered that a compressive stress accelerates a tensile failure. The compression-tension failure envelope obtained was similar to that derived by Kupfer et al [20], and appeared to be independent of the stress path for non-cyclic loading. However, specimens, which had been loaded in compression to about 80% of their uni-axial compressive strength and then unloaded, showed a dramatic decrease in their orthogonal tensile strength when they were subsequently loaded in tension. It is difficult to envisage when such a circumstance could occur in a concrete slab, and use of the bi-axial failure envelope derived from monotonic loadings seems appropriate. Under compressive stresses higher than about 90% of the uni-axial strength, compressive failures resulted.

Maekawa and Okamura found that the value of Young's

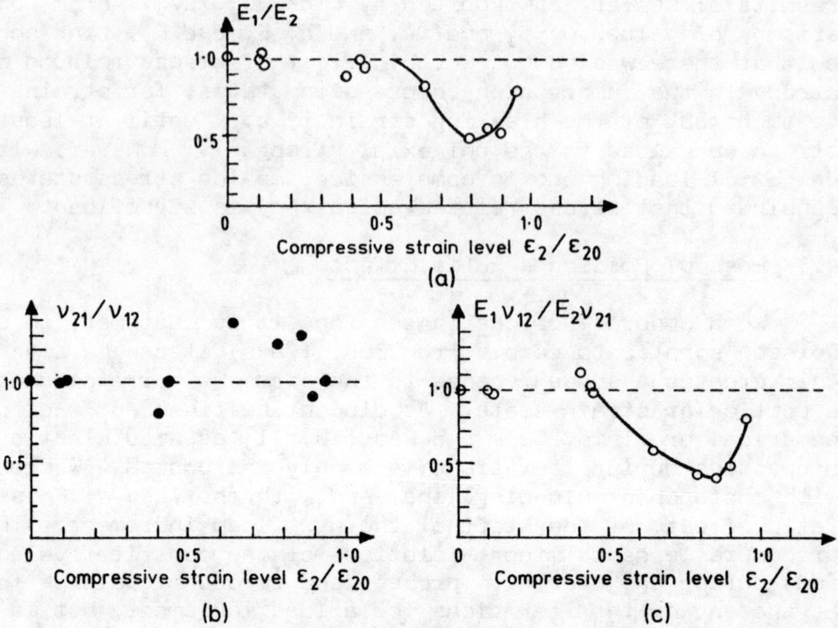

Fig. 3 (a) Ratio of bi-axial moduli;
(b) Ratio of Poisson's ratio;
(c) Ratio of off-diagonal terms.

modulus in the tensile direction decreased almost linearly with increase in the orthogonal compressive strain level. This suggests that use of the above equivalent uni-axial equations will be in error. The value of Poisson's ratio was approximately constant up to a compressive stress of approximately 80% of the uni-axial compressive strength. At higher stress levels, the effective value increased rapidly, to about 0.8.

In the second series of tests, a tensile stress was applied and then held constant while the orthogonal compressive stress was increased. At a particular compressive strain level, the orthogonal tensile stress tended to reduce the value of the compressive Young's modulus. However, the effect was insignificant for compressive stresses of up to about 90% of the uni-axial strength. The value of Poisson's ratio was approximately constant until the compressive stress exceeded about 80% of the uni-axial strength.

Ratios of bi-axial stiffness moduli, Poisson's ratios and of off-diagonal terms of Eq. (5d) were determined, and are shown in Figs. 3a-c. As can be seen, they support the use of a symmetric stiffness matrix for compressive strains of less than half of the strain corresponding to the maximum uni-axial compressive stress. This implies that a symmetric matrix is reasonable for compressive stresses up to 80 - 90% of the uni-axial strength.

5. BI-AXIAL RESPONSE OF REINFORCED CONCRETE

The addition of reinforcement to concrete produces a composite material with greater strength and ductility than plain concrete. In particular, by inhibiting the opening of structural cracks, reinforcement permits additional tensile and shear forces to be transmitted. However, because of their physical presence, reinforcing bars influence compaction conditions locally, and their stiffness can induce micro-cracking and sometimes visible cracking by their restraining action on early thermal contraction. It is assumed that, if average concrete stresses and strains over gauge lengths containing several structural cracks are considered, then it is possible to postulate plane-stress constitutive equations for loading applied to the composite material.

A reinforced concrete slab, relatively lightly loaded normal to its plane, can be considered to act as an infinite number of parallel panels subjected to plane stress conditions. However, the presence of vertical direct stresses, and shear stress components on vertical planes, limit the applicability of such a model. Constitutive equations based on bending test data may be more appropriate than those based on panel test data for plane stress conditions alone. The strain gradient over a slab's thickness, and the different compaction and curing conditions of concrete through the thickness, result in different material properties in the concrete of the constituent "panels". To date, however, this has usually been ignored by analysts.

The response of a reinforced concrete element loaded in uni-axial compression is commonly assumed to be given by superimposing the concrete and steel responses. The concrete behaviour is assumed to be the same as that of a cylinder specimen. In a real structure, the response may be more complicated due to the variation in concrete properties throughout the structure and to the restraining action of reinforcement on transverse cracking. Tests, by for example Bertero and Felippa [29] on square prisms, have shown that the ductility of axially loaded members is significantly affected by the steel ratio of the transverse, confining, reinforcement. Also, when at least part of the loading is sustained, creep of the concrete can affect the proportion of the loading carried by the steel and the concrete.

Tests on a multi-storey building have shown that the assumption of equal strains in steel and concrete in columns can be false. In particular, the reinforcement in lightly loaded columns has been shown to carry considerably more than its anticipated share of the loading [30]. The author is not aware of any similar data for slabs, but it is not unreasonable to suppose that, with time, sustained compressive loading will be similarly transferred from the concrete to any adjacent reinforcement.

The response of a reinforced concrete member loaded in tension is complicated. Prior to cracking, behaviour can be predicted by superposition of the concrete and steel responses. Determination of the cracking load cannot be accomplished with any degree of precision, for the reasons discussed in section 3.1. Cracks in a slab tend to develop gradually and are not usually continuous across a section. When a reinforcement direction does not coincide with the principal tension direction, crack directions can vary locally. Concrete between cracks can carry tension, but there is a tendency for the bond between steel and concrete to become less effective with increase in steel strain and when the load intensity fluctuates.

Tests conducted by Peter [31] on panels with orthogonal, orthotropic reinforcement at various angles to the applied tensile force, showed an initial crack pattern and later, superimposed on it, an inclined failure crack. The initial cracks were oriented approximately normal to the applied tensile force. The inclinations of the failure cracks were dependent on the relative quantities of steel in the two directions and the inclination of the tensile force to those directions.

When a reinforced concrete panel is subjected to tension in one direction, so that cracking occurs, and to compression in the orthogonal direction, failure can be either by crushing of concrete, or by tensile failure of the steel. According to Vecchio and Collins [32], Robinson and Demorieux [33] have shown that increasing tensile strains in the transverse reinforcement can result in significant reductions in the compressive stiffness and strength of the concrete parallel to the cracks.

In the general case, a cracked, reinforced panel can be subjected to both direct forces and to shear forces. Local stress and strain fields are very complicated and reliable test data on which to base a constitutive model of average effects are scarce. After cracking, even in a panel subjected to uniform edge forces, there are considerable variations in the stress and strain fields. Rather than attempt to predict these local effects, only average stresses and strains are considered in analysis. Ideally, the averages are taken over zones large enough to include several cracks. For analysis, it is also assumed that adjacent steel and concrete are subjected to the same strain field. Individual reinforcing bars are not considered, instead, the reinforcement is considered to be uniformly distributed in thin layers that displace no concrete and which possess equivalent, uni-axial stiffnesses in the bar directions. This approach is often called the smeared crack, distributed steel model.

5.1 Data From Panel Tests

Vecchio and Collins [32] provided the first detailed information on the behaviour of panels subjected to general plane-stress loadings. Their carefully conducted study used panels 890 mm square by 70 mm thick, reinforced with two layers of almost elastoplastic wire mesh. Panels were cast flat, to minimise variations in strength due to different compaction conditions. Steps were also taken to ensure a uniform mix through the thickness of the panels. Nevertheless, it is interesting to note their comments that, "Although attempts were made to achieve a reasonably consistent low strength mix, the task proved to be more difficult than expected", and that, "The reinforcing steel also exhibited a wide range of yield strength. This was in spite of the fact that the wire meshes were obtained from the same supplier, and heat treated by the same process".

Most of the tests simulated panel "shear" loading, with no direct forces applied in the orthogonal reinforcement wire directions. With equal steel ratios, the initial cracks formed at approximately 45° to the reinforcement grid and the crack direction did not change with further loading. With unequal steel ratios, however, further loading caused additional cracks to form at a more acute angle to the direction with the greater steel area. These results support those obtained by Peter [31] and mentioned above. Three modes of failure were noted: yielding of wires in both directions; sliding shear failure of the concrete after yielding of the steel in the weaker direction; and sliding shear failure of the concrete prior to any steel yielding.

Those panels that failed after yielding of steel in one direction showed severe crushing and spalling of the concrete, and by a noticeable rotation of the principal strain direction. In general, the sliding shear failure planes avoided crossing the main reinforcing wires. The shear responses of the panels that failed before yielding of any steel were stiffer and more brittle, and an explosive shear failure occurred on a plane approximately parallel to one of the wire directions.

One panel was loaded under combined shear and bi-axial tension, with a bi-axial tension stress to shear stress ratio of 0.32. This panel failed by sliding shear failure before the steel yielded and its behaviour was reported to be similar to that of a similar panel under pure shear. However, its cracking and failure 'shear' loads were, of course, somewhat less than those in the control specimen. The bi-axial tension reduced the shear strength by 9%.

Three panels were tested with combined shear and bi-axial compression. The bi-axial compression to shear stress ratios were 0.39, 0.69 and 0.83. The 'shear' responses were reported to be considerably stiffer, with substantial increases in the loads to cause cracking and failure. The ultimate 'shear-stresses' for the three steel ratios were greater than those of a panel loaded solely by shear forces by 40%, 44% and 25%, respectively. However, the panel with a steel ratio of 0.83 was reported to have failed prematurely, due to the presence of large air voids in the concrete.

Principal, Compressive Stress-Strain Relationship

Vecchio and Collins' results confirmed the findings of Robinson and Demorieux that significant reductions in compressive stiffness and strength are produced by orthogonal cracking and reinforcement stressed in tension. Because of the cracking, it is reasonable to use uni-axial stress-strain relationships and Vecchio and Collins proposed the use of

$$\sigma = f'_c \left(2\left(\frac{\varepsilon}{\varepsilon'_0}\right) - \beta\left(\frac{\varepsilon}{\varepsilon'_0}\right)^2 \right) \qquad (9a)$$

for stresses up to the peak stress and strain (σ_0, ε_0, respectively). Where f'_c is the cylinder crushing strength, ε'_0 is the corresponding strain and β is a function of the principal tensile strain (ε_t) to the principal compressive strain (ε) ratio. The peak stress and strain are given by $\sigma_0 = f'_c/\beta$ and $\varepsilon_0 = \varepsilon'_0/\beta$, respectively. Collins and Vecchio showed that a suitable formula for β is:

$$\beta = 0.85 + 0.27\, \varepsilon_t/\varepsilon \qquad (9b)$$

(both ε_t and ε are given positive values). A comparison of the predicted peak stress with test data is given in Fig. 4a, while Fig. 4b compares the predicted and test values of principal compressive stresses for all of the recorded experimental strain combinations.

Collins and Vecchio suggested the following formula for the post peak stress-strain relationship.

$$\sigma = \sigma_0 \left[1 - \frac{(\varepsilon - \varepsilon_0)^2}{(2\varepsilon'_0 - \varepsilon_0)^2} \right] \qquad (10)$$

whith $\sigma = 0$ for $\varepsilon > 2\,\varepsilon'_0$.

Effect of Confinement

Concrete, forming part of a failing section in a slab, may be surrounded by less highly stressed concrete. In such circumstances, the effect of transverse cracking on the compressive strength may be less than that for a panel, subjected

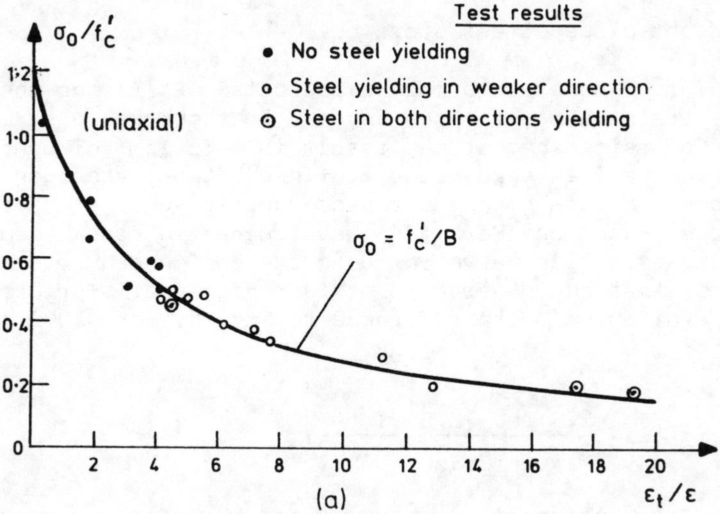

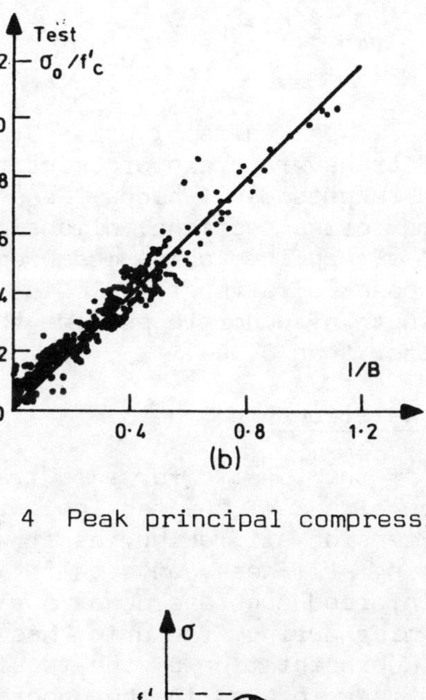

(a) Peak principal compressive stress - principal strain ratio

(b) Comparison of test and predicted principal compressive stresses

Fig. 4 Peak principal compressive stress, based on [32]

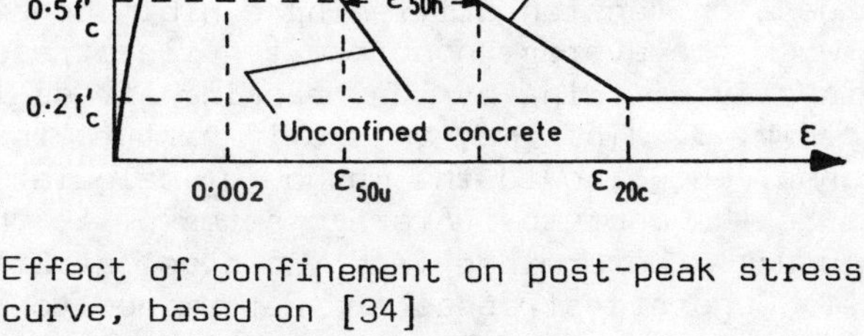

Fig. 5 Effect of confinement on post-peak stress-strain curve, based on [34]

to a constant plane-stress state.

Reinforced concrete bridge slabs often experience their greatest strains close to free edges. In these regions, edge detailing of reinforcement may result in a 'prism' of concrete parallel to the free edge being confined between steel on three faces and less highly stressed concrete on the inner fourth face. Kent and Park [34] have suggested a post peak stress, stress-strain curve for concrete confined by rectangular hoops, that could serve for this region until further test data are available. With reference to Fig. 5, for 0.002 $\leqslant \varepsilon \leqslant \varepsilon_{20c}$.

$$\sigma = f'_c \left(1 - \frac{0.5\,(\varepsilon - 0.002)}{\varepsilon_{50u} + \varepsilon_{50h} - 0.002}\right)$$

$$\varepsilon_{50u} = \frac{3 + 0.002\,f'_c}{f'_c - 1000} \quad \text{and} \quad \varepsilon_{50h} = 0.75\rho_s \sqrt{\frac{b^{11}}{s_h}} \qquad (10a)$$

The stress terms in Eq. (10a) are in psi units (1 psi = 0.00689 N/mm^2). ρ_s = ratio of volume of transverse reinforcement to volume of concrete core measured to outside of hoops; s_h = spacing of "hoops"; b^{11} = minimum cross-sectional dimension (probably depth wise for a slab) of confined core, measured to outside of "hoops". For the purposes of strength calculations, Kent and Park neglected the strength of concrete outside the confining steel for strains greater than 0.004.

Principal Tensile Stress-Strain Relationship

When concrete cracks, some of the tensile force is transferred to the reinforcement. The effect of the tension retained in the concrete is called tension-stiffening, as it was found to increase the post cracking stiffness above that assumed by early theories of reinforced concrete behaviour [1]. The simulation of tension stiffening during fracturing has been discussed in section 3.1. Representation of the tension stiffening caused by tensile stresses induced in the concrete teeth between cracks requires a further idealisation of structural response and is discussed below.

The test data on tension stiffening exhibit considerable scatter, even when the average concrete tensile stress values σ_t are normalised by dividing by the cracking stress for a panel, f'_t, see Fig. 6. This may be due in part to the nature of the phenomenon, but also to the method of calculating the tensile stress in the cncrete. Average stresses in the reinforcement are usually determined from the steel strains in the bar directions. Concrete stresses are determined to be the difference between the applied loading and the steel resist-

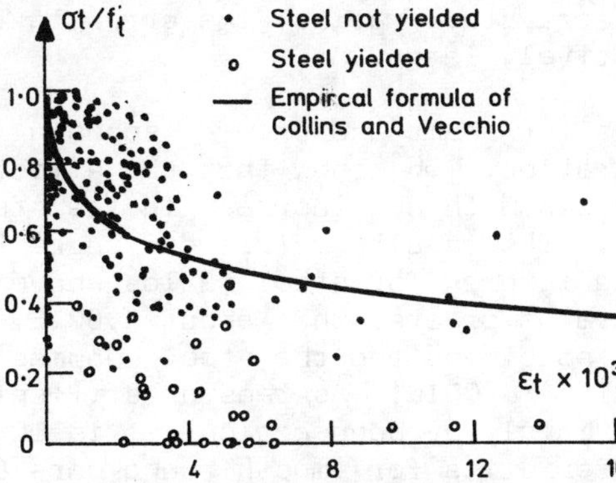

Fig. 6 Principal tensile stress strain data, based on [32]

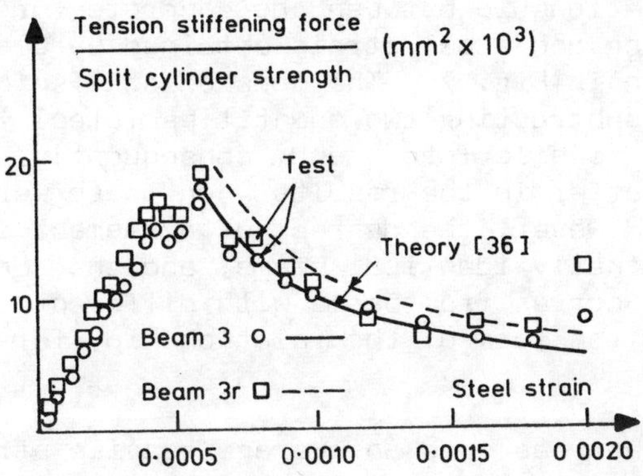

Fig. 7 Tension stiffening force-steel strain, based on beam tests with A_s/bd = 0.79% [36]

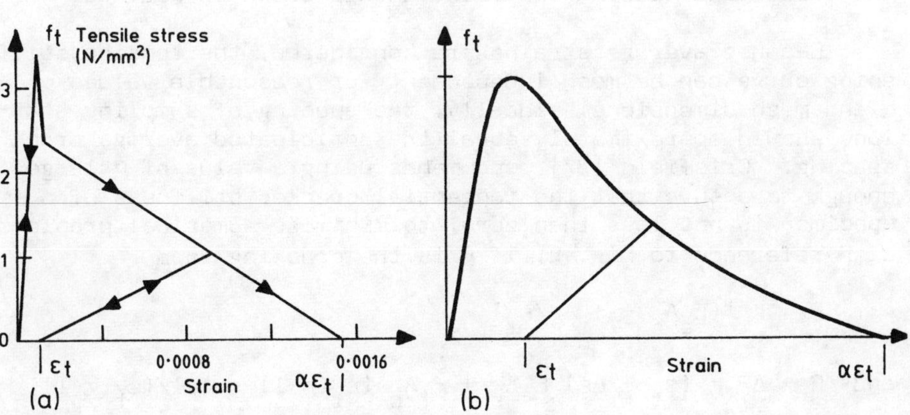

Fig. 8 Analytical tensile stress-strain relationships

ance. As those two quantities can be of a similar magnitude, for relatively large steel ratios, the error in their difference can be relatively large.

Test results for beams and one way spanning slabs, with different steel ratios, [36] show that the effect of tension stiffening decreases with an increase in the steel ratio and the steel strain. The results referenced suggest that tension stiffening can be ignored for steel ratios above about 1.5% and for steel strains greater than about 0.0016. (Tor bar was used in the test specimens and the steel becomes non-linear at a strain of 0.0014 - 0.0016. As tension stiffening between cracks is caused mainly by bond action, it is likely to have different characteristics for smooth round bars and for welded members.)

The relationship between the concrete force due to tension stiffening and steel strain obtained by Clark and Speirs [36] is shown in Fig. 7. The tension stiffening forces were obtained by subtracting two quantities (steel force and calculated total tensile force) and, consequently, there was considerable scatter in the results when these were of a similar magnitude. However, the degrees of agreement obtained from tests on nominally identical beams, and the consistancy in the shape of the curves from beams with different steel ratios, suggest that the form of the relationship, for first loading, is as shown.

The author has used an average tensile stress-strain curve devised to give good predictions of the results of Clark and Speirs for slab analyses [11, 12]. Details of the curve are given in [12] and its general shape is illustrated in Fig. 8a. Today, the author would use a curve of the form shown in Fig. 8b, to minimise the possibility of numerical problems due to the discontinuities in the relationship shown in Fig. 8a.

Because average strains are considered, the tension stiffening curve can be mesh independent for reasonable values of α and mesh dimensions. Ideally, the spacing of sampling stations should approximately equal the anticipated average crack spacing. Crisfield [37] recommends using a value of α large enough to ensure that the tangential cracked stiffness of a specimen is not less than zero, to minimise numerical problems. With reference to Fig. 9, if P is the cracking load

$$P = \varepsilon_t \; (E_s A_s + f_t A_c / \varepsilon_t)$$

and $$P + \Delta P = (\varepsilon_t + \Delta\varepsilon) \; E_s A_s + f_t A_c \; (\varepsilon_t(\alpha-1) - \Delta\varepsilon)/(\varepsilon_t(\alpha-1))$$

Hence $$\Delta P = (E_s A_s - f_t A_c/(\varepsilon_t(\alpha-1))) \; \Delta\varepsilon,$$ and

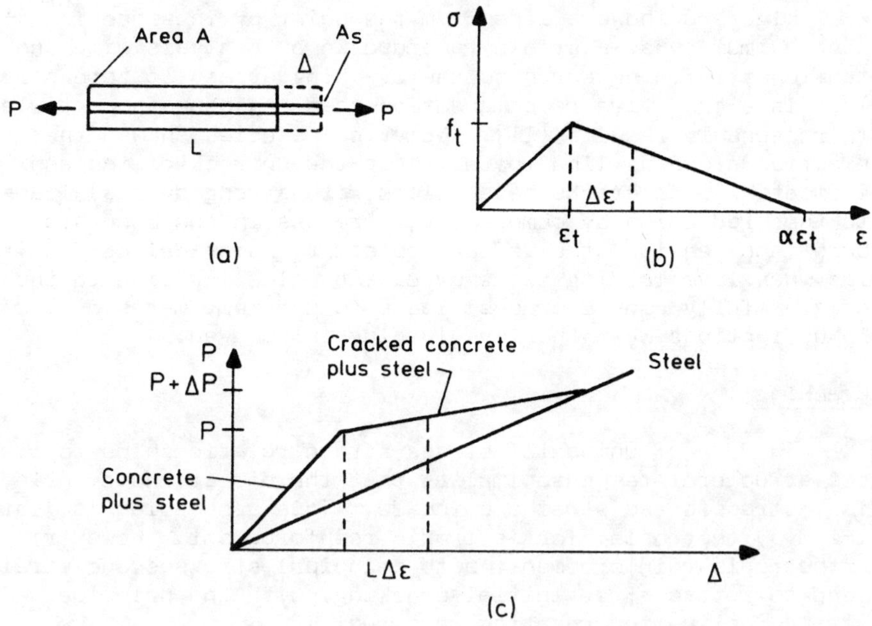

Fig. 9 (a) Tension specimen; (b) Concrete tensile stress-strain curve; (c) Load-displacement lines for specimen

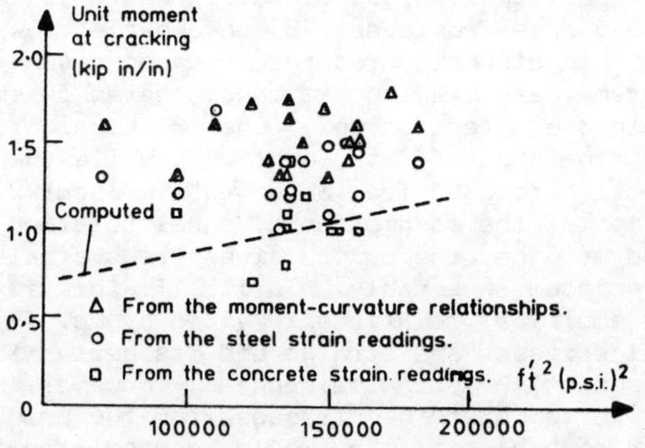

Fig. 10 Moment at cracking-split cylinder strength squared, based on [39]

for $\Delta P > 0, \quad \alpha > f_t A_c/(E_s A_s \varepsilon_t) + 1$

For short term effects $f_t/(E_s\varepsilon_t) \simeq 1/7$. For a relatively low steel percentage of 0.5%, and taking A_c for a slab to be one half of the cross-sectional area, $\alpha \geqslant 15$. The suggested value of 15 for α should not, therefore, lead to numerical difficulties for slabs with a realistic degree of reinforcement.

Figure 8 shows a line from the point of intersection of the maximum tensile strain recorded in an analysis with the tension stiffening curve to the cracking strain at zero stress. This is a tentative recommendation that has been found to lead to reasonable results [11]. Numerical studies showed that an elastic unloading line (parallel to the uncracked line and indicating that cracks have filled with strong debris) leads to the prediction of compressive stresses in the concrete below the tension steel of a slab and to numerical difficulties when live loading is removed. An unloading line to the origin (fully closed cracks) leads to underestimates of locked in deflections when the live loading is removed.

Cracking

The use of uni-axial stress-strain relationships to predict structural response implies that the directions of principal stresses and strains coincide. This assumption is justified by test data, for isotropic reinforcement. However, for orthotropic reinforcement, both principal stresses and strains tend to rotate after initial cracking, with the principal strain inclination rotating more rapidly.

For the panel tested by Vecchio and Collins in shear, with steel areas in the orthogonal directions of 1% and 1.8%, for example, the principal stress direction was calculated from the test data to change from about 45° to about 43°, while the principal strain direction rotated from about 40° to about 35°. For the extreme case examined, a panel tested in shear with 1.8% steel in one direction only, the principal stress direction rotated from about 44° to about 28°, while the principal strain direction rotated from about 45° to about 22°. Despite this violation of the assumption of equal rotation of principal stresses and strains, the compressive stress-strain relationship was predicted reasonably. In itself, this is not surprising, as the empirical relationship given by Eq. (9) is based on principal stresses and strains and its derivation did not take into account the principal angles. However, if the formulae were used to determine stresses from the principal strains predicted by an analysis, they would be predicted to act in the wrong directions.

Although use of Eq. (9) should lead to reasonable predictions for unconfined concrete, the fact that the reduction factor β is expressed as a function of the principal strains should not be taken as implying that these are the only important deformation parameters. For example, as most of the panels were tested in shear, an alternative reduction factor could be deduced in terms of the principal compressive strain and the associated maximum shear strain [35]. The latter approach may well be advantageous in a design environment, but Eq. (9) is preferred for non-linear analysis.

5.2 Data from Slab Panel Tests

Lenschow and Sozen [39] have provided useful data on the post-cracking behaviour of orthogonally-rinforced slab panels loaded monotonically to failure, in either bending or torsion. They reported that deflection measurements did not show a significant rate of increase at the onset of cracking, which they describe as hair cracks traced by a magnifying glass. This supports the use of tension-stiffening models in analysis. The initial cracking directions were approximately perpendicular to the maximum principal moment (and hence tensile stress) direction in all specimens. The spacing of cracks varied from 1" to 3" and in the case with reinforcing bars parallel to the crack direction had a tendency to equal that of the bars. A large variation of concrete strength ($2700 < f'_c < 6610$ psi) was reflected in the cracking moment, but not in the strength and stiffness of the element. This observation confirms the limited influence of concrete material properties on the response of most slabs. The large scatter in tensile strengths and the difficulty in predicting the start of cracking is illustrated in Fig. 10.

From bending tests on slabs with isotropic reinforcement, Lenschow and Sozen concluded that cracking and principal curvatures remained in approximately the same directions at all loading stages, irrespective of the orientation of the steel bars. However, from torsion tests on slabs with isotropic reinforcement, they observed that although all failed at 45°, "it should be mentioned that the failure line at collapse did not always follow the direction of the yield lines, that were dominant at an earlier stage".

In one study, three panels with orthogonal steel of different quantities, inclined at different angles to the bending axis, were tested in one way bending. The researchers commented that, "In each of the three specimens, the directions of the cracks and the directions of the principal curvatures were changing from the time of cracking almost all the way up to failure . . . However, the change in the direction of the principal curvature below the yield point was only of the order 2° - 7° and had little effect on the stiffness of the element".

To investigate the tangential stiffness coefficients of a slab with inclined cracks, a number of 1200 x 520 x 50 mm panels were tested in flexure and torsion, but with different load histories [41]. The slabs were reinforced with top and bottom 6.5 mm "Tor-bar", placed orthogonally at 50 mm centres, and at a number of inclinations to the applied moment axis. The loading rig used to apply flexure was based on the fully-articulated device of Lenschow and Sozen [39], and the torsion rig applied four-point loading to the central zone of a panel to create pure torsion.

Typical crack patterns are shown in Figs. 11a-c. Figure 11a shows two slab panels cast from the same batch of concrete and with 'isotropic' reinforcement oriented parallel to the sides. The slab on the left was subject to several cycles of flexure, followed by several cycles of torsion and then by several cycles of flexure. (Loading was usually cycled five times.) The slab on the right was loaded by torsion, then flexure and finally by torsion again. From the left-hand picture of cracking it is clear that the existing flexural cracks inhibited formation of torsional cracks. From the right-hand picture, it can be seen that existing torsional cracks permitted widespread, but discontinuous flexural cracking.

Figure 11b shows similar results for a slab panel with the reinforcing bars oriented at 45° to the slab sides. The left-hand picture again shows that the existing flexural cracks inhibited the formation and extension of torsional cracking. The right-hand picture shows that the existing torsional cracks tend to change the inclination of the flexural cracks.

Figure 11c shows a similar slab, but with reinforcing bars at 30° and 60° to the slab sides. The left-hand picture shows results consistent with previous trends, as does the right-hand picture, which clearly shows flexural cracks running in different directions in previously-cracked and uncracked concrete areas. It is emphasised that the steel was not yielding in these tests. Maximum crack widths, were of the order of 0.1 mm in one test, but were generally of the order of 0.05 mm.

The average crack spacing for all of the slabs lay in the range 45 - 55 mm. Surface strains were measured using Demec gauges over 100, 200 and 400 mm gauge lengths. In general, the strains measured in four directions over 100 mm gauge lengths did not fit a Mohr's circle. Acceptable errors were, however, obtained using the 200 mm gauge length. Local strain distributions over distances equivalent to 2 - 3 slab depths are very discontinuous and application of the smeared crack approach would seem to be reasonable only after a distributed crack pattern has formed and when relatively large gauge lengths are used.

From the test data, it was deduced that the reinforcing steel orientation had negligible effect on the principal strain directions for the first loading pattern. For those panels tested first in flexure, the flexural cracks were subjected mainly to shear during the subsequent torsion tests. It was found that the applied moment and principal strain directions almost coincided, so the in-plane shear stiffness could not have been degraded substantially. When these panels were subsequently subjected to flexure, the effect of the torsional cracks on the tangential stiffness was negligible.

Although the flexural crack patterns of the slab panels

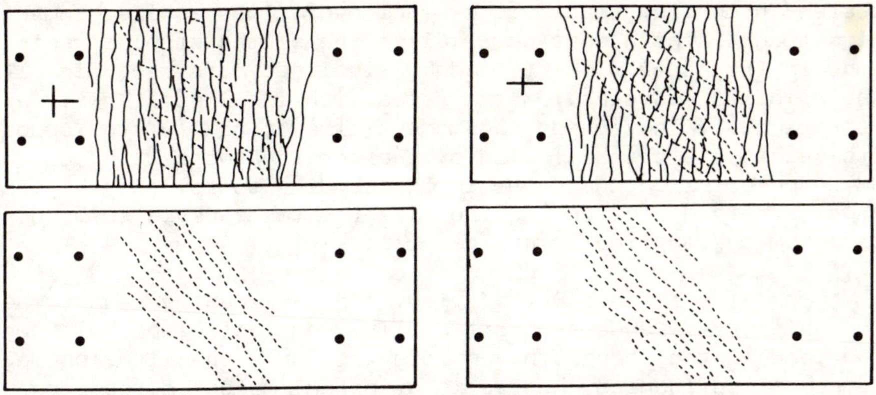

(a) Crack patterns with reinforcement parallel to the sides

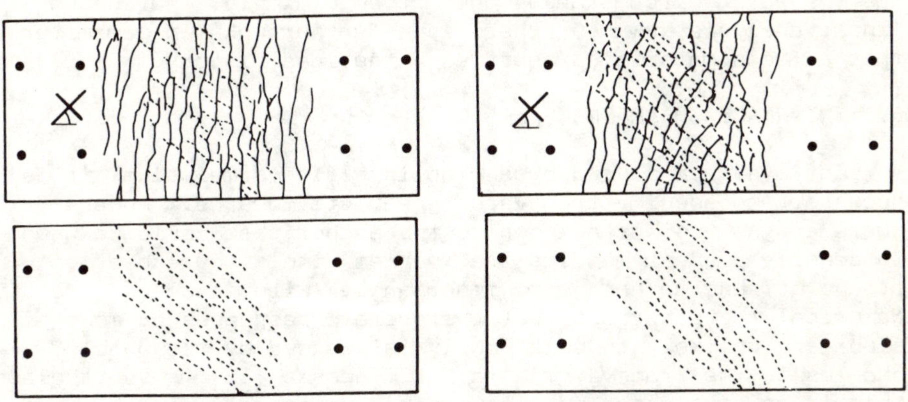

(b) Crack patterns with reinforcement at 45° to the sides

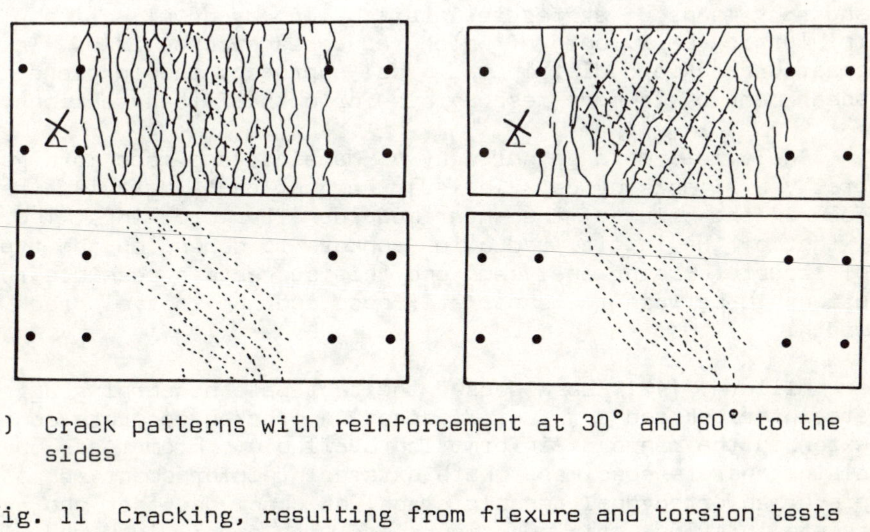

(c) Crack patterns with reinforcement at 30° and 60° to the sides

Fig. 11 Cracking, resulting from flexure and torsion tests

tested in torsion tended to be inclined to the applied moment direction (being almost parallel to one of the steel directions in the panel with 30° and 60° inclined bars, see Fig. 11c), the principal strains, on the faces with inclined cracks, were approximately in the applied moment directions. On the opposite faces, the cracks formed by the torsion tests were subjected to compression and shear in the bending tests, and the principal strain directions in the flexural tests deviated by up to 15° from the applied moment direction [41].

From a detailed point of view, cracking can be seen to be load history dependent. Fortunately, for most slabs, although the load history cannot be anticipated, only the live loading component influences changes in principal moment directions. The extreme swing studied in the tests is unlikely to occur in practice, even for slab bridges, and in design, one can envisage the worst circumstances and act accordingly. For a nonlinear analysis, however, there is a further unknown parameter that affects material properties - the load history.

6. INTERFACE SHEAR TRANSFER

For a slab analysed by a Mindlin plate formulation, it is necessary to specify transverse shear stiffnesses. However, there are no test data known to the author that could be used to specify a shear modulus that varied with either shear rotations or, and perhaps more importantly, with in-plane strains. Numerical tests suggest that the overall responses of most slabs are unlikely to be strongly influenced by reasonable changes to the transverse shear stiffnesses [12]. Nevertheless, this is an area where more test data are needed.

After cracking, reinforced concrete transfers in-plane shear by means of aggregate interlock and by dowel action of reinforcing bars, both of which reduce as cracks widen. In a smeared crack and distributed steel approach, the in-plane shear modulus is made responsible for both of these actions.

Analytical data gathered by Al-Mahaidi, and made more generally available by Gerstle [42], suggest that there is a rapid fall in the value of the shear modulus after cracking. The extent of the initial reduction appears to depend on the type of structure being analysed, and this suggests a possible relationship between the shear stiffness and the initial crack width.

Millard, [43] has assessed the test data concerning aggregate interlock and dowel stiffnesses in cracked concrete and has extended the range of information available. From tests on plain concrete specimens, with external reinforcement and a preformed orthogonal tensile crack, he concluded that the aggregate interlock stiffness decreases with crack width and with increasing shear slip (which also increases the crack width), but it is relatively insensitive to the concrete strength, in the normal range. From tests on dowel action alone, Millard and Johnson [44] concluded that the shear stiffness is similarly influenced by the same parameters as aggregate interlock stiffness. For the normal range of steel ratios, they suggest that the dowel action stiffness is between 25% and 50% of the aggregate interlock stiffness. However, from the results of reinforced concrete tests, Millard and Johnson [45] have shown that the breakdown of local bond in the vicinity of a crack causes a reduction in the effective normal stiffness provided by the bars which control the crack width. Hence, the shear stiffness of reinforced concrete is more sensitive to variations in concrete strength than is indicated by tests on solely aggregate interlock, or dowel action specimens.

At a detailed level, interface shear transfer is clearly a complex phenomenon. In view of the inherent uncertainties in knowledge of material properties, an overall representation of shear stiffness, that omits secondary effects, is highly desirable. As with other material models based on averaging of effects, the suitability of a shear transfer relationship can only be assessed by comparing analytical and test results for particular structural types.

Bazant [46] has suggested the use of an assymmetric, incremental stress-strain relationship. Whilst such expressions are justified for use at a detailed level by test data, and may need to be used for certain structures in the nuclear industry, their use cannot be justified for common-place reinforced concrete slabs. In non-linear analysis it would, in any case, be easier to use an approximate, symmetric, tangent stiffness matrix and to use asymmetric expressions to determine the out-of-balance forces. However, even with an approximate, symmetric matrix there is a problem with the signs of the off-diagonal terms. For example, the incremental, normal, compressive stress across a crack increases with a shear strain of either sign. Thus, the sign of the coefficient relating normal stress to shearing strain cannot be set, until the sign of the shearing strain is known. Setting the wrong sign, could lead to an alternative load path being followed.

For slabs with up to 45° of skew, reasonable analytical results have been obtained by ignoring off-diagonal terms of the stress-strain matrix after cracking, and by calculating released forces using an in-plane shear modulus that reduces with the average normal strain across the cracked zone [11]. This suggests that, for the analysis of most slabs, the incremental compressive stresses induced by shearing strains and the reduction of shear modulus with shear strains, are of secondary importance.

Most of the test data concerning interface shear transfer have been obtained from specimens containing a single, straight,

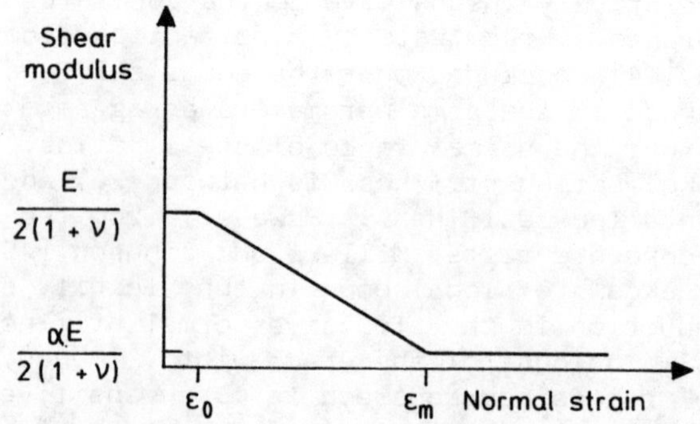

Fig. 12 Secant shear modulus for a cracked station

well-defined crack. Cracking in slabs under serviceability level loadings, however, tends to consist of short, discontinuous cracks, that are only approximately parallel to one another [11]. It is, therefore, likely, that the cracked in-plane shear stiffness of a slab would be greater than that found in a test specimen for the same average crack width. For this reason, a continuous degradation of the secant modulus with the average normal strain has been suggested [11], see Fig. 12. From comparative studies of 45° skew slabs with different steel arrangements, it has been found that acceptable results are obtained with α set to 0.1 and ε_m to $15 \times \varepsilon_o$. However, more test data are required to improve modelling of interface shear transfer. Three of the under-researched areas are: shear stiffness with discontinuous cracks; the effects of repeated loadings; and shear transfer in high strength concretes.

7. FLEXURE OF REINFORCED CONCRETE SLABS

Constitutive models for reinforced concrete slabs can be formed from stress-strain equations. Use of stress-strain relationships, and integration over volume, automatically provides coupled moment-curvature and in-plane force-deformation equations. If it is assumed that plane stress conditions prevail on horizontal planes of a slab, the constitutive equations discussed for panels could be used. However, it should be borne in mind that reinforced concrete slabs built on site are constructed with much less care than laboratory specimens, and the variability of concrete throughout a structure is likely to be much greater than that between specimens. Also, the detailed load history of most slab structures and the extent of cracking and self-equilibrating stresses induced during curing are unquantifiable, complicating factors.

The main trends in the variations of concrete strength with depth in slabs are known. For relatively thin, suspended slabs constructed on site, Bungey, [47] has indicated that there is likely to be an approximately linear decrease in strength from the soffit of about 10%, over seven-eighths of the thickness, with a further 15% fall over the top one-eighth of the thickness. The latter trend may, however, be reversed by the effects of surface trowelling. In deeper members, such as a slab bridge, the reduction in strength over the thickness is approximately linear, with the top concrete having a strength of only 60% of that in the soffit.

The variation of concrete strength with depth has an, as yet, unquantifiable effect on cracking. Because top bars are surrounded by relatively weak concrete, bond is less effective than for bars placed near the soffit. This factor can have important consequences for slabs which are called upon to carry tensile forces via top reinforcement. Clark and Thorogood [48] have reported that the shear capacity is considerably reduced, and that ". . . cracks propagate more rapidly when the bars are positioned in the top . . .".

Variations of strength in plan are likely to be random in nature, and in-situ strengths are usually less than those of standard specimens. Bungey [47] has estimated that the average equivalent wet cylinder strength of the concrete in a thin slab is only about 50% of that of a standard specimen, whereas the equivalent percentage strength for a thick slab is likely to be of the order of 75%. The relatively low in-situ concrete strengths of prototype slabs may not, however, be present in a thin slab constructed under laboratory conditions [12]. This factor complicates the assessment of non-linear analytical procedures, that have been validated against model test results, for use with real, in-situ slabs.

7.1 Cracking

Because a reinforced concrete slab is a complicated statically-indeterminate structure, stress redistribution occurs after cracking and principal stress and strain directions rotate. This implies that the largest tensile strains can be inclined to crack normals, and cracks can be subjected to shearing stresses.

If material property axes for a cracked portion of concrete are fixed by the direction of initial cracking [48], Eqs. (5c) and (5d) are assumed to apply. The directions 1 and 2 are positioned normal and parallel to the crack. Poisson's ratio is set to zero and equivalent uni-axial direct stress-strain relationships are used. A relationship such as that shown in Fig. 14 can be used for the in-plane shear modulus. This approach is referred to below as the fixed material property axis direction (FMP) model.

Fixing the direction of the material property axes makes it difficult for an analytical model of a slab to follow the gradual rotation of principal axes and the eventual formation of an inclined failure crack. To replicate this phenomenon, Cope and Rao [10] permitted the material property axes to follow the rotation of the principal axes. A shear modulus is not required for this model, as the formulation always uses principal stresses. The approach is referred to below as the rotating material property axis direction (RMP) model. The RMP model has been shown to give reasonable predictions of behaviour for: plane stress problems [10, 50]; for slabs [11]; and for shells [51]. It does not, however, meet with the approval of material scientists [52].

A crack, inclined to an initial crack, does not usually form until relatively large strains have been attained, and this may not occur until the steel bars in one direction start to yield. In a hybrid approach, Cope et al [40] suggested that the FMP model be used until the rotation and magnitude of the principal tensile strain exceed prescribed values. These would need to be fixed by reference to test data. The RMP model would then be triggered to determine the new direction of the FMP model in the active crack direction.

De Borst and Nauta [53] have proposed an alternative approach to the problem of rotation of the principal strain direction that promises to be more acceptable to the material scientists. In their method, each strain increment is decomposed into a concrete strain increment and a number of 'crack strain' increments. The latter are fixed in the directions of cracks, and each one has only two, non-zero components; the 'crack strain' normal to the crack and the associated 'crack shearing strain'. Although these two 'crack strain' components define only a part of the total strain field at a point, they are used, without reference to the components of other crack strains, to determine the post-cracking stress components. For this reason, the constitutive equations to use with the approach are not obvious.

8. SOME NUMERICAL CONSIDERATIONS

As there is a significant degree of unavoidable uncertainty concerning the values obtained from constitutive equations, it seems reasonable to fashion stress-strain relationships with numerical procedures in mind. In general, it is advantageous to avoid discontinuities, and to use smooth functions whenever possible.

8.1 Numerical Integration in Plan

Investigators of bending behaviour using plane-stress analyses with a reduced, or selective integration scheme, have warned of the dangers of triggering a zero energy mode after cracking has reduced the rank of the stiffness matrix [37, 53]. This problem has not, to the author's knowledge, been reported with slab analyses. The reason is probably that, because stresses are integrated through the thickness, none of the contributions to the stiffness matrix from a plan integration station are zero.

Presentations of crack patterns for slabs often indicate widespread cracking, with a sampling (integration) station shown cracked after the tensile strength has been exceeded. This can give a misleading representation of the structural action, as physical cracking is only visible after a relatively large strain has been exceeded. Also, such a presentation does not indicate which cracks are opening.

The predicted widespread 'cracking' in a non-linear analysis may be due to the practice of allowing any number of sampling stations to 'crack' simultaneously. In practice, cracks tend to open relatively slowly, at fewer locations, and with stress relief in the concrete between the growing cracks.

Although many presentations of crack patterns and steel yielding seem to indicate failure of a slab, the load-deflection plots continue to show a positive gradient. A reason for this behaviour may be that system failure has more to do with strength and stiffness at nodal points, than at integration stations [55]. If so, this would imply that the constitutive models should function best, close to slab collapse, with fine meshes of low order elements. However, caution should be exercised when using fine meshes, as plane stress analyses of plain concrete specimens, using reduced integration, have shown that the stiffness of a node in a particular direction can become zero, and this can lead to the prediction of unrealistic local displacements before overall failure occurs [57].

Even with fine meshes, linear finite element strain predictions for slabs can be grossly in error near point supports and loads. To prevent spurious damage predictions in a non-linear analysis, Cope and Rao [11] have suggested suppressing damage checking at the plan sampling station adjacent to point supports. An alternative procedure, with a standard program, would be to stiffen a structure locally.

8.2 Numerical Integration Through Depth

There can be large non-linearities in the concrete stress distribution over a section, due both to the compressive stress-strain relationship and to the assumed tensile stress-strain relationship used to simulate tension stiffening effects. To determine the in-plane forces at the mid-plane, and the magnitudes of the corresponding couples, a numerical integration scheme is used. Integration schemes, such as the Trapezoidal Rule and Simpson's Rule, integrate low order polynomials passing through few sampling stations exactly. Other procedures, such as the higher order Newton-Cotes, Gauss-Legendre and Lobbato formulae are designed to integrate higher order polynomials giving smooth curves through a larger number of sampling stations. When one of these procedures is used over the depth, the constitutive equations are effectively replaced by relationships that would provide the stresses defined by the polynomial being integrated.

Ganaba and May [56] have investigated the accuracies of a number of integration schemes. Their studies were restricted to predictions of the concrete force and couple of relatively heavily reinforced sections close to failure, and they indicated unacceptably large errors with the integration schemes commonly in use. Their findings supplement those of Bergen [55] and both sets of investigators recommend adjusting the ranges of the integration according to the crack height. However, Ganaba and May did not obtain results to support Bergen's recommendations for the use of a modified Lobbato rule.

9. CONCLUSIONS

Using the available material models, it is possible to replicate the main features of a tested reinforced concrete slab loaded monotonically to failure. However, because of the inherent variability of material properties, it is less certain that a reliable prediction of behaviour can always be given.

More test data are required before reliable models of reinforced concrete behaviour can be established. This is especially the case when intersecting cracks are involved and when concrete properties dominate the structural response. In the latter case, specified concrete properties should be a function of the location of the material in a slab and of the restraint offered by transverse reinforcement and adjacent concrete.

A wholly satisfactory material model of cracking has yet to be developed. For slabs in which the principal directions remain approximately constant, either the fixed material property axis direction model or the rotating material property axis model can be used. For other cases, the latter can be used to predict effects under serviceability level loadings and to provide a lower bound on the ultimate limit state. If suitable constitutive models can be derived, the strain decomposition approach [53] appears to offer an attractive route of development.

Non-linear analytical procedures are very complicated. Errors can be introduced via: the material models; the spacing of sampling stations; the use of numerical integration; and inadequacies in the non-linear equation solver. The state-of-the-art requires all of these components to be subjected, separately, to critical analysis, if objective solutions are to be obtained.

10. REFERENCES

1. Cope, R. J. and Clark, L. A., Concrete Slabs: Analysis and Design, Elsevier Applied Science, London, 1984.

2. Popovics, S., "A review of stress-strain relationships for concrete ACI Journal, March, 243-248, 1970.

3. Mirza, S. A., Hatzinikdas, M., MacGregor, J. G., "Statistical Descriptions of Strength in Concrete", J. of the Struct. Div. Proc. ASCE, ST6, June, 1021-1037, 1979.

4. Popovics, S., "A numerical approach to the complete stress-strain curve for concrete" Cement and Concrete Research, Vol. 3, No. 5, 583-602, 1973.

5. Gerstle, K. H. et al, "Behaviour of concrete under multiaxial stress states" Proc. ASCE, Vol. 106, No. EM6, 1383-1403, 1980.

6. Hognestad, E., Hanson, N. W. and McHenry, D., "Concrete stress distribution in Ultimate Strength Design" ACI Journal, Vol. 52, No. 6, Dec., 455-479, 1955.

7. Sturman, G. H. Shah, S. P. and Winter, G., "Effect of Flexural Strain Gradients on Micro-cracking and Stress-Strain Behaviour of Concrete" ACI Journal, Vol. 62, No. 7, July, 805-822, 1965.

8. Blume, J. A., Newmark, N. M., Corning, L. H., "Design of Multistorey Reinforced Concrete Buildings for Earthquake Motions", Portland Cement Assoc., Chicago, 1961.

9. Clarke, J. L. and Pmeroy, C. D., "Concrete opportunities for the structural engineer (a review of modified concretes for use in structures)" The Structural Engineer, Vol. 63A, No. 2, 45-53, 1985.

10. Cope, R. J. and Rao, P. V., "Non-linear finite element analysis of concrete slab structures", Proc. Instn. Civ. Engrs., Vol. 63, 159-179, 1977.

11. Cope, R. J. and Rao, P. V., "Moment redistribution in skewed slab bridges", Proc. Instn. of Civ. Engrs., Vol. 75, 419-452, 1983.

12. Cope, R. J., "Material modelling of real, reinforced concrete slabs", Proc. Int. Conf. on Computer Aided Analyses and Design of Concrete Structures, Pineridge Press, Swansea, 85-117, 1984.

13. Hillerborg, A., "The influence of Tensile Toughness of Concrete on the behaviour of Reinforced Concrete Structures" Proc. 9th Congress of the FIP, Stockholm, Vol. 1, 157-164, 1982.

14. Petersson, F. E., Report TUBM - 1006, Div. of Building Materials, University of Lund, Sweden, 1981.

15. Nallathambi, P., Karihaloo, B. L. and Heaton, B. S., "Effect of specimen and crack sizes, water/ cement ratio and coarse aggregate texture upon fracture toughness of concrete". Mag. of Concrete Research, Vol. 36, No. 129, 227-236, 1984.

16. Bazant, Z. P. and Oh, B. H., "Spacing of cracks in reinforced concrete". J. of Struct. Engg. ASCE, Vol. 109, No. 9, 2066-2085, 1983.

17. Bazant, Z. P., "Numerical simulation of progressive fracture in concrete structures: recent developments". Proc. of the Int. Conf. on Computer Aided Analysis and Design of Concrete Structures. Pineridge Press, Swansea, 1-18, 1984.

18. Liu, T. C. Y., Nilson, A. H. and Slate, F. O., Biaxial Stress-Strain relations for concrete. Proc. ASCE Vol. 98, No. ST5, 1025-1034, 1972.

19. Cope, R. J. and Edwards, K. R., "Nonlinear finite element analysis of eccentrically stiffened bridge decks", in FEICOM 85, Pergamon Press, 1985.

20. Kupfer, H., Hilsdorf, H. K. and Rusch, H., "Behaviour of Concrete under Biaxial Stresses", ACI Journal, 656-666, Aug, 1969.

21. Tasuji, M. E., Nilson, A. H. and Slate, F. O., "Biaxial stress-strain relationships for concrete", Mag. of Conc. Research, Vol. 31, No. 9, 217-224, 1979.

22. Saenz, L. P., Discussion of: Equation for the stress-strain curve of concrete by Desayi and Krishnam. ACI Journal, Vol. 61, No. 9, 1229-1235, 1964.

23. Kupfer, H. B. and Gerstle, K. H., "Behaviour of Concrete under Biaxial Stresses", J. Eng. Mech., ASCE, Vol. 99, No. EM4, 853-866, 1973.

24. Kotsovos, M. D. and Newman, J. B., "Generalised Stress-Strain Relations for Concrete", J. Eng. Mech., ASCE, Vol. 104, No. EM4, 845-856, 1978.

25. Cedolin, L. and Mulaş, M. G., "Biaxial Stress-Strain Relation for Concrete", J. of Eng. Mech., ASCE, Vol. 110, No. 2, 187-207, 1984.

26. Gerstle, K. H., "Material behaviour under various types of loading", in High Strength Concrete, University of Illinois at Chicago Circle, 43-78, 1979.

27. Link, J., "Eine Formulierung des zweiaxialen Verformungs- und Bruchverhaltens von Beton und deren Anwendung auf die wirklichkeitsnähe Berechnung von Stahlbetonplatten", Deutscher Ausschuss für Stahlbeton, Heft 270, Berlin 1-119, 1976.

28. Maekawa, K. and Okamura, H., "The deformational behaviour and constitutive equation of concrete, using the elasto-plastic and fracture model". J. of the Faculty of Engineering, Vol. XXXVII, No. 2, Uni. of Tokyo, 253-327, 1983.

29. Bertero, V. V. and Felippa, C., Discussion of 'Ductility of Concrete' by E. H. Roy and M. A. Sozen - Proc. Int. Symp. on Flexural Mechanics of Reinforced Concrete, ASCE-ACI, Miami, p 227-234, 1964.

30. Swamy, R. N. and Arumugasaamy, P., "Structural behaviour of Reinforced Concrete Columns in Service" Structural Engineer, Vol. 56A, No. 11, 319-329, 1978.

31. Peter, J., "The reinforcement of plates and shells for principal stresses acting obliquely to the reinforcement direction" Cement and Concrete Ass. Translation T 621, 1968.

32. Vecchio, F. and Collins, M. P., The response of reinforced concrete to in-plane shear and normal stresses Publication No. 82-03, Univ. of Toronto, Canada, 1982.

33. Robinson, J. R. and Demorieux, J. M., "Essaie de Traction-Compression sur Modeles d'amede poutre en béton armé" Institute de Recherches Appliquées du Béton Arme (IRABA), Paris, June 1968 and May 1972.

34. Kent, D. C. and Park, R., "Flexural members with confined concrete", J. of the Struct. Div., ASCE, Vol. 97, ST7, July 1969-1990, 1971.

35. Vecchio, F. and Collins, M. P., "Stress-strain characteristics of reinforced concrete in pure shear", IABSE Colloquium, Delft, 233-248, 1981.

36. Clark, L. A. and Speirs, D. M., "Tension stiffening in reinforced concrete beams and slabs under short term loads", Tech. Rept. 42.521, Cement and Concrete Ass., London, 1978.

37. Crisfield, M. A., "Difficulties with current numerical models for reinforced concrete and some tentative solutions", Proc. of the Int. Conf. on Computer Aided Analysis and Design of Concrete Structures, Pineridge Press, Swansea, 331-358, 1984.

38. Gilbert, R. I. and Warner, R. F., "Tension stiffening in reinforced concrete slabs", J. of the Struct. Div., ASCE, Vol. 104, No. ST12, 1885-1889, 1978.

39. Lenschow, R. J. and Sozen, M. A., A yield criterion for reinforced concrete under biaxial moments and forces Civil Eng'g. Studies, SRS311, Univ. of Illinois, Urbana, Illinois, 1966.

40. Cope, R. J., Rao, P. V., Clark, L. A. and Norris, P, "Modelling of reinforced concrete behaviour for finite element analysis of bridge slabs" Numerical Methods for Non-linear Problems, Vol. 1, Pineridge Press Ltd., Swansea 457-470, 1980.

41. Norris, P., Post cracking characteristics of reinforced concrete slabs, M. Eng. Dissertation, University of Liverpool, 1981.

42. Gerstle, K. H., "Material modelling of reinforced concrete", IABSE Colloquium, Delft, 41-62, 1981.

43. Millard, S. G., Shear transfer in cracked reinforced concrete, Ph.D thesis, University of Warwick, 1983.

44. Millard, S. G. and Johnson, R. P., "Shear transfer across cracks in reinforced concrete due to aggregate interlock and to dowel action", Mag. of Concrete Research, Vol. 36, No. 126, 9-21, 1984.

45. Millard, S. G. and Johnson, R. P., "Shear transfer in cracked reinforced concrete", Mag. of Concrete Research, Vol. 37, No. 130, 3-15, 1985.

46. Bazant, Z. P., "Advances in Deformation and Failure Models for Concrete" IABSE Colloquium, Delft, 9-39, 1981.

47. Bungey, J. H., The testing of Concrete in Structures, Surrey Univ. Press, 1982.

48. Clark, L. A. and Thorogood, R., "Shear in R. C. slabs: analytical and experimental studies" Report to the Dept. of Transport, Vol. 1, 1983.

49. Hand, F. R., Pecknold, D. A. and Schnobrich, W. C., "A layered finite element nonlinear analysis of reinforced plates and shells", Civil Eng'g. Studies, SRS 389, Univ.

49. Cont'd..
of Illinois, Urbana, Illinois, 1972.

50. Gupta, A. K. and Akbar, H., "Cracking in Reinforced Concrete Analysis", J. of Struct. Engg., ASCE, Vol. 110, No. 8, 1735-1746, 1984.

51. Milford, R. V. and Schnobrich, W. C., "Nonlinear behaviour of reinforced concrete cooling towers", Civil Engg. Studies, SRS, 514 Univ. of Illinois, Urbana, Illinois, 1984.

52. Bazant, Z. P., "Discussion of Session 2, Structural modelling for numerical analysis" In: Advanced mechanics of reinforced concrete, IABSE, Final Report, Delft, p 482, 1981.

53. De Borst, R. and Nauta, P., "Smeared crack analysis of reinforced concrete beams and slabs failing in shear", Proc. Int. Conf. on Computer Aided Analysis and Design of Concrete Structures, Pineridge Press, Swansea, 261-274, 1984.

54. Dodds, R. H., Darwin, D., Smith, J. L. and Leibengood, L.D., "Grid size effects with smeared cracking in finite element analysis of reinforced concrete" SM Report No. 6, University of Kansas, 1-118, 1982.

55. Bergen, P. A., "Some aspects of interpolation and integration in nonlinear finite element analysis of reinforced concrete structures", Proc. Int. Conf. on Computer Aided Analysis and Design of Concrete Structures, Pineridge Press, Swansea, 301-316, 1984.

56. Ganaba, T. H., May, I. M., "An improved numerical integration over the thickness of plates and shells in nonlinear finite element analysis", Report No. 8, University of Warwick, 1984.

57. Bedard, C., Non-linear finite element analysis of concrete structures, Ph.D thesis, University of London, 1983.

CHAPTER 2

COMPUTATIONAL ASPECTS OF SMEARED CRACK ANALYSIS

R. de Borst

Institute TNO for Building Materials and Building Structures
Software Engineering Department (Section DIANA)

ABSTRACT

This paper describes a numerical model which permits different inelastic processes to occur simultaneously in a smeared crack approach. The salient feature is a rigorous decomposition of the strain increment into several contributions. Algorithms for multiple, non-orthogonal smeared cracking and for the combination of smeared crack formation and other non-linear phenomena (e.g. shrinkage, creep, plasticity) are presented. Several examples are included to demonstrate the versatility of the approach. Attention is also drawn to the ability of numerical techniques to correctly predict failure and post-failure behaviour of concrete structures. Although severe difficulties are still encountered, recently developed techniques are in principle capable of tracing post-peak behaviour as is shown by the example of mixed-mode crack propagation in an unreinforced notched beam.

1. INTRODUCTION

In spite of the progress which has been made in nearly twenty years of extensive research [13,26,45,46,47,50,62], numerical analyses of concrete structures still meet with significant difficulties. This is perhaps not so surprising in the light of the many complexities which are inherent in the mechanical behaviour of concrete, but it has led to a growing discrepancy in the development of constitutive models for concrete and the ability to obtain reliable, converged solutions in numerical analyses. Indeed, the past two decades have witnessed the development of very sophisticated models for the description of concrete and also of related materials like soils and rocks, but only a few of these models have been employed in engineering analyses and most finite element codes employ much less refined models.

Ideally, the development of constitutive models and their implementation and use in finite element codes should go hand in hand with each other. This implies that the numerical algorithms which are used for the constitutive models should be as accurate as possible, that the finite elements which we employ should behave well and not exhibit spurious zero-energy modes when using particular types of constitutive models, that the iterative procedure to solve the resulting set of non-linear equations should be as robust as possible and so on. In fact, the above-mentioned arguments hold for any engineering material, but owing to their mechanical properties, which involve softening, cracking and all other sorts of 'undesirable' phenomena, this is particularly true for concrete and materials like soils and rock.

In this paper we will describe some constitutive models and numerical algorithms for these models which appear to yield reliable results in numerical calculations. As it is yet unknown how the various inelastic processes in concrete interact, we have adopted a rigorous resolution of the strain increment into several components (e.g. due to cracking, creep, shrinkage). Such a decomposition is essential for smeared crack analyses if various inelastic processes are considered. If it is not assumed, a troublesome and fallacious interaction between different phenomena may be introduced.

As argued, the principal difficulty in analysing concrete structures lies in the fact that for realistic constitutive models of concrete, uniqueness and stability are guaranteed only below a certain load level. This means that the numerical algorithms should be very accurate as otherwise numerical errors may easily trigger potential physical instabilities thus leading to an underestimation of the failure load. Moreover, we can not suffice by calculating the peak strength of a concrete structure, but we also have to assess the residual strength. It is for this reason that we shall give particular attention to the ability of the finite element method to correctly predict failure loads and post-failure behaviour of concrete structures. In practice, structural failure is often associated with a divergence of the iterative procedure or with failure of the iterative procedure to convergence within a prescribed number of iterations. It is the author's opinion that this is not correct as such a divergence may be caused by a variety of reasons, ranging from an attempt to analyse snap-back behaviour under displacement control to ordinary program errors. Hence, a proper assessment of the peak and eventually of the residual strength of concrete structures can only be made if we trace the entire load-deformation response of concrete structures including the post-peak behaviour.

2. SMEARED CRACK MODEL

In this section, we will outline the fundamentals of a smeared crack model which is capable of properly combining crack formation and the non-linear behaviour of the concrete between the cracks and of handling secondary cracking owing to rotation of the principal stress axes after primary crack formation. In the present approach, a secondary crack is allowed if the major principal stress exceeds the tensile strength and if the angle between the primary crack and the secondary crack exceeds a threshold angle α. This threshold angle need not be equal to 90°, so that the model permits non-orthogonal cracks.

2.1. Non-orthogonal cracks

The fundamental feature of the smeared crack model is a decomposition of the total strain increment into a concrete strain increment $\boldsymbol{\Delta\varepsilon}^{co}$ and into a crack strain increment $\boldsymbol{\Delta\varepsilon}^{cr}$:

$$\boldsymbol{\Delta\varepsilon}=\boldsymbol{\Delta\varepsilon}^{co}+\boldsymbol{\Delta\varepsilon}^{cr} \tag{1}$$

Such a decomposition is not new and has also been advocated by some other researchers [11,42]. Moreover, a similar decomposition, namely into an elastic strain increment $\boldsymbol{\Delta\varepsilon}^{e}$ and into a plastic strain increment $\boldsymbol{\Delta\varepsilon}^{p}$ is also assumed in elasto-plasticity. Indeed, there is nothing which prevents us to employ an elasto-plastic model for the concrete and to split the concrete strain increment into an elastic part and into a plastic part. In doing so, equation (1) becomes

$$\boldsymbol{\Delta\varepsilon}=\boldsymbol{\Delta\varepsilon}^{e}+\boldsymbol{\Delta\varepsilon}^{p}+\boldsymbol{\Delta\varepsilon}^{cr} \tag{2}$$

On the other hand, the crack strain increment $\boldsymbol{\Delta\varepsilon}^{cr}$ may also be decomposed into several contributions:

$$\boldsymbol{\Delta\varepsilon}^{cr}=\boldsymbol{\Delta\varepsilon}_1^{cr}+\boldsymbol{\Delta\varepsilon}_2^{cr}+\dots\dots\dots\dots \tag{3}$$

where $\boldsymbol{\Delta\varepsilon}_1^{cr}$ is the strain increment owing to a primary crack, $\boldsymbol{\Delta\varepsilon}_2^{cr}$ is the strain increment owing to a secondary crack and so on. Combining equations (1) and (3), we obtain

$$\boldsymbol{\Delta\varepsilon}=\boldsymbol{\Delta\varepsilon}^{co}+\boldsymbol{\Delta\varepsilon}_1^{cr}+\boldsymbol{\Delta\varepsilon}_2^{cr}+\dots\dots\dots\dots \tag{4}$$

The relation between the crack strain increment of a particular crack (either primary or secondary) and the stress increment is conveniently defined in the coordinate system which is aligned with the crack. This necessitates a transformation between the crack strain increment $\boldsymbol{\Delta\varepsilon}_n^{cr}$ of crack n in the global x,y,z-coordinates and a crack strain increment $\boldsymbol{\Delta e}_n^{cr}$ which is expressed in local coordinates. Restricting the treatment to a two-dimensional configuration (which is not essential), we observe that a crack only has a normal strain increment Δe_n^{cr} and a shear strain increment $\Delta\gamma_n^{cr}$, so that

$$\boldsymbol{\Delta e}_n^{cr}=(\Delta e_n^{cr}\ \Delta\gamma_n^{cr})^T \tag{5}$$

where the superscript T denotes a transpose. The relation between $\boldsymbol{\Delta\varepsilon}_n^{cr}$ and $\boldsymbol{\Delta e}_n^{cr}$ reads

$$\boldsymbol{\Delta\varepsilon}_n^{cr}=\boldsymbol{N}_n\boldsymbol{\Delta e}_n^{cr} \tag{6}$$

with

$$\boldsymbol{N}_n=\begin{bmatrix}\cos^2\vartheta_n & -\sin\vartheta_n\cos\vartheta_n\\ \sin^2\vartheta_n & \sin\vartheta_n\cos\vartheta_n\\ 2\sin\vartheta_n\cos\vartheta_n & \cos^2\vartheta_n-\sin^2\vartheta_n\end{bmatrix} \tag{7}$$

where ϑ_n is the inclination angle of the normal of crack n with the x-axis.

Substitution of equation (6) in equation (3) gives for multiple cracks

$$\mathbf{\Delta\varepsilon}^{cr} = \mathbf{N}_1 \mathbf{\Delta e}_1^{cr} + \mathbf{N}_2 \mathbf{\Delta e}_2^{cr} + \ldots\ldots\ldots\ldots \tag{8}$$

For the derivation of the stress-strain law of the system of cracks and concrete, it is convenient to assemble all the crack strain increments which are expressed in their own local coordinate system in a vector $\mathbf{\Delta e}^{cr}$,

$$\mathbf{\Delta e}^{cr} = (\Delta e_1^{cr}\ \Delta\gamma_1^{cr}\ \Delta e_2^{cr}\ \Delta\gamma_2^{cr} \ldots\ldots\ldots\ldots)^T \tag{9}$$

When we also introduce the matrix $\mathbf{N}$,

$$\mathbf{N} = \begin{bmatrix}\mathbf{N}_1\ \mathbf{N}_2\ \ldots\ldots\ldots\ldots\end{bmatrix} \tag{10}$$

we observe that we can rewrite equation (8) as

$$\mathbf{\Delta\varepsilon}^{cr} = \mathbf{N}\,\mathbf{\Delta e}^{cr} \tag{11}$$

In a similar way, we can define a vector $\mathbf{\Delta s}_n$

$$\mathbf{\Delta s}_n = (\Delta s_1\ \Delta t_1)^T \tag{12}$$

with Δs_n the normal and Δt_n the shear stress increment in crack n of the integration point. The vector $\mathbf{\Delta s}$ which assembles all the stress increments with respect to their own local coordinate system then reads:

$$\mathbf{\Delta s} = (\Delta s_1\ \Delta t_1\ \Delta s_2\ \Delta t_2 \ldots\ldots\ldots\ldots\ldots\ldots)^T \tag{13}$$

The relation between the stress increment in the global coordinate system $\mathbf{\Delta\sigma}$ and the stress vector $\mathbf{\Delta s}$ can subsequently be derived to be

$$\mathbf{\Delta s} = \mathbf{N}^T \mathbf{\Delta\sigma} \tag{14}$$

To complete the system of equations, we need a constitutive model for the intact concrete and a stress-strain relation for the smeared cracks. For the concrete between the cracks we assume a relationship which has the following structure

$$\mathbf{\Delta\sigma} = \mathbf{D}^{co} (\mathbf{\Delta\varepsilon}^{co} - \mathbf{\Delta\varepsilon}^0) + \mathbf{\Sigma} \tag{15}$$

The matrix $\mathbf{D}^{co}$ contains the instantaneous moduli of the concrete and $\mathbf{\Sigma}$ is a vector function which represents relaxation effects. The autogeneous strain increment $\mathbf{\Delta\varepsilon}^0$ is the part of the concrete strain increment which is independent of the stress (e.g. thermal dilatation or shrinkage). The constitutive law (15) is quite general. It embraces most rate-independent plasticity theories including for instance Besseling's fraction model [14], but also rate-dependent models as visco-plasticity and visco-elasticity with degenerated kernels [29].

In a similar way, we can define a relation between the crack strain increment $\mathbf{\Delta e}_n^{cr}$ of crack n and the stress increment $\mathbf{\Delta s}_n$ in that crack. In this paper, we will assume a relation which formally reads:

$$\mathbf{\Delta s}_n = \mathbf{D}_n^{cr} \mathbf{\Delta e}_n^{cr} \tag{16}$$

with $\mathbf{D}_n^{cr}$ a 2*2 matrix. For the derivation of the stress-strain relation of the cracked concrete, it is again convenient to assemble all the matrices $\boldsymbol{D}_n^{cr}$ in one matrix $\boldsymbol{D}^{cr}$, which is defined as

$$\boldsymbol{D}^{cr} = \begin{bmatrix} \boldsymbol{D}_1^{cr} & \boldsymbol{0} & \cdots\cdots \\ \boldsymbol{0} & \boldsymbol{D}_2^{cr} & \cdots\cdots \\ \cdots\cdots & \cdots\cdots & \cdots\cdots \end{bmatrix} \tag{17}$$

so that the relation between $\boldsymbol{\Delta s}$ and $\boldsymbol{\Delta e}^{cr}$ reads

$$\boldsymbol{\Delta s} = \boldsymbol{D}^{cr} \boldsymbol{\Delta e}^{cr} \tag{18}$$

Using equations (1), (11), (14), (15) and (18) we can develop the stress-strain relation for the cracked concrete. To this end, we substitute (1), (11) and (15) in (14), yielding

$$\boldsymbol{\Delta s} = \boldsymbol{N}^T [\boldsymbol{D}^{co} (\boldsymbol{\Delta \varepsilon} - \boldsymbol{\Delta \varepsilon}^0 - \boldsymbol{N} \boldsymbol{\Delta e}^{cr}) + \boldsymbol{\Sigma}] \tag{19}$$

Substituting this relation in (18) and rearranging gives

$$\boldsymbol{\Delta e}^{cr} = [\boldsymbol{D}^{cr} + \boldsymbol{N}^T \boldsymbol{D}^{co} \boldsymbol{N}]^{-1} \boldsymbol{N}^T [\boldsymbol{D}^{co} (\boldsymbol{\Delta \varepsilon} - \boldsymbol{\Delta \varepsilon}^0) + \boldsymbol{\Sigma}] \tag{20}$$

If we substitute this expression in

$$\boldsymbol{\Delta \sigma} = \boldsymbol{D}^{co} (\boldsymbol{\Delta \varepsilon} - \boldsymbol{\Delta \varepsilon}^0 - \boldsymbol{N} \boldsymbol{\Delta e}^{cr}) + \boldsymbol{\Sigma} \tag{21}$$

we obtain the constitutive relation for the cracked concrete:

$$\boldsymbol{\Delta \sigma} = \left\{ \boldsymbol{I} - \boldsymbol{D}^{co} \boldsymbol{N} [\boldsymbol{D}^{cr} + \boldsymbol{N}^T \boldsymbol{D}^{co} \boldsymbol{N}]^{-1} \boldsymbol{N}^T \right\} \left[\boldsymbol{D}^{co} (\boldsymbol{\Delta \varepsilon} - \boldsymbol{\Delta \varepsilon}^0) + \boldsymbol{\Sigma} \right] \tag{22}$$

It is noted that as long as the constitutive matrices $\boldsymbol{D}^{co}$ and $\boldsymbol{D}^{cr}$ remain symmetric, symmetry is also preserved with regard to the constitutive relation (22) for the cracked concrete.

2.2. Crack stress-strain relation

In the previous sub-section, equation (16) which relates the crack strain increment $\boldsymbol{\Delta e}_n^{cr}$ and the stress increment $\boldsymbol{\Delta s}_n$, was introduced without specifying the matrix $\boldsymbol{D}_n^{cr}$. The structure of this equation is such that it permits incorporation of a relation between the shear crack strain increment and the normal stress increment and of a relation between the normal crack strain increment and the shear stress increment. Such coupling effects for instance occur in crack-dilatancy theories [7,53]. Most of our applications are however restricted to small crack strains and then the off-diagonal terms in $\boldsymbol{D}_n^{cr}$ are less important. Consequently, we have set the off-diagonal terms equal to zero in the sample problems which we will give, so that $\boldsymbol{D}_n^{cr}$ reduces to:

$$\boldsymbol{D}_n^{cr} = \begin{bmatrix} C & 0 \\ 0 & \beta^* G \end{bmatrix} \tag{23}$$

Herein, the tangent modulus C represents the relation between the normal crack strain increment and the normal stress increment (see Figure 1), G is the elastic shear modulus and β^* is a shear stiffness reduction factor.

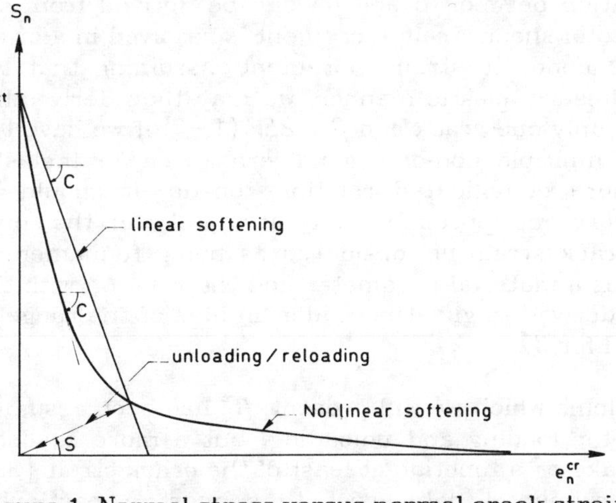

Figure 1. Normal stress versus normal crack strain.

The (softening) modulus C may be determined from a simple tensile test. Recent research however [9] indicates that a straightforward translation from experimental data in a value for C leads to results which are not objective with regard to mesh refinement. To overcome this problem, it has been proposed to consider the fracture energy G_f which is defined as the amount of energy needed to create one unit of area of a continuous crack [9,38] as the fundamental parameter which governs crack propagation. This so-called "tension-softening" model has also been adopted in this study, although it is beginning to emerge that the concept is not entirely free from deficiencies. This is particularly so when we allow for the possibility of multiple cracks. Suppose that a primary crack has been created with a softening modulus C determined from the fracture energy G_f. If upon formation of a secondary crack the same modulus is inserted in the relation for the second crack, the fracture energy will be consumed twice. Possible solutions might be to distribute the fracture energy over both cracks, or to introduce some kind of coupling between the cracks by making the stress increment in a crack a function not only of the strain increment of the same crack, but also of the other cracks. Neither of these solutions however is obvious as a number of problems adhere to both of them and consequently, they have not yet been pursued here.

Other problems with application of fracture energy concepts in smeared crack analyses relate to the application to axisymmetric problem where the integration of the strain over the crack band width entails complications owing to the $1/r$ term and to the shape of the softening branch which may influence the results significantly (see e.g. Section 5 or [58]).

The term $\beta^* G$ gives the relation between the shear stress increment in a crack and the shear crack strain increment and accounts for effects like aggregate interlock. The meaning of the reduction factor β^* differs from the classical shear retention factor β as introduced by Suidan and

Schnobrich [62]. A relation between β and β^* can be derived from the consideration that the total shear strain increment is resolved in a crack strain increment and a concrete strain increment. Assuming that the concrete behaves in a linearly elastic manner, we may then derive that $\beta^*=\beta/(1-\beta)$ if we have only one crack and $\beta^*=2\beta/(1-\beta)$ if we have two orthogonal cracks. For multiple non-orthogonal cracks or for inelastic concrete behaviour, more complicated relations ensue. It should be emphasized that in the present approach, the relation between the shear stress and the shear crack strain is considered as being fundamental. Hence, β^* is conceived as a material parameter and the relation with the more familiar β is only derived to give the reader an idea of the range of values which can be used for β^*.

In the sample problems which we will present, β^* has been assumed to be a constant both for loading and unloading, but a more realistic approach would be to make β^* a function at least of the crack strain [39]. Unfortunately, few experimental data exist to support a particular expression for β^*.

2.3. Closing and opening of cracks

In smeared crack analyses of concrete members we experience a lot of what may be referred to as spurious cracking. Here we mean that there are quite a number of sampling points which crack, but show only small crack strains. This partly causes the diffuse crack pattern of smeared crack analyses. In fact, we often observe that only a limited number of cracks really open and lead to failure (strain-localization). Yet, these sampling points with small crack strains pose a problem as a number of them show unloading, even close and sometimes open again in a later stage of the loading process.

It is therefore important to carefully handle closing and eventually reopening of cracks. For the unloading branch we have adopted a secant approach as is shown in Figure 1. So when a crack starts unloading, expression (23) is replaced by:

$$\boldsymbol{D}_n^{cr}=\begin{bmatrix} S & 0 \\ 0 & \beta^* G \end{bmatrix} \tag{24}$$

where S is the secant modulus of the unloading branch. This assumption is too simple as in reality we may expect some residual strain upon closing of a crack [54], but current experiences indicate that this procedure is numerically stable.

When a crack fully closes, i.e. when the normal stress in the crack changes from tension into compression, the stiffness of the uncracked concrete is inserted again, so that $\boldsymbol{\Delta\sigma}=\boldsymbol{D}^{co}\,\boldsymbol{\Delta\varepsilon}$. Although upon closing of a crack, the normal strain and the normal stress both vanish in the present conception, this need not be true for the shear stress and the shear strain. Consequently, these stresses and strains are considered as initial stresses and initial strains upon closing of the crack and the stress after closing formally follows from:

$$\boldsymbol{\sigma}=\boldsymbol{\sigma}_0+\int \boldsymbol{D}^{co}\,\dot{\boldsymbol{\varepsilon}}\,dt \tag{25a}$$

where $\boldsymbol{\sigma}_0$ is the stress state that exists when the crack closes and $\dot{\boldsymbol{\varepsilon}}$ is the strain-rate vector. If we have linear-elastic material behaviour in compression ($\boldsymbol{D}^{co}=\boldsymbol{D}^e$), equation (25a) reduces to:

$$\boldsymbol{\sigma}=\boldsymbol{\sigma}_0+\boldsymbol{D}^e(\boldsymbol{\varepsilon}-\boldsymbol{\varepsilon}_0) \tag{25b}$$

with $\boldsymbol{\varepsilon}_0$ the strain that exists in the sampling point upon closing of the crack. When a crack opens again the shear crack strain is initialized with the value which follows from $\boldsymbol{\varepsilon}_0^{cr}$, i.e. the crack strain which existed when the crack closed. During reloading equation (24) is used for the incremental relation between the crack stresses and the crack strains. If the normal strain in the crack exceeds the previously reached maximum strain, relation (24) is replaced by equation (23) for the softening branch.

2.4. Threshold angle for non-orthogonal cracks

With the crack model as outlined in the preceding sub-sections, it is possible to allow an arbitrary number of cracks in an integration point. Yet, allowing new cracks to form every time that the stresses rotate slightly and violate the tensile strength in the new principal direction, leads to excessive cracking and closing of existing cracks. Moreover, the existence of multiple cracks in an integration point results in a rapid decrease of the shear stiffness of the total system of concrete and cracks, which may lead to ill-conditioning of the stiffness matrix. For this reason, a threshold angle α has been introduced which allows new cracks to form only when the angle between the current direction of the major principal stress and the normal to the existing cracks has exceeded α.

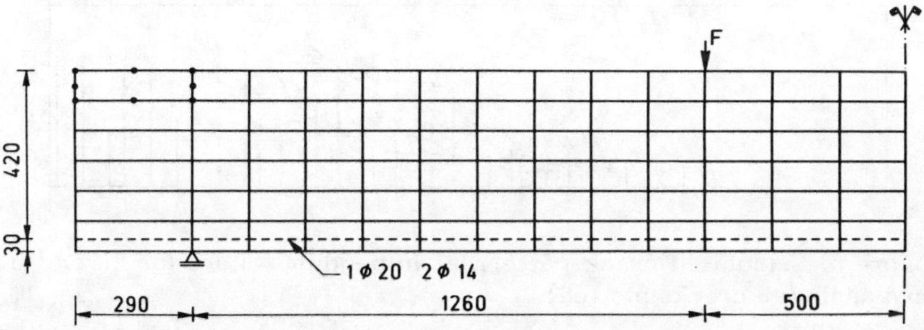

Figure 2. Finite element mesh for moderately deep beam.

The threshold angle α has a similar influence on the results as the shear reduction factor β^* as a low value for α also decreases the shear capacity of the system. For the particular case of two orthogonal cracks ($\alpha=90^o$), an expression between the shear stiffness of the system of concrete and cracks and the shear reduction factor β^* as used here can be derived analytically [20]. For the more general case of non-orthogonal cracks, such a derivation is very cumbersome and the exact way in which the values of α and β^* interact, is difficult to assess. We will therefore demonstrate the possible impact of the value of α on the computational results by means of a moderately deep shear-critical beam [67].

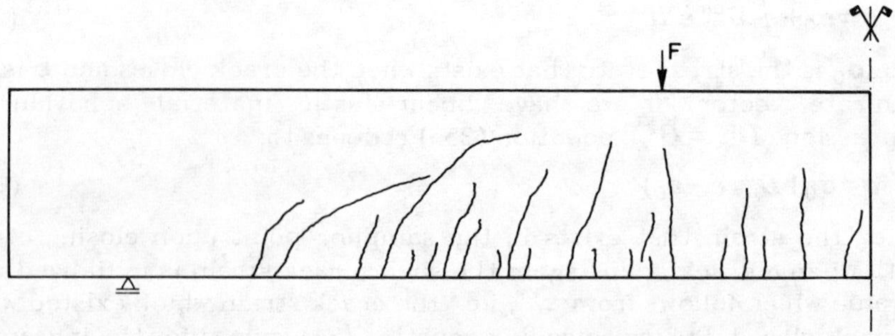

Figure 3. Experimentally recorded crack pattern at impending failure.

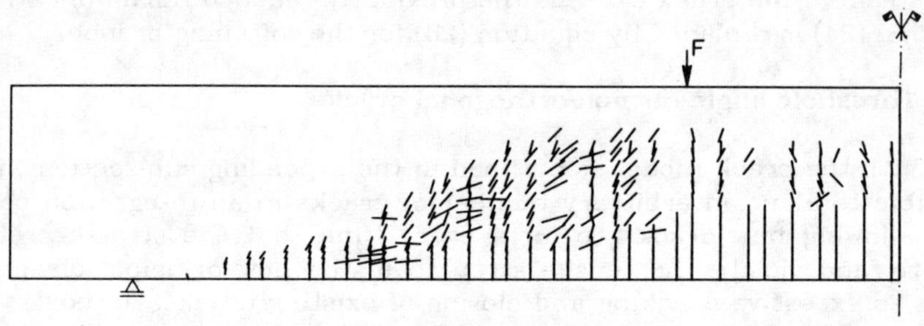

Figure 4. Calculated crack pattern at impending failure for high threshold angle (all cracks plotted).

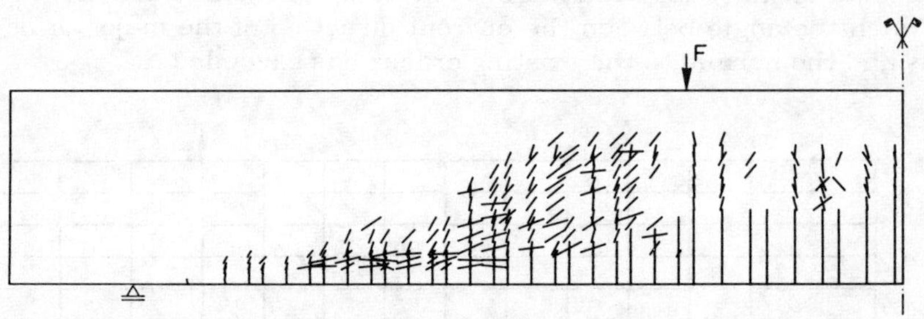

Figure 5. Calculated crack pattern at impending failure for lower threshold angle (all cracks plotted).

The beam which we will consider is shown in Figure 2 and has been analysed using the DIANA finite element program [18], in which the procedures described in the preceding have been implemented. The analysis was carried out using 8-noded plane stress elements with 9-point Gaussian integration to reduce the possibility of the occurrence of spurious zero-energy modes, an issue to which we will return in a subsequent section. The properties of the concrete were assumed to be: Young's modulus E_c=28000 N/mm^2, Poisson's ratio ν=0.2, tensile strength f_{ct}=2.5 N/mm^2, shear reduction factor β^*=0.087 and fracture energy G_f=0.06 N/mm. The beam has a thickness of 200 mm and has no shear reinforcement. The reinforcement at the bottom of the beam has a Young's modulus E_s=210000 N/mm^2 and a yield stress σ_y=440 N/mm^2.

Numerical analyses have been carried out for two different threshold angles, namely for $\alpha=60^o$ and $\alpha=30^o$. In both analyses we observe that vertical cracks due to bending arise first. On subsequent loading the stresses rotate and new, non-vertical crack form in the region between the point load and the support (see Figure 3). Now we observe a marked difference between the results for both threshold angles. The diagonal cracks in the analysis for the high threshold angle tend to run rather steep (Figure 4), whereas for the lower threshold angle ($\alpha=30^o$), the shear capacity at a cross-section is exhausted more quickly. The formation of cracks between the load and the support remains more or less near the reinforcement and the initiation of diagonal cracking is forced to move to the centre of the beam (Figure 5). A similar trend has been observed for another beam upon reduction of the shear reduction factor β^*, as in that case inclined cracks tended to propagate along the reinforcement for very low values of β^* [57].

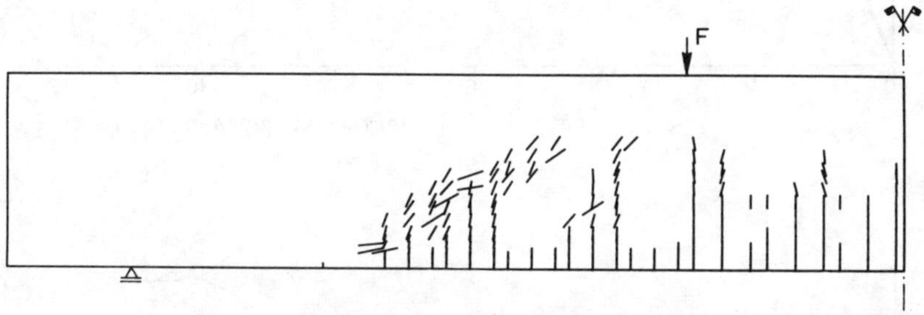

Figure 6. Cracks that transfer no normal stress at impending failure for high threshold angle.

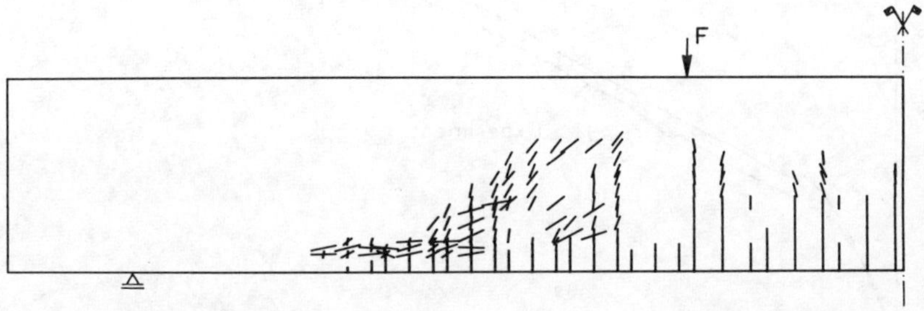

Figure 7. Cracks that transfer no normal stress at impending failure for lower threshold angle.

The difference between the results for both threshold angle becomes even more apparent when only those cracks are plotted which are out of the softening branch, which may be interpreted as micro-cracks which have coalesced into one macro-crack (Figures 6 and 7). These figures also serve the purpose of demonstrating that the somewhat diffuse crack pattern which is observed in most smeared crack analyses largely disappears when we plot only those cracks which really open. Indeed, the crack patterns of Figures 6 and 7 clearly show strain-localization including the initiation of a diagonal crack.

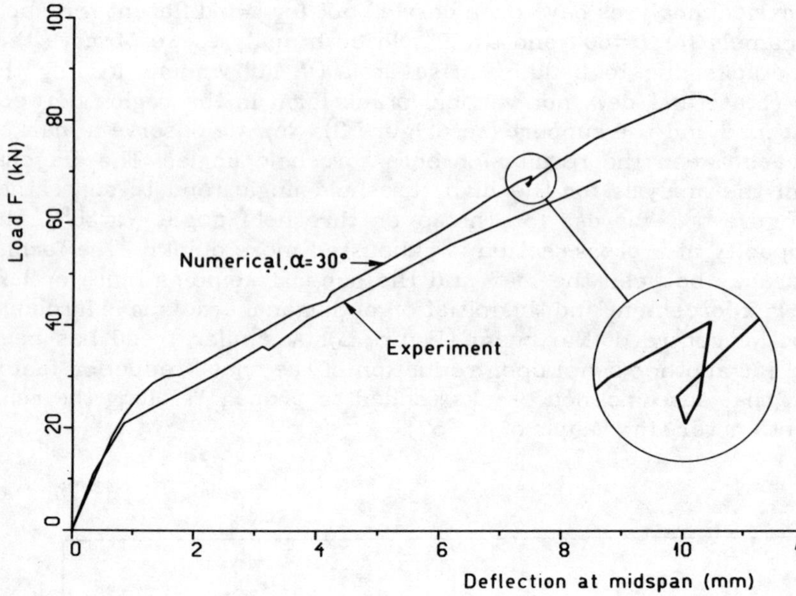

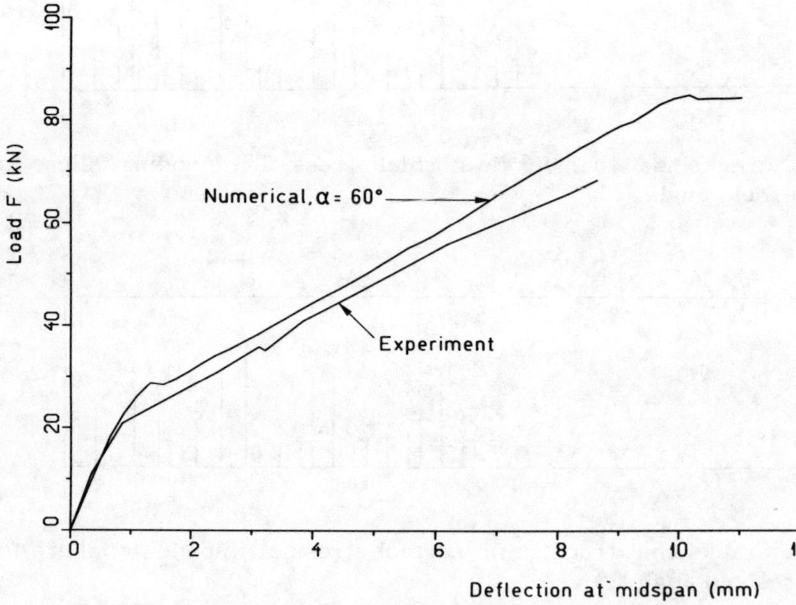

Figure 8. Load versus deflection at midspan for the beam problem (a) $\alpha = 30^o$, (b) $\alpha = 60^o$.

The impact of using a different threshold angle on the global load-displacement curve of Figures 8a and 8b appeared negligible until a displacement at midspan of 7.2 mm. Then, the analysis with α=30° (Figure 8a) diverged. No convergence could be achieved no matter how small the imposed displacement increments were taken. However, when we decremented displacements we were able to get a converged solution, although a number of new cracks arose. When no more physical changes were observed, we again incremented displacements and we could continue the solution up to and beyond the point where the steel started yielding.

It is debatable whether the "snap-back" behaviour of Figure 8a is physical or merely numerical. Spurious non-physical solutions have for instance been observed in fluid mechanics which appeared to vanish on mesh refinement [4]. This observation would support the conclusion that the observed behaviour is numerical rather than physical. Indeed, when we recalculated this problem with a mesh which was slightly refined near the location of the reinforcement [20], no numerical difficulties were met at this point. Yet, convergence was slow also in this case and small steps had to be taken. Although the behaviour is thought to be non-physical in the present case, snap-backs are possible when employing softening models as will be argued in Section 5 (see also [25,60]).

2.5. Relation with other smeared crack models

In the past, various other smeared crack models have been developed and definite relations can be elaborated with some of these models. The first smeared crack analysis was probably performed by Rashid [50]. In his approach the normal stress was set to zero immediately upon crack formation. Moreover, no shear resistance across a crack was permitted as the shear modulus G was also assumed to vanish upon crack formation. Later, it was recognized that such an approach grossly underestimates the stiffness of a structure after crack formation and a reduced shear stiffness was inserted in the stiffness matrix after crack formation [62]. Also, the sudden drop in normal tensile stress across a crack was replaced by a gradual softening branch. Initially, this softening branch was attributed to the contribution of the stiffness of the concrete between the cracks to the stiffness of the reinforcement, but later [9,38], it was also related to the softening of the concrete itself. Further, Bažant and Oh [9] proposed to start from a compliance approach instead of a stiffness approach, so that off-diagonal terms in the orthotropic stiffness matrix of the cracked concrete were retained. The present approach relates to their approach because for the case of only one crack and elastic concrete behaviour (to which their treatment was restricted), the present model coincides with their model [20].

The introduction of a reduced shear stiffness to represent aggregate interlock and possibly dowel action, and the gradual decrease of the normal tensile stress after crack formation introduced the problem that on subsequent loading, the direction of the current major principal stress may deviate from the normal to the crack and moreover, the current major principal stress may exceed the tensile strength. An approach which is often adopted is to allow a secondary crack to form only orthogonal to a primary crack [47]. Again this approach is recovered as a particular case of the model adopted here if the threshold angle is set equal to 90° [20]. The problem of rotation of principal stress axes after primary crack formation was also recognized by Cope et al. [22,23], who proposed to co-rotate the material axes with the rotation of principal stress axes, either without [22], or with a threshold value for the angle between the current direction of the major principal stress and the original crack direction after which the material axes are rotated [23]. Although this approach was criticized by Bažant because it violates the principle of material frame indifference [8,12,63], recent research has indicated that results obtained using such a 'rotating crack model' are often superior to results obtained using the traditional fixed crack models [44]. Other models which allow for the possibility of multiple crack formation to solve this problem were proposed in [31,40,42]. Nevertheless, these models have limited capabilities in describing the stress-strain relation within a crack, and à priori assume linear-elastic material behaviour of the concrete between the cracks. Owing to the rigorous decomposition of strain increments, such limitations do not have to be imposed in the approach outlined here. Moreover, the model complies with the principle of material frame indifference as will be shown in another publication.

3. CONCRETE SUBJECT TO LONG-TERM LOADING

In this section we shall outline how the crack model can be used in analyses of concrete structures under long-term loading. It is demonstrated that the combination of visco-elasticity and crack formation naturally fits the crack model. An example is included in which we employ a visco-elastic constitutive model with a degenerated kernel for the concrete and a tension-softening model for the cracks.

3.1. Cracking and visco-elasticity

Creep of concrete is often described using visco-elastic models. Use of visco-elastic models however implies that the constitutive law is formulated in a hereditary integral, which is not convenient in numerical programs, either finite element or finite difference. This is because such an approach requires storage of the entire strain history. Therefore, most analysts resort to developing the kernel of the hereditary integral in a series. Then, the inelastic creep process is entirely described by a finite number of so-called internal variables and we can suffice by storing the internal variables rather than the entire strain history [2,5]. As the former description is simply not suitable for numerical applications, we will follow the last-mentioned approach.

Leaving out shrinkage or thermal effects for sake of simplicity, a visco-elastic law with a degenerated kernel can be written in the following incremental form (see e.g. [29]):

$$\boldsymbol{\Delta\sigma} = \boldsymbol{D}^{co}\, \boldsymbol{\Delta\varepsilon}^{co} + \boldsymbol{\Sigma}_{t-\Delta t} \tag{26}$$

We observe that we only need data from time $t-\Delta t$ and from the time increment itself for the calculation of the stress increment from $t-\Delta t$ to t. Equation (26) is identical with equation (15) but for the fact that a term which accounts for shrinkage and thermal dilatation has now been omitted. Hence, a visco-elastic law with a degenerated kernel has precisely the structure which was assumed for the concrete in deriving the crack model in Section 2, so that a visco-elastic law can directly be inserted in equation (22).

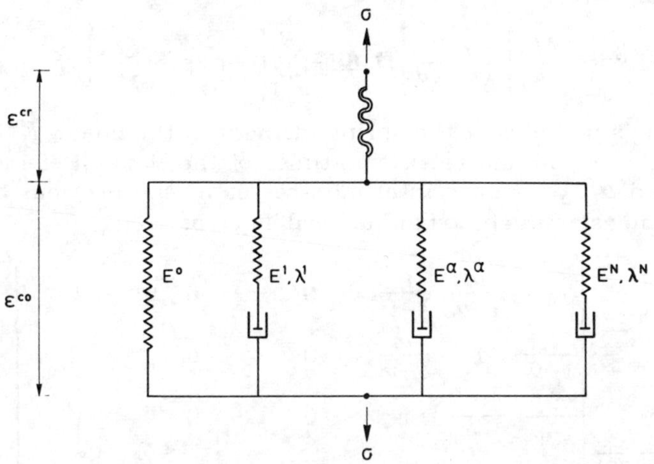

Figure 9. Maxwell chain and crack model linked in series.

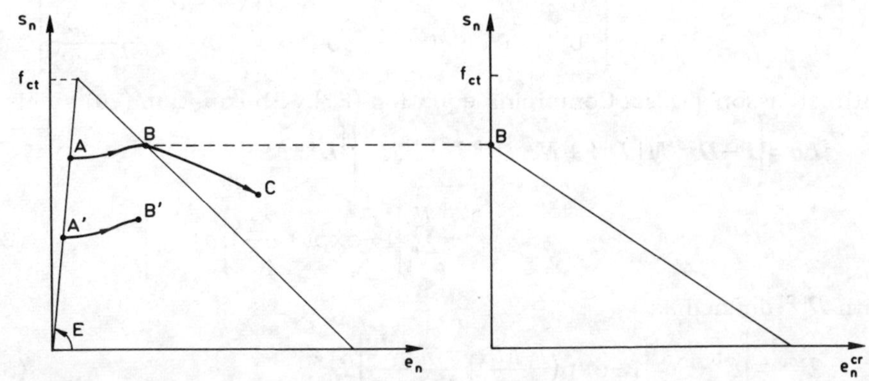

Figure 10. Envelope of stress-strain combinations which bound the uncracked state and stress-normal crack strain relation after crack initiation.

The above statement applies to any visco-elastic model which can be expressed in the form (26), but in this paper we shall limit ourselves to a so-called Maxwell-chain model (see Figure 9). Such a model arises if a relaxation type hereditary integral is developed in a Dirichlet series [5,16] and can mechanically be interpreted as a parallel arrangement of Maxwell chains. In the present conception, the Maxwell chain model is thought to apply only to the concrete and not to the cracks, so that the Maxwell elements all have the same concrete strain increment to which the crack strain increment must be added to obtain the total strain increment. Here again, we profit from the rigorous decomposition of the total strain increment into a concrete and into a crack strain increment, as we can independently specify the constitutive laws for the concrete and for the smeared-out cracks.

For a Maxwell chain model, equation (26) attains the following form [5,16]:

$$\boldsymbol{\Delta\sigma}=\left[E^0+\sum_{\alpha=1}^{N}\left[1-\exp(-\frac{\Delta t}{\lambda^\alpha})\right]\frac{E^\alpha}{\Delta t/\lambda^\alpha}\right]\boldsymbol{D}^*\boldsymbol{\Delta\varepsilon}-\sum_{\alpha=1}^{N}\left[1-\exp(-\frac{\Delta t}{\lambda^\alpha})\right]\boldsymbol{\sigma}^\alpha_{t-\Delta t} \quad (27)$$

where E^0 is Young's modulus of the spring element in the chain, E^α and λ^α are the Young's moduli and relaxation times of the Maxwell elements (see Figure 9) and $\boldsymbol{\sigma}^\alpha_{t-\Delta t}$ are the internal stresses at the previous time $t-\Delta t$. $\boldsymbol{D}^*$ is a dimensionless matrix which is defined as

$$\boldsymbol{D}^*=\frac{1-\nu}{(1+\nu)(1-2\nu)}\begin{bmatrix} 1 & \frac{\nu}{1-\nu} & \frac{\nu}{1-\nu} & 0 & 0 & 0 \\ \frac{\nu}{1-\nu} & 1 & \frac{\nu}{1-\nu} & 0 & 0 & 0 \\ \frac{\nu}{1-\nu} & \frac{\nu}{1-\nu} & 1 & 0 & 0 & 0 \\ 0 & 0 & 0 & \frac{1-2\nu}{2(1-2\nu)} & 0 & 0 \\ 0 & 0 & 0 & 0 & \frac{1-2\nu}{2(1-2\nu)} & 0 \\ 0 & 0 & 0 & 0 & 0 & \frac{1-2\nu}{2(1-2\nu)} \end{bmatrix} \quad (28)$$

with ν Poisson's ratio. Combining equation (27) with equation (20) yields

$$\boldsymbol{\Delta\sigma}=\left\{\boldsymbol{I}-\boldsymbol{D}^{co}\boldsymbol{N}[\boldsymbol{D}^{cr}+\boldsymbol{N}^T\boldsymbol{D}^{co}\boldsymbol{N}]^{-1}\boldsymbol{N}^T\right\}\left[\boldsymbol{D}^{co}\boldsymbol{\Delta\varepsilon}\right.$$

$$\left.-\sum_{\alpha=1}^{N}\left[1-\exp(-\frac{\Delta t}{\lambda^\alpha})\right]\boldsymbol{\sigma}^\alpha_{t-\Delta t}\right] \quad (29)$$

with $\boldsymbol{D}^{co}$ defined as

$$\boldsymbol{D}^{co}=\left[E^0+\sum_{\alpha=1}^{N}\left[1-\exp(-\frac{\Delta t}{\lambda^\alpha})\right]\frac{E^\alpha}{\Delta t/\lambda^\alpha}\right]\boldsymbol{D}^* \quad (30)$$

Maxwell-chain models are recognized to reasonably describe the creep behaviour of plain concrete, although there are some limitations. For instance, such models are only applicable if the stress level remains moderate, say 50 percent of the ultimate stress, and deviations also arise upon stress reversals. For higher stress levels, a combination of visco-elastic and visco-plastic models is perhaps more appropriate. Less is known about the time-dependent stress-strain behaviour of the smeared cracks. For this reason, we have simply neglected any time-dependent effects in the cracks in the example of the next sub-section and we have employed a time-independent tension-softening model as plotted in Figure 1 for the cracks.

While the stress-strain behaviour in the cracks has for the present been assumed to be time-independent, such an assumption appeared to rigorous for the criterion for inception of a crack. Consider for instance a pure creep problem where the stresses are prescribed. For a statically

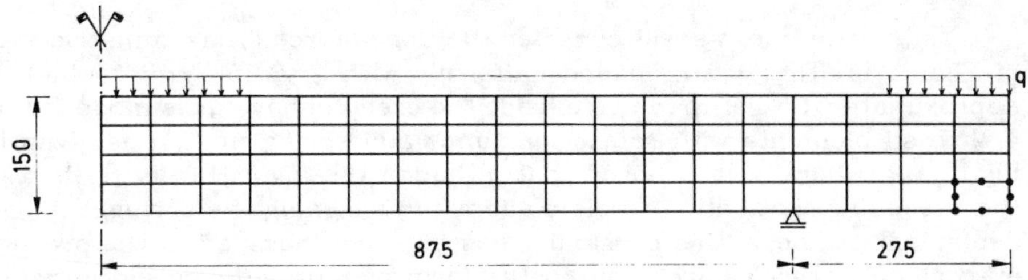

Figure 11. Finite element mesh of axisymmetric slab used in creep analysis.

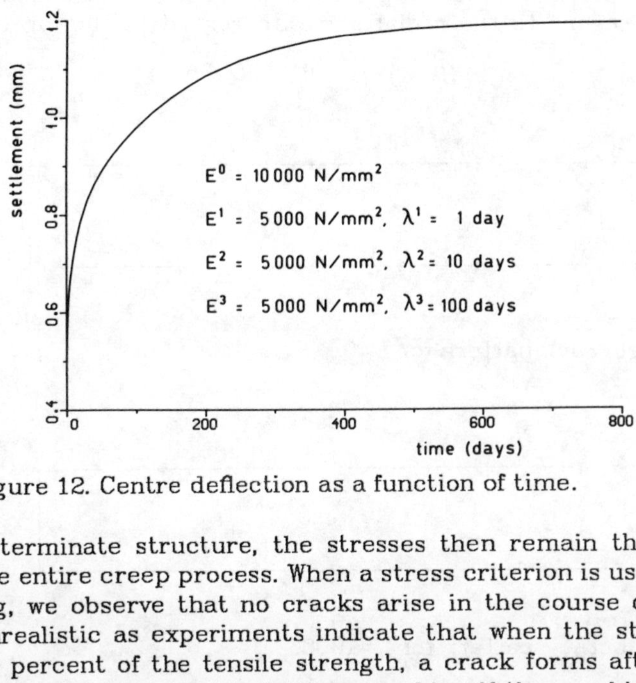

Figure 12. Centre deflection as a function of time.

determinate structure, the stresses then remain the same throughout the entire creep process. When a stress criterion is used to initiate cracking, we observe that no cracks arise in the course of time. This seems unrealistic as experiments indicate that when the stress exceeds about 80 percent of the tensile strength, a crack forms after some time. It is therefore suggested to initiate cracking if the combination of stress and strain in the major principal strain direction reaches the envelope of the left part of Figure 10. This diagram is obtained when we add the concrete strain obtained from a linear-elastic model to the crack strain normal to the crack as given the right part of Figure 10. Suppose that when a structure is loaded, point A of Figure 10 is reached. In the course of time the path AB is traversed and a crack is initiated when point B is reached. Thereafter, the stress-strain law in the crack is again considered to be fundamental and is given by the descending branch of the right part of Figure 10. It is stressed that although the descending branch of the left part of Figure 10 acts as an envelope for possible stress-strain combinations, this does not mean that the stress-total strain curve follows this branch after crack formation. For a visco-elastic model it for instance follows path BC. Figure 10 further shows that for lower stress levels, the crack criterion does not lead to crack formation. This is exemplified by the path A'-B'.

3.2. Creep analysis of an axisymmetric slab

As an example, we will consider the unreinforced axisymmetric slab of Figure 11. The slab is loaded uniformly with q=0.25 N/mm^2 which is approximately 65% of the failure load. The creep behaviour is modelled by 4 Maxwell elements with relaxation times and spring moduli as given in Figure 12, whereby it is noted that although the algorithm described in the preceding section in principle allows for assigning the springs an age-dependent stiffness, this possibility has not been pursued in the present example. The results are given in the form of a time-deflection curve of the centre of the slab (Figure 12) and of crack patterns for t=0 and t=1000 days (Figures 13 to 16). It is interesting to note that owing to the criterion for crack initiation, cracking gradually progresses in the course of time, both with regard to the radial and with regard to the circumferential cracks.

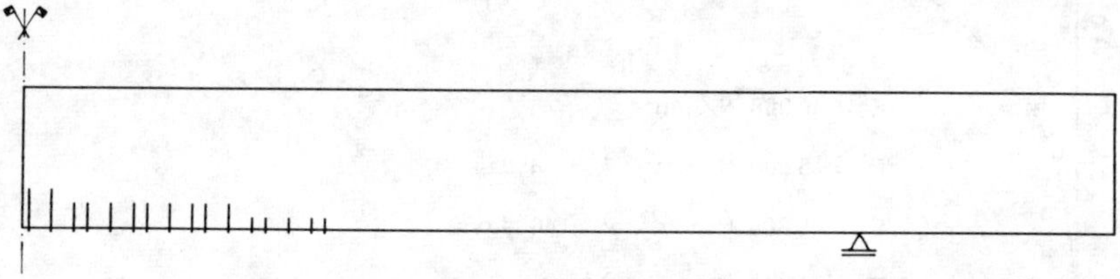

Figure 13. Tangential crack pattern for t=0.

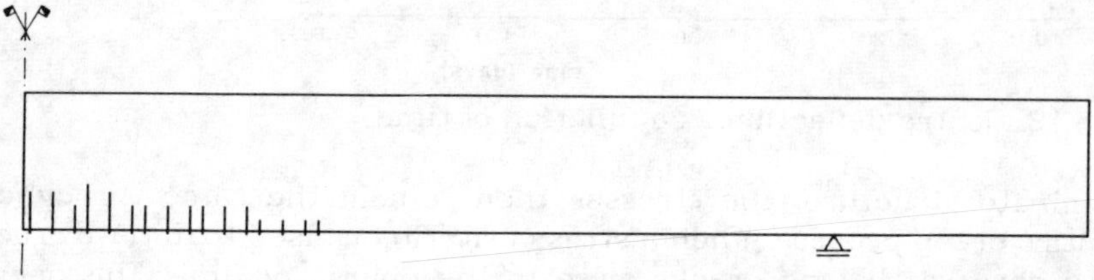

Figure 14. Tangential crack pattern for t=1000.

In the past, other analyses of concrete structures subjected to long-term have been presented, although most analysts employ Kelvin type visco-elastic models [1,2,51]. It seems that neither of them has used a softening model for the cracks, but that all previous analyses have been carried out under the assumption that all stiffness normal to a crack is lost upon crack formation. However, a softening crack stress-strain relation can only be incorporated if a decomposition of the strain increment is employed as starting point of the algorithm. If such a decomposition is not assumed, fracture and creep of the concrete between the cracks can not be distinguished properly. A similar algorithm has recently been proposed by Bažant and Chern [11], and indeed, if we restrict cracking to one crack, their algorithm coincides with the present approach.

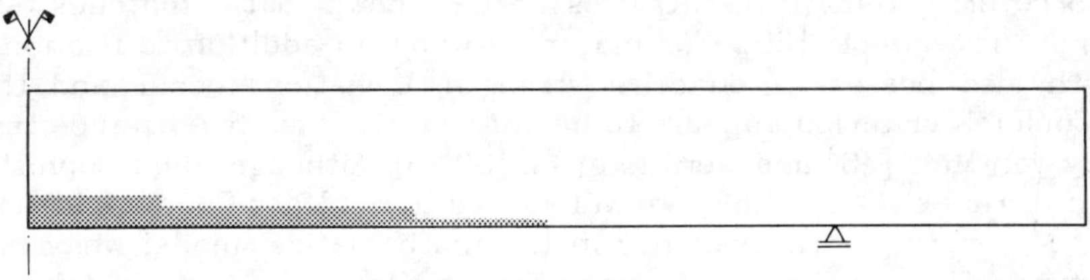

Figure 15. Area in which radial cracks have developed for $t=0$ (shaded).

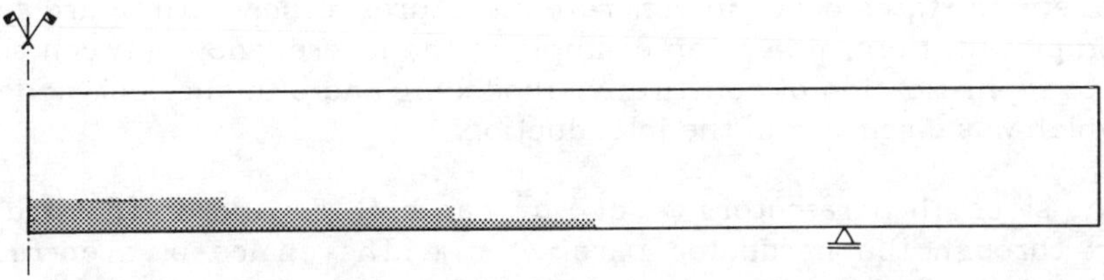

Figure 16. Area in which radial cracks have developed for $t=1000$ (shaded).

4. CONCRETE SUBJECT TO SHORT-TERM COMPRESSIVE LOADINGS

In this section we shall outline a plasticity model which is able to describe some of the basic characteristics which concrete displays in triaxial compression [66]. The treatment involves as salient characteristics hardening on the frictional and softening on the cohesive properties of concrete, while a non-associated flow rule is employed to control the inelastic volume changes. Thereafter, we will indicate how such a plasticity model can be combined with the crack model of the preceding section and a numerical example is included. The example also serves the purpose of demonstrating the danger of employing a low order numerical integration scheme.

4.1. A non-associated plasticity model for concrete

A first observation with regard to the strength properties of materials like concrete, rock and soils is that their strength depends on the first stress invariant. The oldest criterion which implies a dependence of the strength on the stress level, is the Mohr-Coulomb criterion. Let f be the yield function, σ_1 be the major and σ_3 be the minor principal stress (tension being taken as positive in accordance with the sign convention in continuum mechanics). Then, this criterion reads:

$$f=\tfrac{1}{2}(\sigma_1-\sigma_3)+\tfrac{1}{2}(\sigma_1+\sigma_3)\sin\varphi^*-c^* \tag{31}$$

The Mohr-Coulomb criterion has two strength parameters, namely the so-called mobilised friction angle φ^* and a mobilised cohesion c^* (which has a slightly different definition than is normally given to a cohesion as we observe from equation (31)). We have introduced the notation * to indicate that φ^* and c^* are not constants, but depend on the plastic strain history through a hardening parameter κ.

The Mohr-Coulomb criterion is simple and has a clear physical meaning. Moreover, extensive testing on sand has shown that it matches test data quite reasonably [35]. For materials which in addition to frictional strength also possess a cohesive strength, deviations occur and the Mohr-Coulomb criterion appears to be conservative, see for instance test data by van Mier [43] and Gerstle et al. [32,33]. Although more sophisticated criteria exist [21,41,69], we will adhere to the Mohr-Coulomb idea in order to preserve a transparency in the mathematical model which we will develop to describe the inelastic behaviour of concrete in triaxial compression. Another reason is that in actual computations in which triaxial stress states occur in concrete structures, other criteria are seldom employed. Here, we see an example of the discrepancy between the levels of sophistication of constitutive modelling and of numerical modelling, which was discussed in the introduction.

The strength parameters φ^* and c^* can both depend on the strain history through the hardening parameter κ. The choice of a correct definition for κ is difficult as its definition should be such that it attains the same value for all points on the yield surface f. Schreyer and Babcock [61] for instance use a linear combination of the three inelastic strain invariants weighted by the stress level for κ and they found that for a low strength concrete, only the first and the third inelastic strain invariants are of importance. Vermeer [65] on the other hand only uses the second inelastic strain invariant in describing another granular material, namely sand. It seems that a proper choice for κ is still an open question which has to be answered by considering a great deal of experimental data.

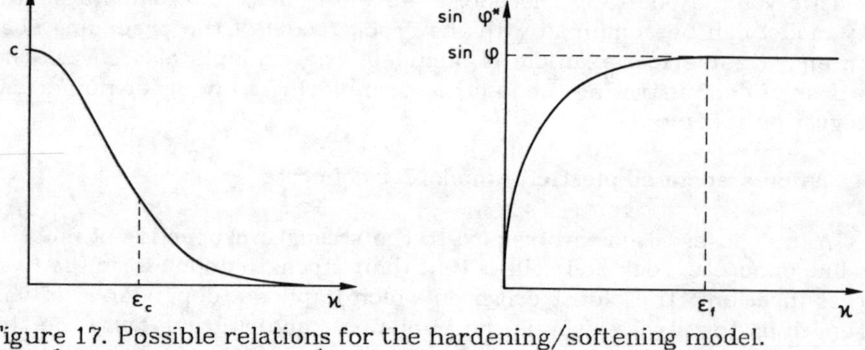

Figure 17. Possible relations for the hardening/softening model.
(a) $c^*-\kappa$ relation (b) $\sin\varphi^*-\kappa$ relation.

In spite of the paucity of experimental data to support a definition of κ, we are able to say something about the dependence of $\sin\varphi^*$ and c^* on κ, for $\sin\varphi^*$ should generally be an ascending function of κ, while c^* may expected to be a descending function of κ. The rationale behind the latter assumption is that during loading of a specimen of originally intact concrete, micro-cracks gradually develop. So the cementation gradually vanishes at continued loading thus leading to a degradation of the cohesive strength. A possible choice which expresses such a softening on the cohesive strength mathematically, is:

$$c^{*}=c\exp\left[-\left(\frac{\kappa}{\varepsilon_c}\right)^2\right] \tag{32}$$

with c the cohesion of the intact material. The meaning of ε_c is shown in Figure 17a, in which relation (32) has been plotted. It is noted that the softening on the cohesion as formulated here implies that softening is conceived as a material property. Lately, there has been much debate whether softening is indeed a material property or whether it is a structural property (e.g. [52]) and we will return to this issue in the next section.

At continued loading, concrete becomes more and more cracked and the initially cemented material very much begins to resemble a particulate material with only frictional resistance. The frictional resistance gradually increases in the loading process as the asperities protruding from the faces of the micro-cracks offer more and more resistance to subsequent sliding. This phenomenon may be modelled by applying hardening on the friction angle in addition to the softening on the cohesion. In [66], it has been suggested to use the relationship (see also Figure 17b):

$$\begin{cases} \sin\varphi^{*}=2\dfrac{\sqrt{\kappa\varepsilon_f}}{\kappa+\varepsilon_f}\sin\varphi & \kappa\leq\varepsilon_f \\ \sin\varphi^{*}=\sin\varphi & \kappa>\varepsilon_f \end{cases} \tag{33}$$

for the $\sin\varphi^{*}-\kappa$ relation. In it, ε_f is the value which κ attains if the frictional strength has been mobilised fully ($\varphi^{*}=\varphi$).

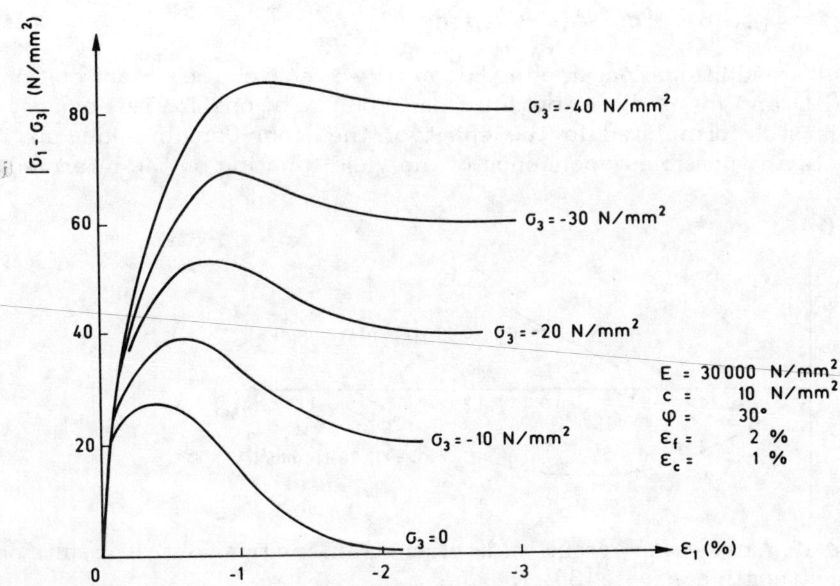

Figure 18. Computed response of concrete in triaxial tests.

Let us now explore the implications of the combination of friction hardening and cohesive softening. This is most easily done by simulating a triaxial test for different levels of confining pressure, the results of which have been plotted in Figure 18. Two important observations can be made from these results. First, the model predicts an increased ductility with higher confining pressure, and secondly, the amount of post-peak softening gradually vanishes at higher load levels. At first sight, especially the latter result seems peculiar, because the cohesion is assumed to vanish entirely for all stress levels. It is however not correct to identify the entire degradation of the cohesion at all stress levels with a vertical shift of the limit surface [72]. Indeed, the softening on the cohesion corresponds to a vertical shift, but the hardening on the friction angle corresponds to a rotation of the yield surface. These effects interact and cause the amount of softening to vanish with increasing stress level, so that for high stress levels we end up with a ductile, non-softening material.

An inherent drawback in adopting a Mohr-Coulomb yield function is that it possesses straight meridians. Hence, the ultimate yield surface (that is the yield surface which is reached when all cohesion has vanished and when the frictional resistance has been mobilised fully), also has straight meridians. For high stress levels, it was argued that the difference between the failure surface (which gives the peak stress states) and the ultimate yield surface vanishes asymptotically, but for low stress levels the ultimate yield surface is clearly inside the failure surface. As the ultimate yield surface is straight, these observations imply that the failure surface is not convex. A possible solution is to slightly modify the yield function, as follows:

$$f = \tfrac{1}{2}(\sigma_1 - \sigma_3) - [c^* - \tfrac{1}{2}(\sigma_1 + \sigma_3)\sin\varphi^*]^m \qquad (34)$$

with m an additional parameter. For $m=1$ we recover the linear relationship (31) and for $m=½$ we obtain a parabolic relationship. Yet, equation (34) is still formulated in the spirit of the Mohr-Coulomb idea as it neglects any possible dependence of the yield function on the intermediate stress.

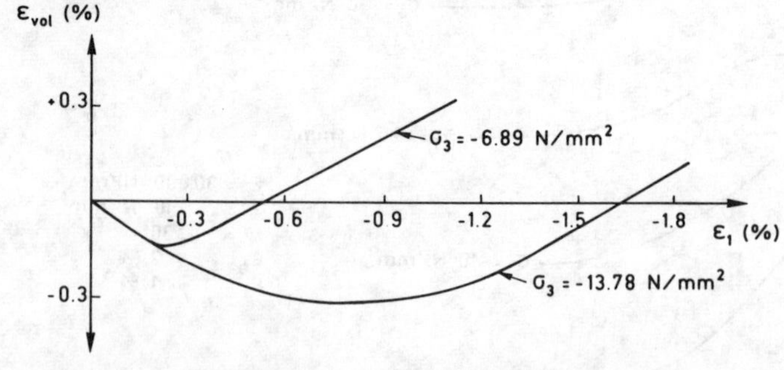

Figure 19. Axial strain versus volumetric strain for triaxial test results on a high-strength concrete [36].

Another salient characteristic of granular materials including concrete is the observed shear-enhanced inelastic volume changes. When a concrete specimen is sheared, we not only measure a shear strain, but also a volumetric strain. In the early stages of the loading process, this volumetric strain appears to be negative (so that we have contraction), but at continued loading the volumetric strain increment becomes positive (see Figure 19). This phenomenon is not captured within classical plasticity theory, as the classical associated flow rule which is based on normality of plastic strain increment and yield surface, predicts positive volumetric strain increments from the onset of inelastic behaviour. Moreover, the predicted rate of volume production greatly exceeds the experimental data, as application of an associated flow rule

$$\Delta \boldsymbol{\varepsilon}^p = \lambda \frac{\partial f}{\partial \boldsymbol{\sigma}} \tag{35}$$

with λ a non-negative multiplier, gives in conjunction with a Mohr-Coulomb yield function the relation

$$\Delta\varepsilon_v^p = \frac{-2\sin\varphi^*}{1-\sin\varphi^*}\Delta\varepsilon_1^p \tag{36}$$

between the inelastic volumetric strain increment and the inelastic axial strain increment. For realistic values of φ^*, however, this leads to a large deviation from the experimental curves of Figure 19. Therefore, φ^* is replaced by the so-called mobilised dilatancy angle ψ^*. Rewriting equation (36) with ψ^* taking the place of φ^*, gives

$$\sin\psi^* = \frac{\Delta\varepsilon_v^p}{-2\Delta\varepsilon_1^p + \Delta\varepsilon_v^p} \tag{37}$$

which may be regarded as the definition for the mobilised dilatancy angle ψ^* and which is valid for general triaxial stress states [66]. Mathematically, this phenomenon can be incorporated within plasticity theory via a non-associated flow rule. Then we must define an independent function g

$$g = \tfrac{1}{2}(\sigma_1-\sigma_3) + \tfrac{1}{2}(\sigma_1+\sigma_3)\sin\psi^* - c^* \tag{38}$$

which differs from the yield function f as the mobilised friction angle φ^* is replaced by the mobilised dilatancy angle ψ^*. Replacing the yield function f in equation (35) by the plastic potential function g, we precisely obtain expression (37) for $\sin\psi^*$. It is noted that for $\varphi^*=\psi^*$, we have $f=g$ and the classical associated flow rule is recovered.

From Figure 19, we observe that $\Delta\varepsilon_v^p/\Delta\varepsilon_1^p$ is initially positive (plastic contraction) and then becomes more and more negative (dilatancy) until a limiting value is attained at and beyond peak stress level. With definition (37), this means that the mobilised dilatancy angle is negative at the onset of loading and becomes more and more positive at continued loading and ultimately results in a limiting value $\sin\psi$. This gradual increase of $\sin\psi^*$ may be modelled by making $\Delta\varepsilon_v^p/\Delta\varepsilon_1^p$ a function of some stress measure. An example thereof is Rowe's [59] stress-dilatancy theory which states that:

$$1-\Delta\varepsilon_v^p/\Delta\varepsilon_1^p = \left\{\frac{\tan(45^o+\varphi^*/2)}{\tan(45^o+\varphi_{cv}/2)}\right\}^2 \tag{39}$$

with φ_{cv} a constant, which is usually referred to as the 'friction angle of constant volume'. This name is logical as the above equation gives $\Delta\varepsilon_v^p = 0$ for $\varphi^* = \varphi_{cv}$. As a matter of fact, the equation predicts negative dilatation for small friction angles ($\varphi^* < \varphi_{cv}$) and positive dilatation for larger values ($\varphi^* > \varphi_{cv}$). Hence, φ_{cv} marks a turning point at which plastic contraction stops and expansion starts. Rowe's stress-dilatancy equation (39) has originally been proposed to describe the dilatant behaviour of soils and it has been proved to be accurate for sand [65], but it has not yet been validated for concrete. It may be expected however, that this or other similar relationships are also capable of representing the observed shear-enhanced inelastic volume changes in concrete.

It is convenient to use equation (39) in a somewhat different form. Eliminating $\Delta\varepsilon_v^p / \Delta\varepsilon_1^p$ from (37) and (39), we obtain the more suitable form

$$\sin\psi^* = \frac{\sin\varphi^* - \sin\varphi_{cv}}{1 - \sin\varphi^* \sin\varphi_{cv}} \tag{40}$$

This is the desired relationship between the mobilised dilatancy angle and the effective strain as φ^* is a function of κ. The constant φ_{cv} is readily calculated from the limit dilatancy angle ψ and the limit friction angle φ. Substituting $\psi^* = \psi$ and $\varphi^* = \varphi$ and rearranging equation (40) we obtain

$$\sin\psi_{cv} = \frac{\sin\varphi - \sin\psi}{1 - \sin\varphi \sin\psi} \tag{41}$$

For the concrete of Figure 19 [36] we have $\varphi = 35^o$, $\psi = 12.6^o$ and consequently $\varphi_{cv} = 24^o$.

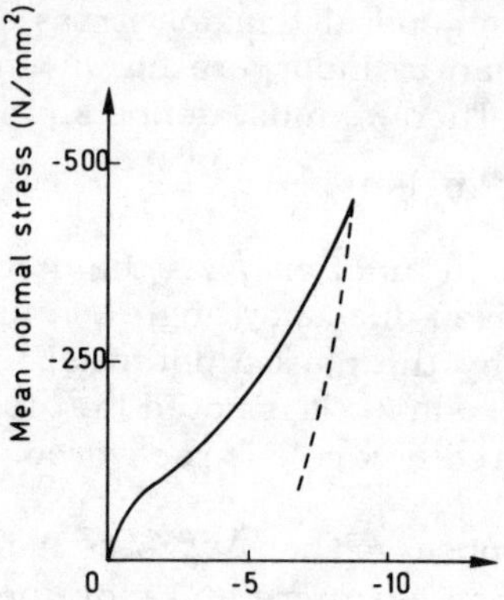

Figure 20. Hydrostatic loading of concrete specimen [36].

The above formulated constitutive model is not able to represent all characteristics of concrete. An example thereof is the inelastic behaviour of concrete in pure hydrostatic loading (see Figure 20). This behaviour may be explained from the consideration that above some threshold stress level, the pores collapse and a softer response to subsequent load increments is obtained. At a later stage in the loading process, a stiffening is again observed. However, the Mohr-Coulomb yield function is open in the hydrostatic direction, so that the response of a concrete specimen will be purely elastic when it is loaded hydrostatically. In order to capture plastic volume changes under hydrostatic stressing, we need a yield cap that closes the Mohr-Coulomb surface (e.g.[65]).

4.2. Cracking and plasticity

For elasto-plasticity, the general stress-strain law for the concrete (15) reduces to:

$$\Delta\boldsymbol{\sigma}=\boldsymbol{D}^{ep}\,\Delta\boldsymbol{\varepsilon} \tag{42}$$

with $\boldsymbol{D}^{ep}$ the elasto-plastic stiffness matrix, defined as

$$\boldsymbol{D}^{ep}=\boldsymbol{D}^{e}-\frac{\boldsymbol{D}^{e}\dfrac{\partial g}{\partial \boldsymbol{\sigma}}\dfrac{\partial f}{\partial \boldsymbol{\sigma}}^{T}\boldsymbol{D}^{e}}{h+\dfrac{\partial f}{\partial \boldsymbol{\sigma}}^{T}\boldsymbol{D}^{e}\dfrac{\partial g}{\partial \boldsymbol{\sigma}}} \tag{43}$$

where h represents the hardening and $\boldsymbol{D}^{e}$ is the elasticity matrix. For the elasto-plastic stress-strain law of equations (42) and (43), equation (22) for the cracked concrete reduces to

$$\Delta\boldsymbol{\sigma}=\left\{\boldsymbol{D}^{ep}-\boldsymbol{D}^{ep}\boldsymbol{N}\left[\boldsymbol{D}^{cr}+\boldsymbol{N}^{T}\boldsymbol{D}^{ep}\boldsymbol{N}\right]^{-1}\boldsymbol{N}^{T}\boldsymbol{D}^{ep}\right\}\Delta\boldsymbol{\varepsilon} \tag{44}$$

As argued, the decomposition of the total strain increment into a concrete and into a crack strain increment allows for the development of an algorithm that yields a stress state which not only complies with the yield function, but also agrees with the fracture function. Yet, straightforward application of equation (44) to integration points which are not only cracked but also show plastic behaviour, generally results in a stress state lying outside of the yield surface. This is because equation (44) is in fact a linearisation around the current stress state, and strictly speaking this equation is only exact for infinitesimal changes but not for finite strain increments. The problem of drifting from the yield surface exists whether we have crack formation or not. A number of methods have been devised that aim at minimizing or eliminating this drifting tendency. A possibility, which has been shown to be accurate for frictional materials like concrete and soils is to apply a correction $\boldsymbol{\delta\sigma}$

$$\boldsymbol{\delta\sigma}=-\frac{f(\boldsymbol{\sigma})}{h+\dfrac{\partial f}{\partial \boldsymbol{\sigma}}^{T}\boldsymbol{D}^{e}\dfrac{\partial g}{\partial \boldsymbol{\sigma}}} \tag{45}$$

to the stress point [17,64]. The stress state after correction however does not necessarily comply with the cracking criterion. Presently, a sophisticated algorithm is being developed which properly solves this problem, but in the calculations presented in the next sub-section, we have applied the following procedure.

Upon violation of the tension cut-off criterion, we first determine the intersection of the stress path with the tension cut-off. The strain increment from the initial state up to the tension cut-off is considered as elastic and the remaining part of the strain increment is treated as elastic-fracturing. A first estimate of the stress increment for the latter part is made using equation (44). If we have no plasticity the elasticity matrix $\boldsymbol{D}^e$ is substituted for $\boldsymbol{D}^{co}$ and we may proceed to the next integration point. If we have encountered plasticity during the previous loading step so that the stress in the beginning of the present step $\boldsymbol{\sigma}^0$ is on the yield locus, the elasto-plastic matrix $\boldsymbol{D}^{ep}$ is substituted for $\boldsymbol{D}^{co}$. The stress state which results from application of equations (43) and (44) is on the softening branch of Figure 1, but a drift from the yield surface may have occurred. We therefore recompute the stress increment using:

$$\Delta\boldsymbol{\sigma}=\boldsymbol{D}^e(\Delta\boldsymbol{\varepsilon}-\Delta\boldsymbol{\varepsilon}^{cr})-\frac{f(\boldsymbol{\sigma})}{h+\frac{\partial f}{\partial\boldsymbol{\sigma}}^T\boldsymbol{D}^e\frac{\partial g}{\partial\boldsymbol{\sigma}}}\boldsymbol{D}^e\frac{\partial g}{\partial\boldsymbol{\sigma}} \tag{46}$$

with

$$\boldsymbol{\sigma}^t=\boldsymbol{\sigma}^0+\boldsymbol{D}^e(\Delta\boldsymbol{\varepsilon}-\Delta\boldsymbol{\varepsilon}^{cr}) \tag{47}$$

The resulting stress state $\boldsymbol{\sigma}^1=\boldsymbol{\sigma}^0+\Delta\boldsymbol{\sigma}$ can be shown to be on the yield surface for the Mohr-Coulomb yield locus, and also for the Drucker- Prager criterion provided that this yield function is written in a particulate form [17]. The crack strain increment follows from the initial estimate using equation (44) and remains unaltered during this corrector step. The stress increment in the crack is changed by the application of equation (46) and consequently, the crack strain increments $\Delta\boldsymbol{e}^{cr}$ and the stress increments $\Delta\boldsymbol{s}$ are no longer related by the crack relations (17), (18) and (23). Conceiving the computed plastic strain increment $\Delta\boldsymbol{e}^p$ as an initial strain increment, we may obtain an improved estimate for the stress increment using:

$$\Delta\boldsymbol{\sigma}=[\boldsymbol{D}^e-\boldsymbol{D}^e\boldsymbol{N}(\boldsymbol{D}^{cr}+\boldsymbol{N}^T\boldsymbol{D}^e\boldsymbol{N})^{-1}\boldsymbol{N}^T\boldsymbol{D}^e](\Delta\boldsymbol{\varepsilon}-\Delta\boldsymbol{\varepsilon}^p) \tag{48}$$

which gives stress increments that are related to the crack strain increment via equations (17), (18) and (23), but the total stress may now again violate the yield function.

After an initial estimate using equation (44) we therefore apply an inner iteration loop using equations (46) to (48). Current experiences indicate that this process always converges to a stress state which complies with the yield function and which agrees with the crack relation, although convergence may be rather slow.

4.3. Axisymmetric slab failing by punching shear

The example that we consider is the axisymmetric slab of Figures 21 and 22. The properties of the concrete are assumed to be: Young's modulus E_c=28000 N/mm^2, Poisson's ratio ν=0.2, tensile strength f_{ct}=2.6 N/mm^2, shear reduction factor β^*=0.25 and fracture energy G_f=0.06 N/mm. A simplified linear elastic-perfectly plastic version of the constitutive model was employed with a cohesion c=9.6 N/mm^2 and a friction angle φ=30°. The slab is reinforced isotropically with a reinforcement ratio of 1% and the properties of the reinforcement are E_s=205000 N/mm^2 and σ_y=465 N/mm^2. The experimental failure mechanism of the slab is ultimately due to punching shear [68].

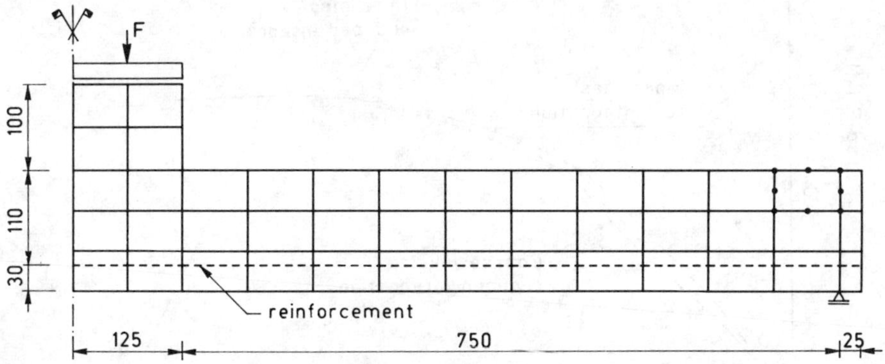

Figure 21. Coarse mesh for axisymmetric slab.

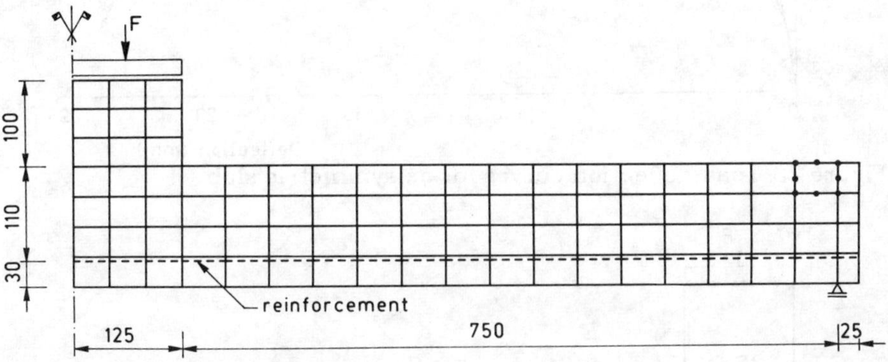

Figure 22. Refined mesh for axisymmetric slab.

The first analysis was carried out for the coarse mesh of Figure 21 and incrementing the displacement up to a deflection of 9.6 mm no convergence appeared possible although very small displacement increments were imposed (see Figure 23). However, when the analysis was repeated with exactly the same mesh and the same material parameters, but with 'full' 9-point integration, displacements could be imposed until and beyond a plateau in the load-displacement curve. The same trend was observed for computations with the refined mesh of Figure 22, i.e. the analysis with 'full' integration could be continued after the plateau had been reached and the analysis with reduced integration diverged at some displacement level (Figure 23). Yet, the latter calculation with 'reduced' integration could be continued much further than the computation with 'reduced' integration for the coarser mesh. This would again indicate that spurious snap-backs and divergence gradually disappear with mesh refinement.

The poor behaviour of reduced integration in conjunction with crack formation has been explained to be caused by the introduction of spurious zero-energy modes upon the formation of cracks in 'reduced' integrated elements [30]. With the formation of a new crack, an extra spurious zero-energy mode is introduced, so that we have four additional zero-energy modes when all four integration points are cracked.

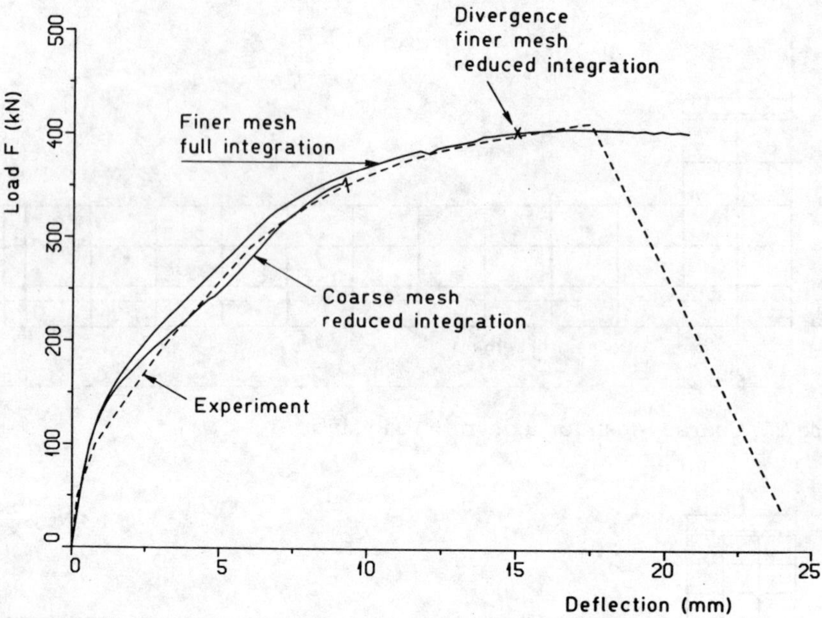

Figure 23. Load-deflection curves for axisymmetric slab.

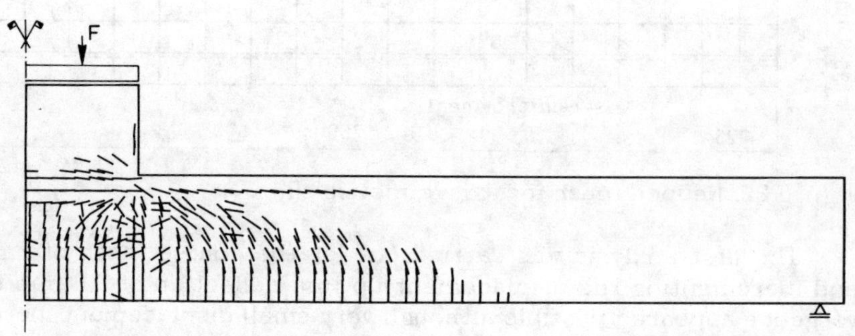

Figure 24. Tangential crack pattern at peak load.

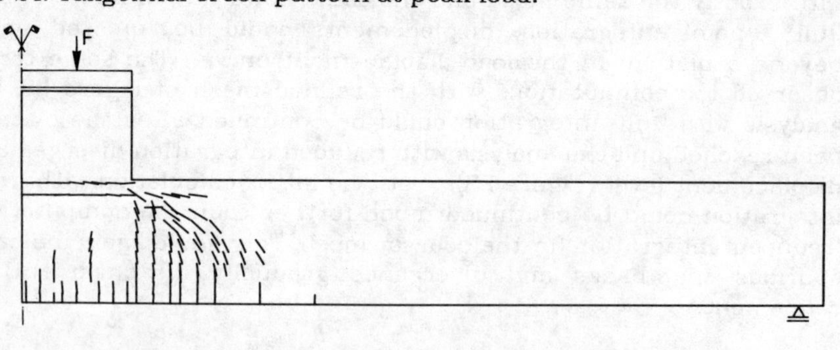

Figure 25. Major cracks (no normal stress transfer) at peak load.

Observations about incorrect predictions of structural behaviour when using reduced integration have also been reported by Crisfield [26,27].

The load-deflection curve of Figure 23 ultimately reaches a plateau. Usually such a plateau is identified with yielding of the reinforcement. In this example this was not the case as yielding started only after a deflection of approximately 20 mm, thus after that a considerable part of the plateau had been traversed. Probably, the plateau is due to yielding of the concrete in the compression zone (see Figure 26). If we had employed the hardening/softening model for the concrete instead of a perfectly-plastic model, we would presumably have obtained a softening behaviour instead of a plateau. Preliminary calculations with the hardening/softening model proved to be rather unstable, and indicated that zero-energy modes may then occur even with 9-point Gaussian integration.

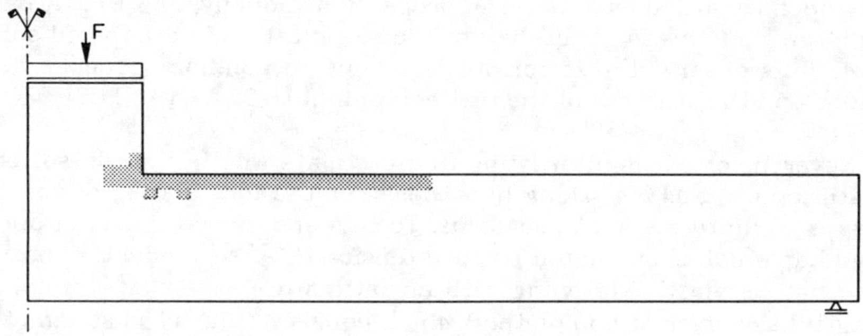

Figure 26. Plastic region at peak load level (shaded).

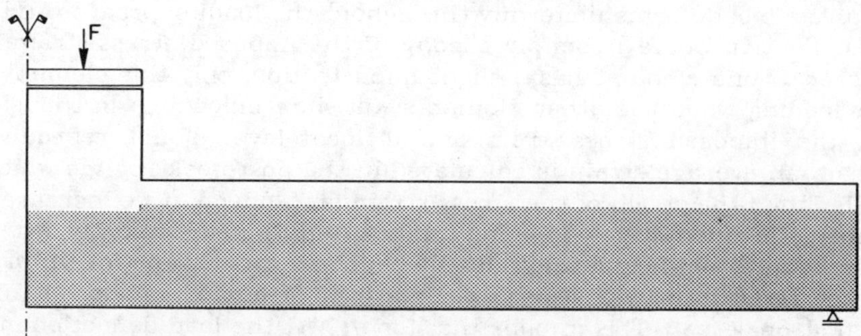

Figure 27. Region in which radial cracks have developed (shaded).

The tangential crack pattern of Figure 24 calculated at peak load is rather shallow. This is in accordance with the experimental observations which also show that tangential cracking spreads over a rather large area [68]. The observation that radial cracks arise almost throughout the slab already in an early stage of the loading process is also in agreement with experimental findings (Figure 27). When we plot only the so-called major cracks which carry no more tensile stress, we again observe a much more localized pattern which reveals the initiation of a diagonal shear crack (see Figure 25).

5. CALCULATION OF LIMIT POINTS AND POST-PEAK BEHAVIOUR

In this section we will argue that use of a local strain-softening constitutive model may lead to 'snap-back' behaviour on structural level. A numerical example is included to demonstrate that numerical methods are in principle capable of tracing such equilibrium paths.

5.1. Some consequences of strain-softening

In the preceding section, we touched on the question whether the softening which is observed experimentally is a material property or whether it is a structural effect. In the past, strain-softening has generally been regarded as a material property. Gradually, however, it begins to emerge that this is generally not true as most and perhaps all softening which is observed in experiments, is due to non-homogeneous deformations and the stiffness of the test equipment [6,33,37,43,52,66].

Nevertheless, it is interesting to investigate whether strain-softening can still be used as a working hypothesis for the formulation of constitutive laws in finite element programs. To this end, we consider an unreinforced bar which is subjected to pure tension [25]. We model the material of the bar as elastic-softening with an ultimate strain ε_u (at which the strength has vanished completely) which equals n times the strain at the tensile strength. The bar is modelled with m elements. If we have a perfect bar, so that all elements have exactly the same tensile strength and so on, the bar deforms uniformly throughout the loading process and the load-deflection curve is simply a copy of the imposed stress-strain law. However, if one element has a slight imperfection, only this element will show loading while the other elements will show unloading. In this situation, the imposed stress-strain law at local level is not reproduced. Instead, an average strain is calculated in the post-peak regime which is smaller than the strain of the stress-strain law. Indeed, if we increase the number of elements while keeping the length of the bar fixed, the average strain at a certain load level gradually becomes smaller and for $m > n$ the average strain in the post-peak regime even becomes smaller than the strain at peak load. This implies that for $m > n$, the load-deflection curve shows a 'snap-back' [25].

The above result implies that computational results for materials with a local softening constitutive law are not objective upon mesh refinement. To remedy this disease, it has been proposed to make the softening modulus dependent on the element size [9,38]. Numerical experiments for tensile loadings have confirmed that numerical results are then objective with regard to mesh refinement (e.g. [9,56,57]). For softening under compressive stresses, numerical studies are not decisive as contradictory results have been reported [48,71]. The problem of making the softening modulus dependent on the element size is that for an arbitrary concrete structure, the spread of the softening region is not known in advance. Consequently, the observation that use of a local softening law may involve snap-back behaviour on structural level also holds when we use a model in which the softening modulus has been adapted to some structural size. An example thereof has been encountered in a previous section and another example is given in the next subsection.

As a local softening law may lead to snap-back behaviour on structural level, it is necessary to explore whether numerical techniques permit tracing such paths. Clearly, post-peak behaviour can not be traced under load control, but if we apply a two-stage procedure and iterate both to displacement increments and to the load increment, tracing post-peak behaviour is possible [24,49]. The magnitude of the load increment is then usually determined from the requirement that the norm of displacement increments remains constant throughout the loading step, so that the loading process is constrained by the condition:

$$\Delta \boldsymbol{a}_j^T \Delta \boldsymbol{a}_j = \Delta l^2 \tag{49}$$

where the subscript j refers to the loading step and Δl is the arc-length of the equilibrium path in the n-dimensional displacement space. The draw-back of this so-called spherical arc-length method [24] is that it yields a quadratic equation for the load increment. To circumvent this problem, one may linearize equation (49), yielding [49]

$$\Delta \boldsymbol{a}_j^T \Delta \boldsymbol{a}_{j-1} = \Delta l^2 \tag{50}$$

This method, known as the updated normal path method, results in a linear equation for the load increment.

Both equations (49) and (50) have been employed very successfully within the realm of geometrically non-linear problems, where snapping and buckling of thin shells can be traced very elegantly. Nevertheless, for physically non-linear problems the method sometimes fails, which may be explained by considering that for physically non-linear problems, failure or bifurcation modes are often highly localized. Hence, only a few nodes contribute to the norm of displacement increments, and failure is not sensed accurately by such a global norm. As straightforward application of equations (49) or (50) is not always successful, we may amend these constraint equations by applying weights to the different degrees of freedom or omitting some of them from the constraint equation. As a consequence, the method loses some of its generality and elegance.

The fact that local softening laws may involve snap-back phenomena and other numerical instabilities, makes it questionable whether such a constitutive law in which the stress increment only depends on the strain increment and the strain history of the same material point, is suitable for describing structural softening. Indeed, some attempts have recently been published [10,60] which aim at describing the behaviour of concrete by a non-local constitutive law, in which the stress increment in a material point also depends on the neighbourhood, either via the strain gradient or via the strain increment and strain history of nearby material points. The difficulty of this approach is that it does not fit the nature of the finite element method.

5.2. Snap-back in mixed mode fracture of a notched beam

The example which we consider, is the notched unreinforced beam of Figure 28. The beam has been analysed using 8-noded plane stress elements and 6-noded triangles have been used in the transition region between the coarse part and the fine part of the mesh. Nine-point Gauss quadrature was applied for the quadrilateral elements. The concrete has been modelled as linearly elastic in compression with a Young's modulus E_c=24800 N/mm^2 and a Poisson's ratio ν=0.18. This approach is justified in this case, because the compressive stresses remain low. In tension, the crack model as outlined in the preceding has been employed. The crack parameters have been taken as: tensile strength f_{ct}=2.8 N/mm^2 and fracture energy G_f=0.055 N/mm. The width of the crack band was assumed to be h=10.167 mm.

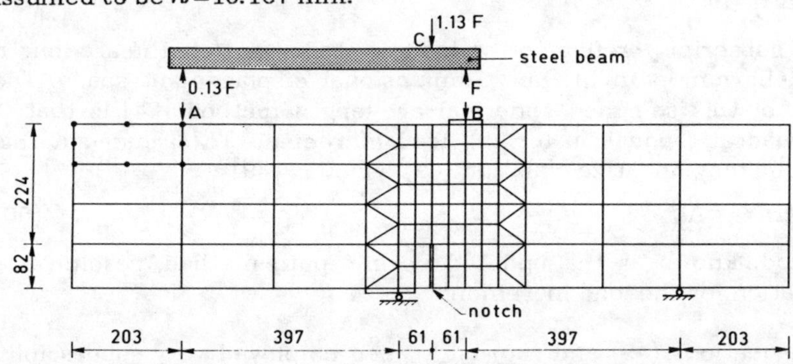

Figure 28. Finite element mesh of notched beam.

Figure 28 also gives the loading configuration which shows that the beam is loaded asymmetrically so that the crack propagating from the notch shows opening as well as sliding. In the experiment [3] the load was applied cyclically at point C of the steel beam AB and was controlled by a feed-back mechanism with the so-called Crack Mouth Sliding Displacement (CMSD) as control parameter. After peak, the envelope of the load cycles falls down sharply (see Figure 29), which is particularly challenging for a numerical simulation.

A number of researchers have endeavoured a numerical analysis of this beam [3,15,28,34,56,57,70], either with the smeared crack approach or with the discrete crack approach. Invariably, however, they adopted displacement control to the point of load application (point C in Figure 28), and all the calculated load-CMSD curve showed far too much ductility in the post-peak regime. Moreover, the post-peak regime was repeatedly reported to be highly unstable and converged equilibrium states could not be obtained. Indeed, we will show that such attempts are deemed to fail.

In the present study, the loading process has been controlled indirectly by a displacement parameter. Initially, a global norm of displacements was employed, but this analysis was not successful as this constraint equation failed near peak load, probably for reasons as argued

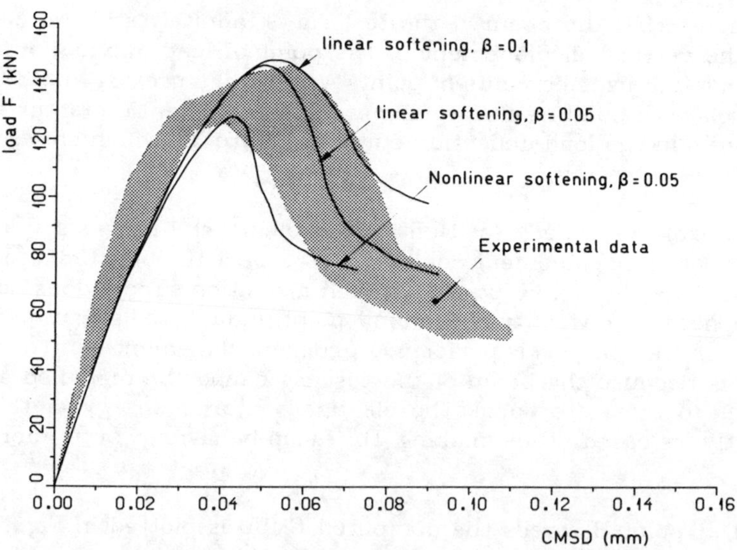

Figure 29. Load versus CMSD computed with CMSD-control. The shaded area denotes the range of experimental results.

in the preceding. Indeed, we observe a strong localization if we plot the eigen displacement mode at peak load (see Figure 30). However, for the present problem, a displacement parameter which can be used to control the loading process is naturally available, namely the CMSD itself. In this way the numerical analysis entirely parallels the experiment. With CMSD-control, the load-increment is in the linearised version determined from the condition

$$\Delta a_j^T \Delta a_{j-1} = \Delta l^2 \tag{51}$$

with Δa_j the increment in CMSD at step j. Using the constraint equation (51), the limit point could be overcome without problem with the same step size where the constraint formulae (49) and (50) failed.

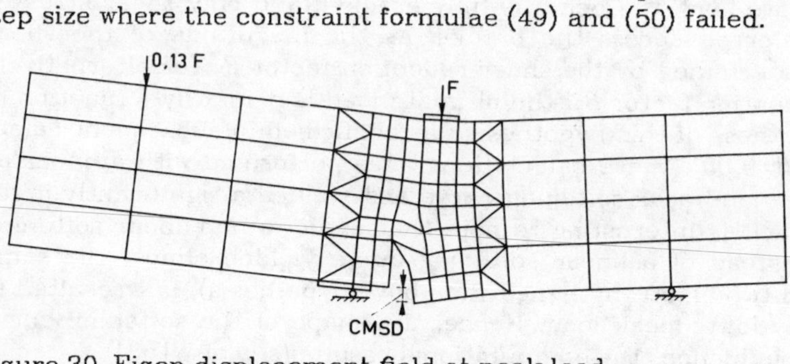

Figure 30. Eigen-displacement field at peak load.

Near the peak load, a negative pivot was encountered upon factorizing the tangent stiffness matrix. An eigenvalue analysis was performed subsequently and this resulted in one negative eigenvalue with the eigenmode of Figure 30. Hence, the peak load is a limit point and not a bifurcation point. After peak, the load was decremented and a genuine equilibrium path could be obtained.

If we assume that the beam of the test rig is infinitely stiff, we can recalculate the vertical displacement of the point of load application C from the calculated displacements of points A and B (Figures 31 and 32). For the analysis with non-linear softening and a shear retention factor of 0.05, this results in the load-deflection curve of Figure 33, which shows a violent snap-back behaviour.

The snap-back after peak load entirely explains why previous solutions which adopted displacement control with respect to point C did not result in a stable post-peak response, as such a solution simply does not exist, at least not in the vicinity of the limit point. In fact, the situation is even worse when the analysis is performed under displacement control of point C. This is because the beam of the test rig cannot be modelled as infinitely stiff, so that after peak the elastically stored energy of this beam is partly released, thus making the snap-behaviour even more violent.

The load at point B versus the computed CMSD is plotted in Figure 29. The most ductile curve corresponds to $\beta^{*}=0.1111$ (so that we have for the shear retention factor $\beta=0.1$) and a linear softening curve. We observe that the computed ultimate load overestimates the experimental values. Yet, Figure 34, which gives the incremental displacements and Figures 35 which show the crack pattern at ultimate load, reveal that the crack arising from the notch has developed fully. Consequently, all stresses which are transferred in this crack, are shear stresses, and these stresses cannot decrease because of the constant shear retention factor and because of the relatively high threshold angle for the formation of secondary cracks ($\alpha=60^{o}$). Indeed, beyond this point, the load again falls down sharply, which appeared to be due to formation of cracks elsewhere in the beam. Because the solution then becomes physically meaningless, this part of the curve has not been plotted.

At ultimate load, the beam is fully cracked and only shear stresses can be transferred across the cracks. As the magnitude of the shear stresses is determined by the shear reduction factor β^{*} (or alternatively the shear retention factor β), the ultimate load is primarily a function of β. The correctness of this hypothesis was confirmed in subsequent calculations with $\beta=0.05$. These calculations were performed with a linear as well as with a non-linear softening curve and yielded a significantly lower ultimate load. It is interesting to note that use of a non-linear softening curve [55] instead of a linear softening curve yielded almost the same ultimate load (which strengthened the above hypothesis), but resulted in a significantly lower peak load. Hence, the shape of the softening curve also strongly influences the computational results (see also [58]).

We finally note that the excellent match of the computational results with an experiment in which we have crack sliding as well as crack opening (mixed-mode fracture) casts doubts as to the necessity of refining the stress-strain relation in the crack very much. Replacement of the constant shear retention factor by an expression which makes it a function of the crack strain seems essential, but it is doubted whether the off-

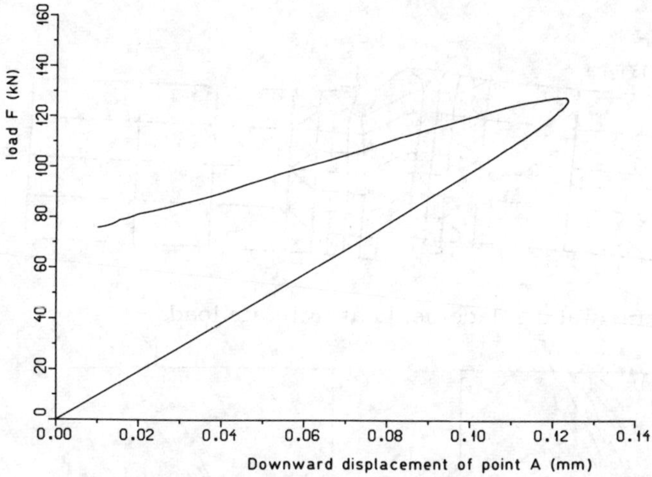

Figure 31. Load-deflection curve of point A.

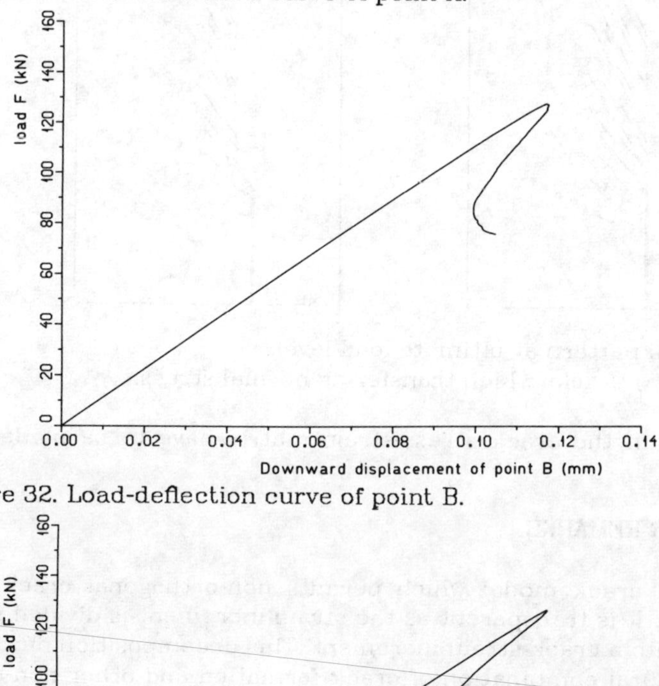

Figure 32. Load-deflection curve of point B.

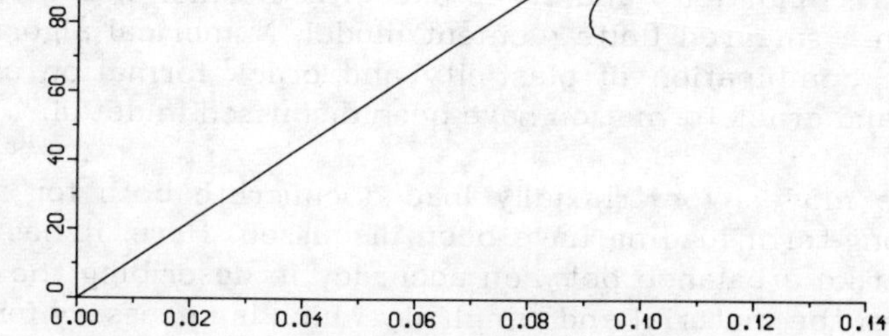

Figure 33. Load-deflection curve of point C.

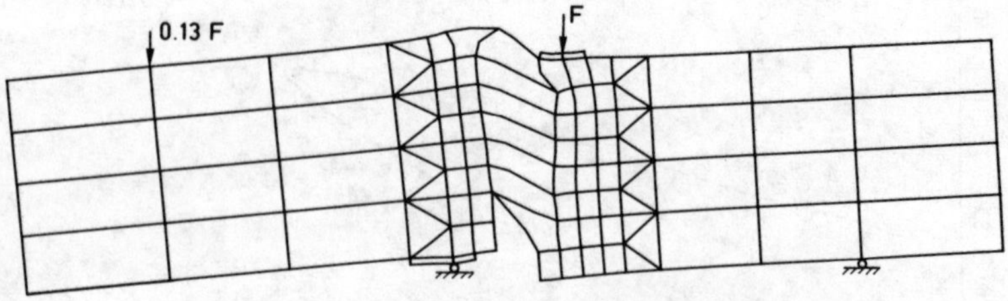

Figure 34. Incremental displacements at ultimate load.

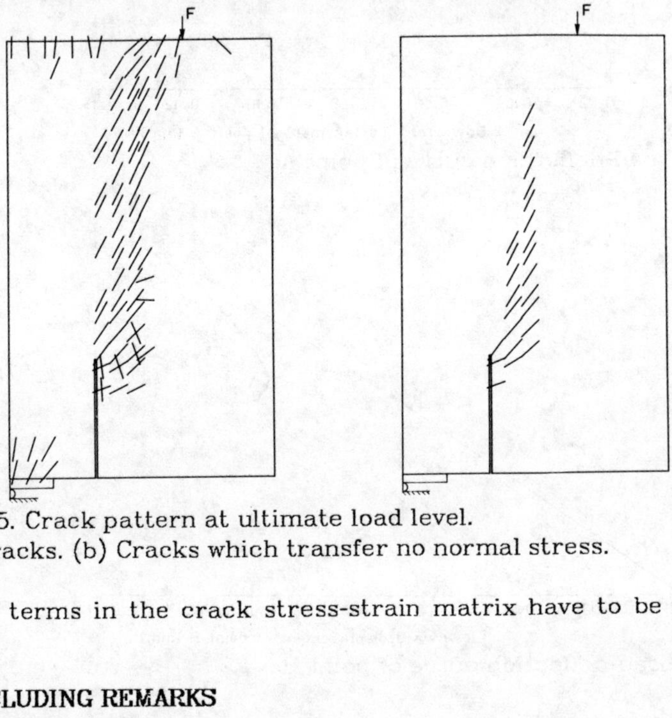

Figure 35. Crack pattern at ultimate load level.
(a) All cracks. (b) Cracks which transfer no normal stress.

diagonal terms in the crack stress-strain matrix have to be made non-zero.

6. CONCLUDING REMARKS

A smeared crack model which permits non-orthogonal cracks has been described. It is transparent as the strain increment is divided into a concrete and into a crack strain increment. This decomposition moreover allows for a natural combination of crack formation and other non-linear phenomena such as plasticity and creep and with thermal dilatation and shrinkage within a smeared finite element model. Numerical algorithms that handle the combination of plasticity and crack formation and of visco-elasticity and crack formation have been discussed in detail.

Constitutive models for triaxially loaded concrete both for short-term and for long-term loading have been discussed. Here, it has been attempted to strike a balance between accuracy in describing the basic characteristics of the material and simplicity which is necessary for successful application in numerical programs.

The use of so-called reduced integration in conjunction with crack formation may lead to erroneous results and even to divergence of the iterative procedure. When using 8-noded elements, 9-point or 'full' Gauss integration is more reliable for concrete structures as this largely prevents the occurrence of spurious zero-energy modes.

Smeared crack approaches are capable of predicting dominant diagonal cracks which occur in shear failures. Reasonable crack patterns were obtained for the sample problems which have been discussed despite a certain degree of diffusion. So, strain-localization can to a certain extent be represented by smeared crack models. Here, the situation is similar to non-associated plasticity, where the localization of deformation in thin shear bands can also be predicted by smeared representations [19].

It is incorrect to identify structural failure with divergence of the iterative procedure as is sometimes done for brittle failures. For a proper assessment of the structural behaviour it is necessary to trace the post-failure response also. This may involve analysing snap-through and snap-back behaviour for which the arc-length method, or modifications on it, seems very suitable also for concrete structures. Examples thereof have been discussed.

ACKNOWLEDGEMENTS

The author wishes to thank his colleagues of the Section 'DIANA' for their continuous support, in particular Mr. P. Nauta for his important contributions in implementing the described procedures in the DIANA-finite element package, Mr. J.G. Rots for the valuable discussions on the calculation of the notched beam and Mr. P. van den Berg for the discussions on creep problems. The research described in this paper was assisted financially by CUR-Committee A26 'Concrete Mechanics' and by the Section 'Bouwspeurwerk' of the Department of Public Works.

REFERENCES

1. Anderson, C.A. "Numerical creep analysis of structures", in: Creep and Shrinkage in Concrete Structures, (eds. Z.P. Bažant and F.H. Wittmann), Wiley, Chichester, 259-303, 1982.
2. Argyris, J.H., Pister, K.S., Szimmat J. and Willam, K.J. "Unified concepts of constitutive modelling and numerical solution methods for concrete creep problems", Comp. Meth. Appl. Mech. Eng., 10, 199-246, 1977.
3. Arrea, M. and Ingraffea, A.R. "Mixed-mode crack propagation in mortar and concrete", Report No. 81-13, Department of Structural Engineering, Cornell University, Ithaca, New York, 1981.
4. Azmy, Y.Y. and Dorning, J.J. "Arc-length continuation of nodal integral method solutions to the non-linear Navier-Stokes equations", in Numerical Methods for Non-linear Problems, (eds. C. Taylor, E. Hinton, D.R.J. Owen and E. Oñate), Pineridge Press, Swansea, Vol. 2, 672-687, 1984.

5. Bažant, Z.P. and Wu, S.T. "Rate-type creep law for aging concrete based on Maxwell chain", RILEM Materials and Structures, 7, 45-60, 1974.
6. Bažant, Z.P. "Instability, ductility and size effect in strain-softening concrete", ASCE J. Eng. Mech. Div., 102, 331-344, 1976.
7. Bažant, Z.P. and Gambarova, P. "Rough cracks in reinforced concrete", ASCE J. Struct. Div., 106, 819-842, 1980.
8. Bažant, Z.P. "Discussion on Session 2, Structural modelling for numerical analysis", in Final Report IABSE Colloquium on Advanced Mechanics of Reinforced Concrete, Delft University Press, Delft, 482, 1981.
9. Bažant, Z.P. and Oh, B. "Crack band theory for fracture of concrete", RILEM Materials and Structures, 16, 155-177, 1983.
10. Bažant, Z.P., Belytschko, T.B. and Chang, T.-P. "Continuum theory for strain-softening", ASCE J. Eng. Mech. Div., 110, 1666-1692, 1984.
11. Bažant, Z.P. and Chern, J.C. "Strain softening with creep and exponential algorithm", ASCE J. Eng. Mech. Div., 111, 391-415, 1985.
12. Becker, E. and Bürger, W. Kontinuumsmechanik, Teubner, Stuttgart, 1975.
13. Bergan, P.G. and Holand, I. "Nonlinear finite element analysis of concrete structures", Comp. Meth. Appl. Mech. Eng., 17/18, 443-467, 1979.
14. Besseling, J.F. "A theory of elastic, plastic and creep deformation of an initially isotropic material showing strain hardening, creep recovery and secondary creep", J. Appl. Mech., 25, 529-536, 1958.
15. Blaauwendraad, J., van den Berg, F.J.M. and Merks, P.J.G. "The use of the rough crack model of Walraven and the fictitious crack model of Hillerborg in F.E. analysis", in Preprints William Prager Symp. on Mechanics of Geomaterials: Rocks, Concretes, Soils, (ed. Z.P. Bažant), Northwestern University, Evanston, 643-646, 1983.
16. de Borst, R. "Thermomechanical analysis of aging visco-elastic solids", TNO-IBBC Report No. BI-84-35, Rijswijk, 1984.
17. de Borst, R. "Implicit integration of stress-strain laws in thermoplasticity", TNO-IBBC Report No. BI-84-43, Rijswijk, 1984.
18. de Borst, R., Kusters, G.M.A., Nauta, P. and de Witte, F.C. "DIANA, a three-dimensional, nonlinear finite element package on a microcomputer", in Engineering Sofware for Microcomputers, (eds. B.A. Schrefler, R.W. Lewis and S.A. Odorizzi), Pineridge Press, Swansea, 435-446, 1984.
19. de Borst, R. and Vermeer, P.A. "Possibilities and limitations of finite elements for limit analysis", Géotechnique, 34, 199-210, 1984.
20. de Borst, R. and Nauta, P. "Non-orthogonal cracks in a smeared finite element model", Eng. Comput., 2, 35-46, 1985.
21. Chen, W.F. Plasticity in reinforced concrete, McGraw-Hill, New York, 1982.
22. Cope, R.J. and Rao, P.V. "Non-linear finite element analysis of concrete slab structures", Proc. Instn. Civ. Engrs., 63, 159-179, 1977.
23. Cope, R.J., Rao, P.V., Clark, L.A. and Norris, P. "Modelling of reinforced concrete behaviour for finite element analysis of bridge slabs", in Numerical Methods for Non-Linear Problems, (C. Taylor, E. Hinton and D.R.J. Owen), Pineridge Press, Swansea, Vol. 1, 457-470,

1980.
24. Crisfield M.A., "A fast incremental/iterative procedure that handles snap-through", Computers & Structures, 13, 55-62, 1981.
25. Crisfield, M.A. "Local instabilities in the non-linear analysis of reinforced concrete beams and slabs", Proc. Instn. Civ. Engrs., Part 2, 73, 135-145, 1982.
26. Crisfield, M.A. "Difficulties of current numerical models for reinforced concrete and some tentative solutions", in Proc. Int. Conf. Computer Aided Analysis and Design of Concrete Structures, (eds. F. Damjanić et al.), Pineridge Press, Swansea, Part 1, 331-358, 1984.
27. Crisfield, M.A., "Some recent research on numerical techniques for structural analysis", in Numerical Methods in Engineering: Theory & Applications, (eds. J. Middleton and G. Pande), Balkema, Rotterdam, Vol. 2, 565-575, 1985.
28. Darwin, D., "Summary report: Bench-mark problems", in Concrete Mechanics, Cooperative Research between Institutions in The Netherlands and the USA, (eds. P. Gergely, R.N. White, J.W. Frénay and H.W. Reinhardt), 66-80, 1984.
29. Dill, E.H. "Rate formulations of nonlinear solid mechanics for the finite element method", Computers & Structures, 19, 51-56, 1984.
30. Dodds, R.H., Darwin, D., Smith, J.L. and Leibengood, L.D. "Grid size effects with smeared cracking in finite element analysis of reinforced concrete", SM Report No. 6, University of Kansas, Lawrence, Kansas, 1982.
31. Ebbinghaus, P. Herleitung eines Verfahrens zur Berechnung von Stahlbetonscheiben unter Berücksichtigung der Rissentwicklung, Dissertation, Rheinisch-Westfälische Technische Hochschule, Aachen, 1975.
32. Gerstle, K.H. et al. "Strength of concrete under multiaxial stress states", in Proc. Douglas-McHenry Symp. on Concrete and Concrete Structures, ACI Specialty Publication SP-55, 103-131, 1978.
33. Gerstle, K.H. et al. "Behaviour of concrete under multiaxial stress states", ASCE J. Eng. Mech. Div., 106, 1383-1403, 1980.
34. Glemberg, R., "Dynamic analysis of concrete structures", Publication 84:1, Department of Structural Mechanics, Chalmers Institute of Technology, Göteborg, 1984.
35. Goldscheider, M. "True triaxial tests on dense sands", in Constitutive Relations for Soils (eds. G. Gudehus, F. Darve and I. Vardoulakis), Balkema, Rotterdam, 1984.
36. Green, S.J. and Swanson, S.R. "Static constitutive relations for concrete", Rep. AWFL-TR-72-244, Kirtland Air Force Base, New Mexico, 1973.
37. Hettler, A. and Vardoulakis, I. "Behaviour of dry sand tested in a large triaxial apparatus", Géotechnique, 34, 183-198, 1984.
38. Hillerborg, A., Modeer, M. and Petersson, P.E. "Analysis of crack formation and crack growth in concrete by means of fracture mechanics and finite elements", Cement and Concrete Research, 6, 773-782, 1976.
39. Kolmar, W. and Mehlhorn, G. "Comparison of shear stiffness formulation for cracked reinforced concrete elements", in Proc. Int. Conf. Computer Aided Design and Analysis of Concrete Structures, (eds. F.

Damjanić et al.), Pineridge Press, Swansea, Vol. 1, 133-147, 1984.

40. Kristjansson, R. Physikalisch und geometrisch nichtlineare Berechnung von Stahlbetonplatten mit Hilfe Finiter Elemente, Dissertation, Technische Hochschule Darmstadt, Darmstadt, 1977.
41. Lade, P.V. "Three-parameter failure criterion for concrete", ASCE J. Eng. Mech. Div., 108, 850-863, 1983.
42. Litton, R.W. A contribution to the analysis of concrete structures under cyclic loading, Dissertation, University of California, Berkeley, 1976.
43. van Mier, J.G.M. Strain-softening of concrete under multiaxial loading conditions, Dissertation, Eindhoven University of Technology, Eindhoven, 1984.
44. Milford R.V. and Schnobrich, W.C. "Numerical model for cracked reinforced concrete", in Proc. Int. Conf. Computer Aided Analysis and Design of Concrete Structure (eds. F. Damjanić et al.), Pineridge Press, Swansea, Part 1, 71-84, 1984.
45. Ngo, D. and Scordelis, A.C. "Finite element analysis of reinforced concrete beams", J. Am. Concrete Inst., 64, 152-163, 1967.
46. Nilson, A.H. "Non-linear analysis of concrete by the finite element method", J. Am. Concrete Inst., 65, 757-766, 1968.
47. Nilson, A.H. et al. ASCE State-of-the-art Report on Finite Element Analysis of Reinforced Concrete, ASCE, 1982.
48. Pietruszczak, St. and Mróz, Z. "Finite element analysis of deformation of strain softening materials", Int. J. Num. Meth. Eng., 17, 327-334, 1981.
49. Ramm, E. "Strategies for tracing the nonlinear response near limit points", in Nonlinear Finite Element Analysis in Structural Mechanics (eds. W. Wunderlich, E. Stein & K.-J. Bathe), Springer Verlag, Berlin, 63-83, 1981.
50. Rashid, Y.R. "Analysis of prestressed concrete pressure vessels", Nucl. Eng. Design, 7, 334-344, 1968.
51. Rashid, Y.R. "Nonlinear analysis of two-dimensional problems in concrete creep", J. Appl. Mech., 39, 475-482, 1972.
52. Read, H.E. and Hegemier, G.A. "Strain softening of rock, soil and concrete - A review article", Mechanics of Materials, 3, 271-294, 1984.
53. Reinhardt, H.W. and Walraven, J.C. "Cracks in concrete subject to shear", ASCE J. Struct. Div., 108, 207-224, 1982.
54. Reinhardt, H.W. and Cornelissen, H.A.W. "Post-peak cyclic behaviour of concrete in uniaxial tensile and alternating tensile and compressive loading", Cement and Concrete Research, 14, 263-270, 1984.
55. Reinhardt, H.W. Personal Communication, 1985.
56. Rots, J.G., Kusters, G.M.A. and Blaauwendraad, J. "The need for fracture mechanics options in finite element models for concrete structures", in Proc. Int. Conf. Computer Aided Analysis and Design of Concrete Structures (eds. F. Damjanić et al.), Pineridge Press, Swansea, Part 1, 19-32, 1984.
57. Rots, J.G., Nauta, P., Kusters, G.M.A. and Blaauwendraad, J. "Smeared crack approach and fracture localization in concrete", Heron, 30, No. 1, 1-48, 1985.

58. Rots, J.G. "Strain-softening analysis of concrete fracture specimens", in Preprints Int. Conf. on Fracture Mechanics of Concrete (ed. F.H. Wittmann), Elsevier Science Publ, Amsterdam, 1985.

59. Rowe, P.W. "Theoretical meaning and observed values of deformation parameters for soil", in Proc. Roscoe Mem. Symp. on Stress-Strain Behaviour of Soils, Foulis, Henley-on-Thames, 143-194, 1972.

60. Schreyer, H.L. "Modelling of soil, concrete and interfaces", AWFL-TR-85-05, Kirtland Air Force Base, New Mexico, 1984.

61. Schreyer, H.L. and Babcock, S.M. "A third-invariant plasticity theory for a low strength concrete", ASCE J. Eng. Mech. Div., Vol, 111, 545-558, 1985.

62. Suidan, M. and Schnobrich, W.C. "Finite element analysis of reinforced concrete", ASCE J. Struct. Div., 99, 2109-2122, 1973.

63. Truesdell, C. and Noll, W. The non-linear field theories of mechanics, The encyclopedia of physics, Vol. III/3, Springer-Verlag, Berlin, 1965.

64. Vermeer, P.A. "A modified initial strain method for plasticity problems", in Proc. Third Int. Conf. Num. Meth. Geomech., (ed. W. Wittke), Balkema, Rotterdam, 377-387, 1979.

65. Vermeer, P.A. "A five-constant model unifying well-established concepts", in Constitutive Relations for Soils, (eds. G. Gudehus, F. Darve and I. Vardoulakis), Balkema, Rotterdam, 175-197, 1984.

66. Vermeer, P.A. and de Borst, R. "Non-associated plasticity for soils, concrete and rock", Heron, 29, No. 3, 1-64, 1984.

67. Walraven, J.C. "The influence of depth on the shear strength of lightweight concrete beams without shear reinforcement", Report 5-78-4, Stevin Laboratory, Delft University of Technology, Delft, 1978.

68. Walraven, J.C. Personal Communication, 1984.

69. Willam, K.J. and Warnke, E.P. "Constitutive model for the triaxial behaviour of concrete", Colloquium on Concrete Structures subjected for triaxial stresses, ISMES Bergamo, IABSE Report Vol. 19, 1974.

70. Willam, K.J., "Computational aspects of concrete mechanics", in Concrete Mechanics, Cooperative Research between Institutions in The Netherlands and the USA, (eds. P. Gergely, R.N. White, J.W. Frénay and H.W. Reinhardt), 81-109, 1984.

71. Willam, K.J. "Experimental and computational aspects of concrete fracture", Proc. Int. Conf. Computer Aided Analysis and Design of Concrete Structures, (eds. F. Damjanić et al.), Pineridge Press, Swansea, Part 1, 33-70, 1984.

72. Willam, K.J., Hurlbut, B. and Sture, S. "Experimental, constitutive and computational aspects of concrete failure", in Preprints US-Japan Seminar on Finite Element Analysis of Reinforced Concrete Structures, 149-172, 1985.

CHAPTER 3

A GENERAL CONSTITUTIVE MODEL FOR CONCRETE STRUCTURES

R. Glemberg*, M. Oldenburg**, L. Nilsson**, and A. Samuelsson*

* Dep. of Structural Mechanics, Chalmers Univ. of Technology, Gothenburg, Sweden
** Div. of Computer Aided Analysis and Design, Luleå Univ. of Technology, Luleå, Sweden

SUMMARY

A constitutive model for concrete is presented. The constitutive properties are represented by an elasto-plastic model, which in compression is either strain hardening or strain softening. In tension a smeared crack approach is used in which the physical crack opening is related to a crack strain consistent with a measure of the fracture energy developed in non-linear fracture mechanics. The objectivity of the crack model is checked for a uniaxial tension test using a few types of finite elements.

The constitutive model is used in the numerical study of the behavior of a simply supported reinforced beam. The correlation between experimental and calculated results is good.

1. INTRODUCTION

A constitutive model for reinforced concrete under monotonic loading is developed, see also [8], [2], [13]. The material behavior is described by use of two surfaces in the principal stress space, see Fig. 1. The outer surface called the *failure surface* defines the onset of unstable fracture propagation either as crushing or as cracking. The inner surface called the *plastic loading surface* defines the state of stresses initiating the stable fracture propagation which is the limit of the elastic region for the material.

The plastic loading surface expands with increasing inelastic strains, so called *strain hardening*. Additional straining of the material for stresses at the failure surface will cause stress releases, so called *strain softening*. The inelastic strains are in both cases assumed to be parallel to the gradients of the current surface.

No discontinuities are introduced due to cracks in the structure. Instead their effect is treated under the assumption of continuity. It is assumed that a fully developed crack causes permanent anisotropy with a small stiffness perpendicular to the crack surface, and that the stresses are released in such a way that the energy dissipation in the real crack is simulated.

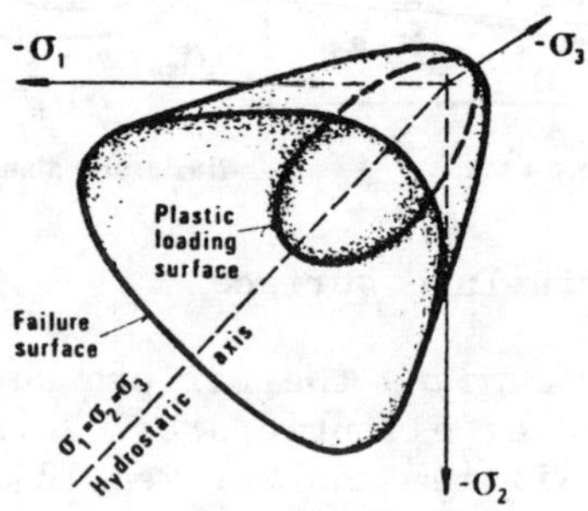

Figure 1. Schematic failure and plastic loading surfaces in the three dimensional stress space

2. FAILURE SURFACE

Failure surfaces mostly are defined by the stress states for which the unstable fracture propagation starts or cracking is initiated, see Kotsovos and Newman [5]. These states of stress also represent the ultimate strength for long-time loadings. The short-time loadings, which are studied here, allow additional loadings of about 20%. The failure surface is here proposed to connect stress states of ultimate strength in short-time loadings instead of those causing onset of unstable fracture propagation. The initiation of the volume increase observed at the onset of unstable fracture propagation is also proposed to be associated with the ultimate stress states.

The concrete is assumed to be initially isotropic. Thus, the shape of the failure surface in the deviatoric plane must possess a threefold symmetry indicated by the form of the surface in Fig. 1. Therefore, it is necessary to explore only the sector $\theta = 0^o$ to $\theta = 60^o$, while other sectors are known by the symmetry conditions, see Fig. 2. The two extreme meridian planes corresponding to $\theta = 0^o$ and $\theta = 60^o$ are called the tensile and the compressive meridians respectively.

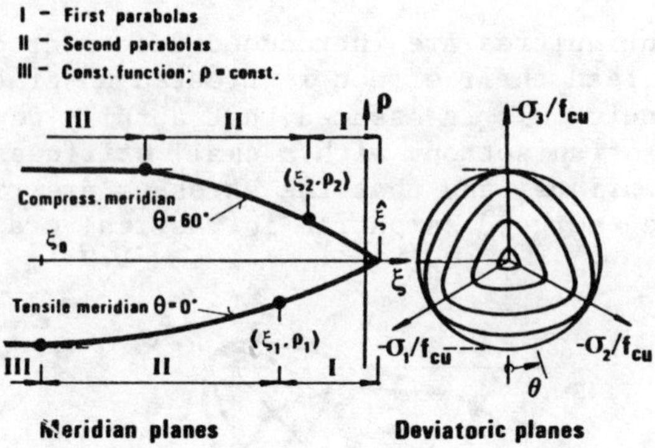

Figure 2. Trace of failure surface

In the *deviatoric plane* the failure surface is assumed to have a trace composed of elliptic arcs. This approach has been used for example by William and Warnke [12], Kotsovos [6], and Nilsson [8]. Experiments show, however, that the trace should change from a nearly triangular shape for zero mean stress to a more circular form for high compressive mean stresses.

In the *meridian plane* parabolas, suggested by William and Warnke [12], can be made to fit experimental data quite well. However, these parabolas are not open in the direction of compressive mean stresses and it is therefore difficult to accurately fit experimental data in the complete stress space. It is here suggested to use two parabolas and a constant function for each meridian.

The failure surface $F^f = 0$ predicts failure if the state of stress satisfies the condition

$$F^f = \rho/\rho^f(\xi,\theta) - 1 = 0 \tag{1}$$

Notations, stress and strain measures are briefly explained in Appendix.

The tensile ρ_t and compressive ρ_c meridians of the surface ($\theta = 0^o$ and $\theta = 60^o$) are expressed by two second-order parabolas, which are connected in the deviatoric plane by an ellipsoidal surface.

The two meridians are required to intersect the hydrostatic axis at the same point corresponding to hydrostatic tension $\hat{\xi}$ giving $\rho_t(\hat{\xi}) = \rho_c(\hat{\xi}) = 0$. The strength in uniaxial compression f_{cu}, uniaxial tension f_{tu}, and equal biaxial compression f_{bc} together with one point on each of the meridians in the high compressive region, (ξ_1,ρ_1) and (ξ_2,ρ_2), define the failure surface, see Fig. 2.

Better fittings to experimental data are obtained if the single parabolas are replaced by two parabolas and a constant function for each meridian. The connection between the two parabolas in each of the meridians is defined by the two stress points in the high compressive region, see Fig. 2.

The second parabolas in each of the meridians are determined so that the meridians become continuous both in function and first derivative. The connection between the second parabolas and the constant function is here defined by specifing the connection point on the tensile meridian ξ_0. The point on the compressive meridian is calculated from the requirement that the radius of the tensile and compressive meridians should be equal, $\rho_c = \rho_t$, for the constant functions. The derivatives are required to be continuous also at these points.

3. PLASTIC LOADING SURFACE

The initial plastic loading surface is assumed to coincide with the elastic limit surface defined by the onset of stable fracture propagation. According to the theory of plasticity admissible stresses can only be found on or inside of this surface. The surface is symbolically written as a function of stresses and hardening functions denoted H_k

$$F^p(\underset{\sim}{\sigma}, H_k) = 0; \quad k = 1, 2, \ldots n \qquad (2)$$

Additional loading to a stress point situated on the plastic surface produces inelastic strains, which are assumed to obey the associated flow rule.

The plastic loading surface is used to control the stresses between the states of stable fracture propagation and the state of ultimate strength. In case of small mean stresses it is found that the region between these two states is narrow and the constitutive model can be simplified by disregarding the plastic loading surface giving a constitutive model, which is linear elastic-fracturing.

The plastic loading surface is generated by an infinite number of generalized ellipsis. Thus, for every angle θ there is a semi ellipse for which one axis is parallel to the deviatoric axis and the other coincides with the hydrostatic axis. The proposed plastic loading surface is formally given in Eq. 3, while the parameters are defined in Fig. 3.

$$F^p = \left[\left(\frac{\xi - \xi_m(\theta)}{(\xi_m(\theta) - \xi_1)}\right)^2 + \left(\frac{\rho}{\rho^p(\xi_m(\theta))}\right)^2\right]^{1/2} - 1 = 0 \qquad (3)$$

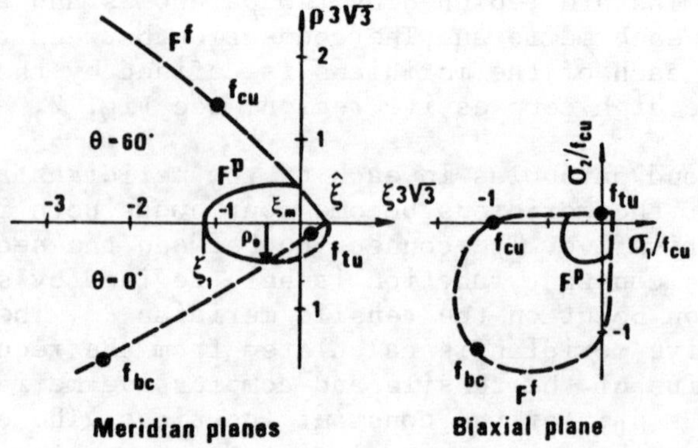

Figure 3. Plastic loading surface F^p and failure surface F^f

4. HARDENING AND SOFTENING

The nonlinear behavior of concrete is simulated within the constitutive model by letting the plastic loading surface and the failure surface be strain hardening and strain softening respectively. The hardening and softening behavior is governed by plastic strain measures.

The hardening and softening are proposed to obey the associated flow rule. For derivation of the inelastic strains and the constitutive tensor at a material point, see [8].

The plastic strain is in the hardening and softening functions represented by the integrated length of the plastic octahedral volume strain ε_0^p, the plastic octahedral shear strain γ_0^p, and the length $\bar{\varepsilon}^p$ of the combined plastic strain vector given by

$$\bar{\varepsilon}^p = ((\varepsilon_0^p)^2 + (\gamma_0^p)^2/4)^{1/2} \tag{4}$$

The parameter ξ_m of the plastic loading surface is chosen to govern the translation of the plastic loading surface according to

$$\xi_m = \xi_m^0 e^{-a\varepsilon_0^p}(1 + b(\bar{\varepsilon}^p)^{2d}) + \int_0^{\gamma_0^p} H(F^f)\,d\gamma_0^p \tag{5}$$

where ξ_m^0 is the initial value of the parameter ξ_m, and a, b, and d are material parameters to be determined and the function $H(F^f)$ is defined below.

The exponential factor of this function makes the hardening unlimited under hydrostatic compression. The term $b(\bar{\varepsilon}^p)$

controls the decrease in the rate of hardening due to increasing plastic strains. Finally, the integral reflects the influence of the failure surface on the hardening. Here, the funtion $H(F^f)$ is assumed to be a parabolic function of the distance between the actual stress state and the failure surface defined by the failure function, Eq. (1).

$$H(F^f) = H_1\alpha + H_2\alpha^2; \quad \alpha = 1 - F^f(\underset{\sim}{\sigma}, \underset{\sim}{\varepsilon}^p) \tag{6}$$

where H_1 and H_2 are material parameters to be determined. The failure surface defines the ultimate strength of the concrete and additional loadings will cause the material to fail by crushing. The function of the failure surface is rewritten to allow softening by including the function f^f.

$$F^f = (\rho/\rho_f)f^f(\underset{\sim}{\sigma}, \underset{\sim}{\varepsilon}^p) - 1 = 0 \tag{7}$$

The rapid increase of the strain in the direction of the maximal principal stress is the most significant characteristic of concrete when loaded beyond the ultimate strength. The stress release function, which governs the value of the present failure surface, is here proposed to be a function of the maximal principal strain and the mean stress according to

$$f^f = e^{-(\varepsilon_1^m/g(\sigma_0))^2}, \quad \varepsilon_1^m = \varepsilon_1^p - \varepsilon_1^{p,max}; \quad \varepsilon_1^p > 0 \tag{8}$$

where $\varepsilon^{p,max}$ is the maximal principal strain at the ultimate strength. The function $g(\sigma_0)$ is defined by

$$g = c(1 - \sigma_0/f_{cu}); \quad \sigma_0 < 0 \tag{9}$$

where c is a material parameter to be determined from experimental data.

5. CRACKING

5.1 Proposed crack model

The tensile cracking of concrete creates physical discontinuities. These cracks can not be treated directly within the context of continuum mechanics. However, the constitutive model can be arranged to include an objective measure, the fracture energy G_f, see Fig. 4b, which is assumed to be a material parameter for small and moderate values of the rate of crack opening, cf Hillerborg et al [3]. The approach was proposed by Nilsson and Oldenburg [9] and has been used by Jiang [4], and by Glemberg [2].

In the vicinity of a crack tip one finds a region, see Fig. 4, where microcracking starts to form the crack but stresses still can be transformed between the two crack

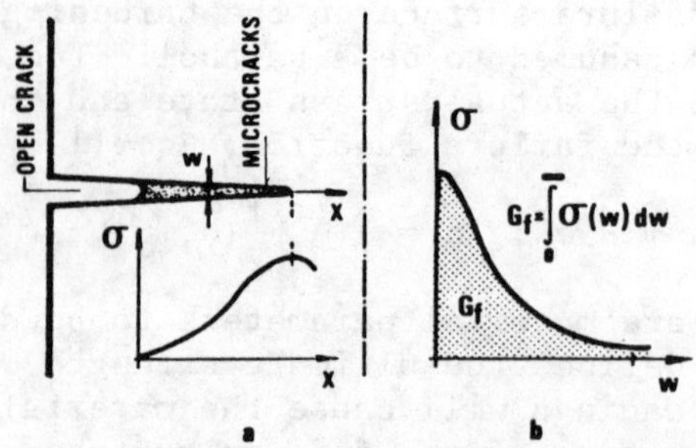

Figure 4. Stress distribution in the vicinity of a crack tip

surfaces. In the following only normal stresses perpendicular to the crack surfaces are considered, so called cracks of Modus I. Shear crack modes, Modus II and Modus III, can be treated with the same type of approach as applied for Modus I.

The continuous crack approach replaces the discrete crack opening w in a control volume V, often chosen to be the finite element, with an equivalent crack strain ε^c causing the same energy dissipation in the volume as the crack over the crack surfaces S. That is

$$\int_S \sigma\dot{w}\,dA = \int_V \underset{\sim}{\sigma}:\underset{\sim}{\dot{\varepsilon}}^c\,dV$$

leading under constant conditions within the control volume to

$$w = (V/S)\varepsilon^c = l_c \cdot \varepsilon^c \qquad (10)$$

The ratio V/S can be identified as a characteristic length l_c of the control volume versus the crack area.

A maximal stress function can now be defined as a plane, $F^c = 0$, in the stress space with the normal in the direction of the axis of the first principal stress. The following explicit function, proposed by Nilsson and Oldenburg [9], is used in the present analysis

$$F^c = \sigma_1 - \mu e^{-\varepsilon^c/a_c} = 0; \quad a_c = G_f/(\mu l_c) \qquad (11)$$

Here μ is the maximal principal stress at the onset of the crack. The fracture energy G_f is determined experimentally and the characteristic length l_c can be calculated exactly or approximately as the square root of the area associated with each integration point.

A crack opens at a material point of the structure if the stress reaches the failure surface F^f and the maximal principal stress is greater than half the strength in uniaxial tension.

Additional loading to this stress point will take place on the surface defined by the maximal stress function Eq. (11), giving an appropriate stress release perpendicular to the crack surface.

The use of the crack strain ε^c instead of the physical crack opening w makes it possible to treat the stress release with use of normal strain measures. It is assumed that the crack preserves its direction during the rest of the analysis. The stiffness perpendicular to the crack surface can be established from the maximal stress function, Eq. (11), in accordance with the theory of plasticity and associated flow.

The constitutive tensor has to be modified for each new crack that develops in the stress point. For simplicity, each new crack in a point is assumed to be perpendicular to the already excisting cracks and for every crack in the point there exists a maximal stress surface of the form given in Eq. (11). Rotation of the principle strain direction is thus disregarded.

For a closing crack it is assumed that the orthotropy of the cracked point is kept unchanged and that the remaining E-modulus is a function of the already developed crack strain. The calculation of the remaining E-modulus is given in Fig. 5.

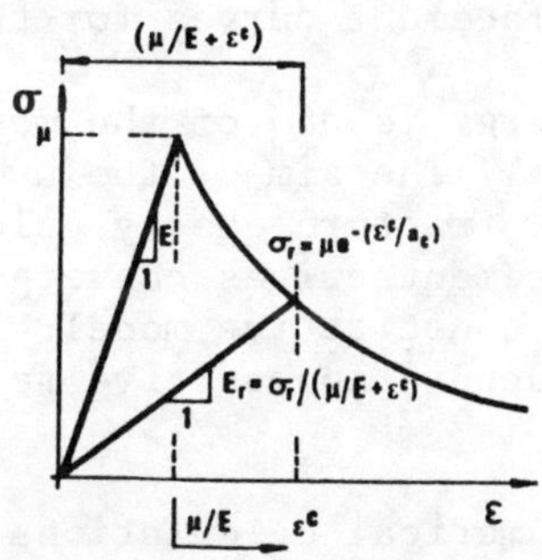

Figure 5. Calculation of the remaining E-modulus in a closing crack

It is here proposed that the shear modulus in the crack is reduced in the same manner as the E-modulus.

$$G_r = (E_r/E)G \tag{12}$$

The proposed reduction of the shear modulus makes the shear stiffness neutral with respect to the loading direction. The same shear stiffness is supposed both for opening and closing cracks.

5.2 Verification of the crack model

The objectivity of the proposed crack model with respect to the size of the finite elements is demonstrated by analysing a necked specimen subjected to tensile loading. The specimen was tested by Petersson [11] for determination of the value of the fracture energy G_f. The value of the fracture energy is obtained as the ratio of the area under the load displacement curve and the crack surface area, see Fig. 6. This figure also shows the characteristics of the used specimen.

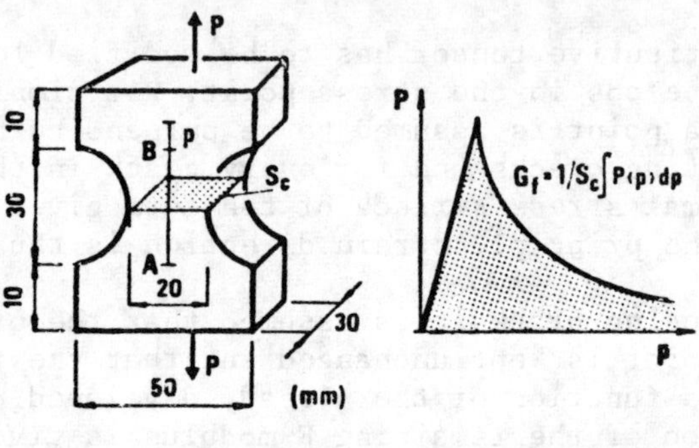

Figure 6. Load-displacement curves for the necked specimen

The fracture energy is one of the material parameters in the constitutive model. The aim of the numerical analysis is to investigate if the fracture energy calculated from the obtained load-displacement curves converges to the value given as input data to the constitutive model. If this is true the proposed crack approach is insensitive to the size of the finite elements.

Two series of numerical calculations are performed. In one series [2] half the specimen is modelled with 8-node isoparametric elements for four different element meshes, see Fig. 7. The given value of G_f is 125 N/m while the calculated values are listed in the figure.

The calculated load-displacement curves for the four element meshes are also plotted. The calculations converge monotonically to the given value of G_f, thus proving objectivity.

In the other series [13] a quarter of the specimen is modelled with 4 node quadrilaterals and triangular constant strain elements. The results displayed in Fig. 8 and Fig. 9 show convergence to the given value of G_f = 130 N/m for the quadrilateral mesh but no convergence for the triangular one.

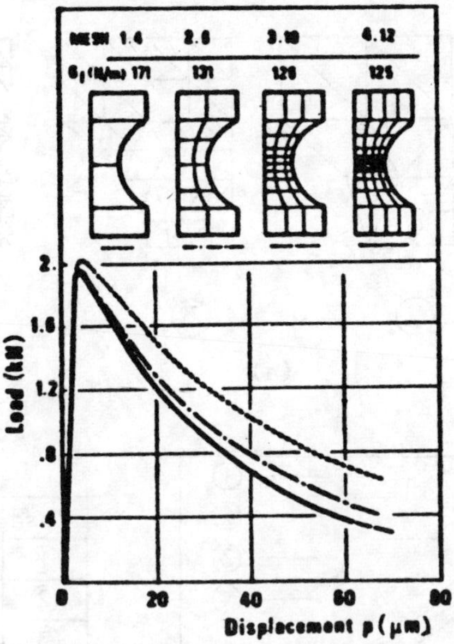

Figure 7. Calculated load displacement curves and fracture energies for 8-node elements

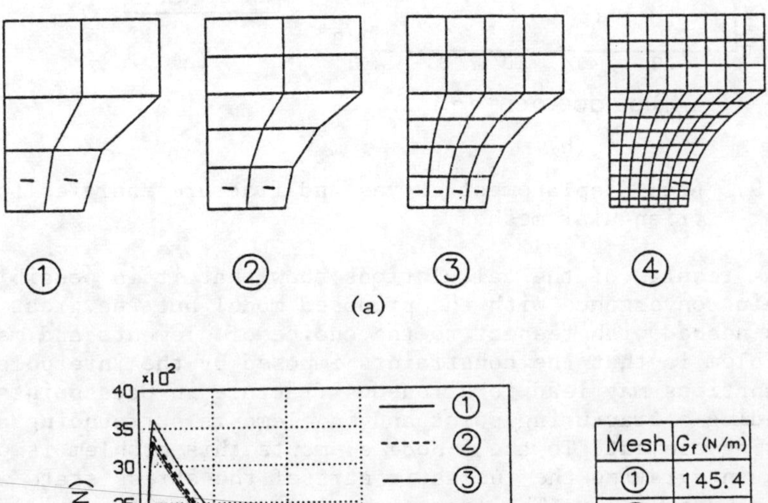

(a)

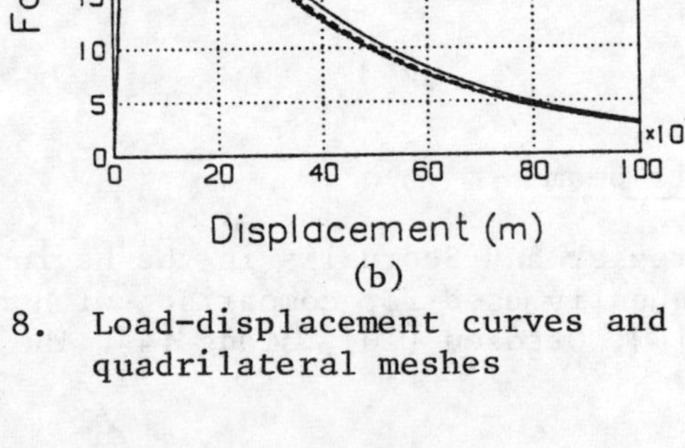

(b)

Mesh	G_f (N/m)
①	145.4
②	136.9
③	134.1
④	131.1
Exact	130.0

(c)

Figure 8. Load-displacement curves and fracture energies for quadrilateral meshes

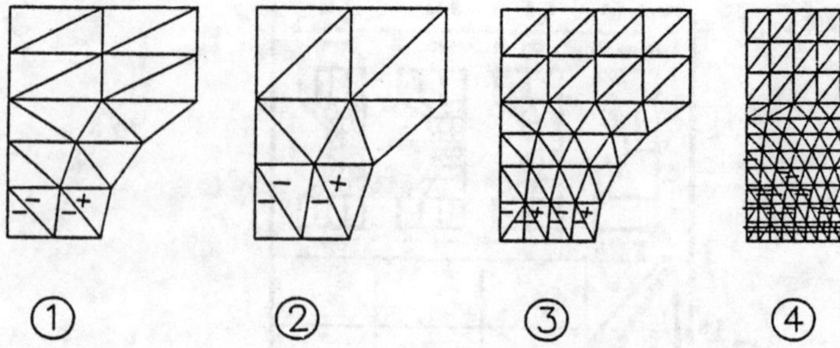

(a)

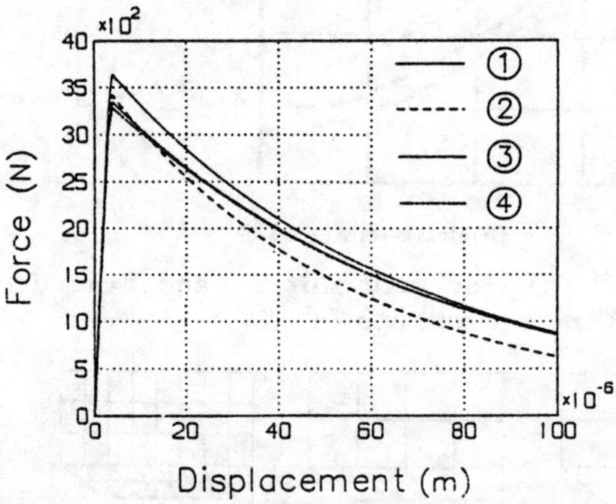

(b)

Mesh	G_f (N/m)
①	252.0
②	199.3
③	244.1
④	241.5
Exact	130.0

(c)

Figure 9. Load-displacement curves and fracture energies for triangular meshes

The results of the calculations show that it is possible to obtain convergence with the proposed model but the great care is needed with respect to the choice of elements and mesh. One problem is that the constraints imposed by the interpolation functions may lead to erroneous fracture in Gausspoints surrounding a fracturing point and in elements surrounding a fracturing element. To the 4-node elements this problem is overcome by integrating the inelastic part of the stress state with only one Gauss point. The lack of convergence for the triangular mesh is mainly due to the appearance of two rows of cracks, thus overestimating the fracture energy.

6. NUMERICAL RESULTS

6.1 A Bresler-Scordelis beam

Beams tested by Bresler and Scordelis in the beginning of the 1960's [1] are frequently used for comparison with numerical results, e.g. Ngo [7], Ottosen [10], Jiang [4]. The analysis discussed in this subsection conserns the OA-2 beams of the Bresler and Scordelis experiments.

Preliminary results indicate that the behavior close to failure is sensitive to effects such as bond-slip, dowel action, changes in geometry etc. To obtain some information about at least one of these effects the calculation includes the modelling of bond-slip by a 6-node bond-slip element. The OA-2 beam is reinforced with longitudinal reinforcement only. The principal dimensions of the beam are given in Fig. 10.

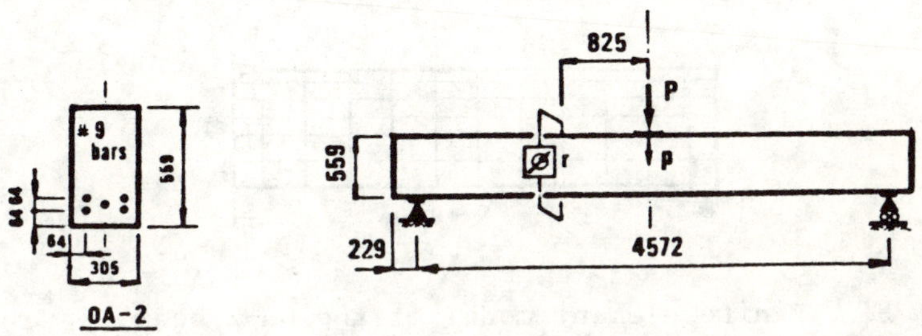

Figure 10. Dimensions of the OA-2 beam

The set of material parameters used in the constitutive model are listen in Table 1.

Table 1. Material parameters used in the constitutive model

Parameter	Value	Parameter	Value	Parameter	Value
E	21.3 [GPa]	ξ_1	-1.275	a	48.0
ν	0.18	ϱ_1	1.97	b	22.0
f_{cu}	24.2 [MPa]	ξ_2	-0.995	d	0.125
f_{cb}	1.16 f_{cu}	ϱ_2	0.818	$\xi^0{}_m$	0.240
f_{tu}	0.08 f_{cu}	ξ_0	138.6	c	0.01
G_f	[N/m]			H_1	1200.
				H_2	1500.

The material properties of the reinforcement was modelled by polygonial stress-strain relations determined from the experimentally obtained stress-strain curves, [1].

Two calculations are discussed here, one with rigid connection between reinforcement and concrete and one with bond-slip, which is assumed to obey the bilinear relation given in Fig. 11.

The beam is modelled with 8-node isoparametric membrane elements and the corresponding 3-node bar or 6-node bond and bar element for the reinforcement. The symmetry of the beam is utilized giving the element mesh shown in Fig. 12. The steel plates at the support and the loading point are modelled by membrane elements.

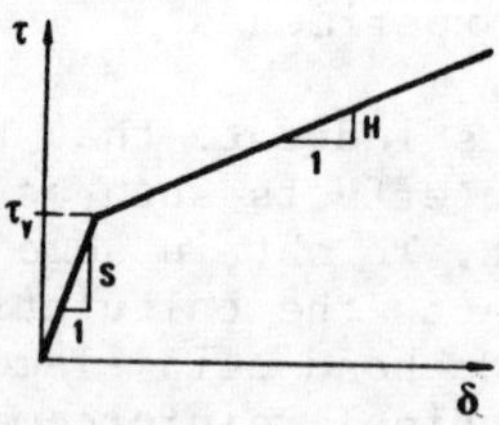

Figure 11. Bilinear τ-δ relation

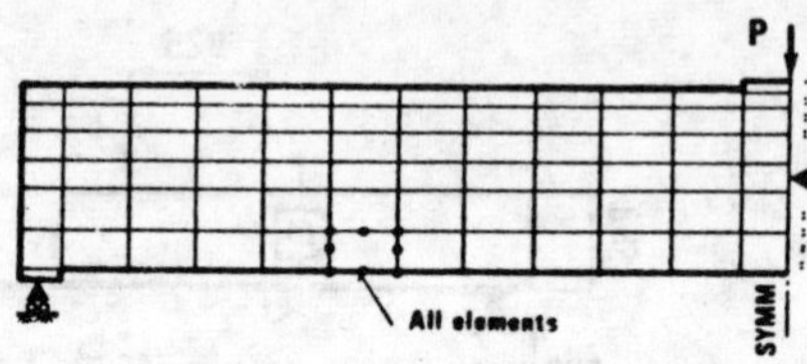

Figure 12. Finite element model of the OA-2 beam

6.2 Results

The load displacement curves for the OA-2 beam obtained with and without considering the bond-slip are plotted in Fig. 14. The failure was initiated by crushing in the compressive zone of the beam.

The numerical failure was in all cases connected to uncontrolable increase in crack strain for a number of "cracked" integration points. These final crack increases seemed to be initiated by the force redistribution caused by the crushing at the loading point. Thus, the compressive strength is critical for the ultimate load. The visible shear crack, observed in the experiments, Fig. 13, did probably not initiate the failure but it was a consequence of the softening in the compressive zone. This behavior is also mentioned by Ottosen [10].

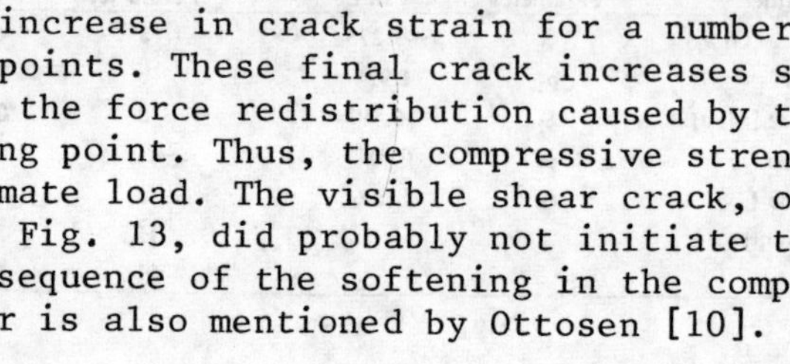

Figure 13. Experimentally observed crack patterns for the OA-2 beam [1] at failure load P = 363 kN

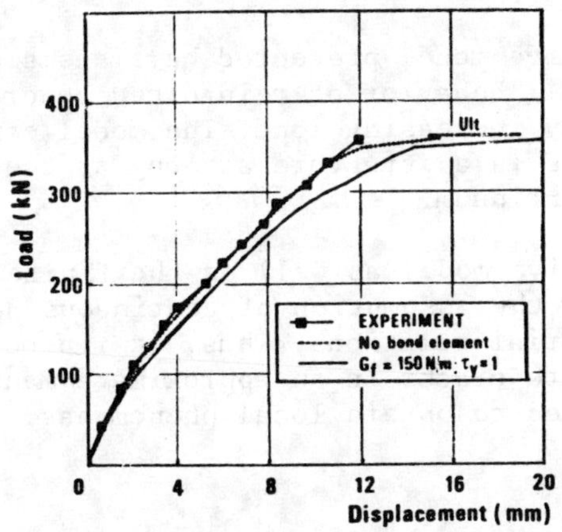

Figure 14. Load-displacement curves for the OA-2 beam

As expected, the bond-slip has no effect on the ultimate load. However, the slope of the load-displacement curve is sensitive to this effect, which can be seen in Fig. 14. The influence of the bond-slip is also seen in the calculated crack patterns, which are plotted for the Gauss points of the elements in Fig. 15.

The calculated crack widths at failure load vary from about 0.1 mm for the rigid bond to about 0.3 mm for the bond-slip.

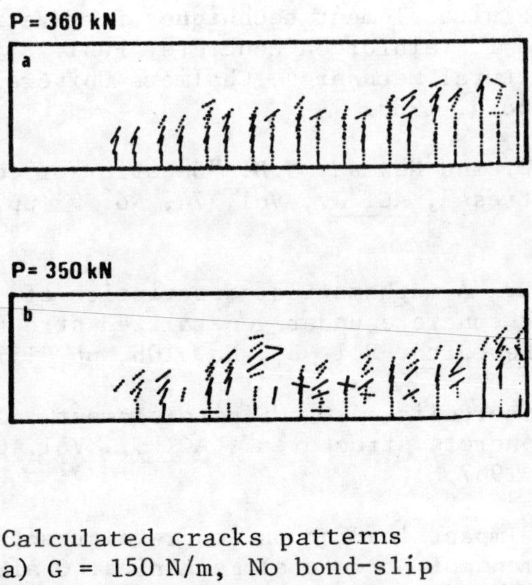

Figure 15. Calculated cracks patterns
a) G = 150 N/m, No bond-slip
b) G = 150 N/m, τ_y = 1 MPa, S = 400 GN/m^3

7. CONCLUSIONS

The constitutive model presented here seems to accurately predict the over-all behavior of reinforced concrete structures under monotonically increasing load. The model can follow the progressive failure in a structure as long as the structure can be treated as a continuum.

The constitutive model as well as the finite element method is developed under the assumption of continuous behavior in both kinematic and material relations. Thus, it reproduces the singularities of discrete cracks in an approximate manner, and a dense mesh is needed to obtain local phenomenas.

8. REFERENCES

1. Bresler, B. and Scordelis, A.C. "Shear strength of reinforced concrete beams", Proc., ACI J., Vol. 60, No. 1, pp. 51-74, 1963.

2. Glemberg, R. Dynamic analysis of concrete structures, Constitutive modelling, Dynamic substructuring, Computer Implementation, Publ. 84:1, Department of Structural Mechanics, Chalmers University of Technology, Göteborg, 1984.

3. Hillerborg, A., Modeer, M., and Petersson, P-E. "Analysis of crack formation and crack growth in concrete by means of fracture mechanics and finite elements", Cement and Concrete Research, 6, pp. 773-782, 1976.

4. Jiang, J.J Finite element techniques for statical analysis of structures in reinforced concrete, Publ. 83:2, Department of Structural Mechanics, Chalmers University of Technology, Göteborg, 1983.

5. Kotsovos, M.D. and Newman, J.B. "Behavior of concrete under multiaxial stress", ACI J., Vol. 74, No. 9, pp. 443-446, 1977.

6. Kotsovos, M.D. "A mathematical description of the strength properties of concrete under generalized stress", Magazine of Concrete Research, Vol. 31, No. 108, pp. 151-158, 1979.

7. Ngo, D. and Scordelis, A.C. "Finite element analysis of reinforced concrete structures", ACI J., Vol. 64, No. 3, pp. 152-163, 1967.

8. Nilsson, L. Impact loading on concrete structures, Publ. 79:1, Department of Structural Mechanics, Chalmers University of Technology, Göteborg, 1979.

9. Nilsson, L. and Oldenburg, M. Nonlinear wave propagation in plastic fracturing materials - A constitutive modelling and finite element analysis, IUTAM, Symposium Nonlinear Deformation Waves, Tallin 22-28 Aug. 1982, Springer, 1983.

10. Ottosen, N.S. "2-D Finite element analysis of massive RC structures", ASCE, Vol. 108, No. ST8, pp. 1874-1893, 1982.

11. Petersson, P-E. Crack growth and development of fracture zones in plain concrete and similar materials, Division of Building Materials, Lund Institute of Technology, Report TVBM-1006, Lund, 1981.

12. Willam, K.J. and Warnke, E.P. "Constitutive models for triaxial behavior of concrete subjected to triaxial stresses", Int. Assoc. Bridge Struct. Eng. Proc., Vol. 19, pp. 1-30, 1975.

13. Oldenburg, M. Finite element analysis of tensile fracturing structures, Lic.thesis, Division of Computer Aided Analysis and Design, Luleå University of Technology, Luleå, 1985.

APPENDIX

The most important stress and strain measures used in the paper are summarized together with the formulas defining the failure surface.

Stress measures

$$\sigma_0 = (\sigma_1 + \sigma_2 + \sigma_3)/3 \tag{13}$$

$$\tau_0 = [(\sigma_1 - \sigma_2)^2 + (\sigma_2 - \sigma_3)^2 + (\sigma_3 - \sigma_1)^2]^{1/2}/3 \tag{14}$$

$$\cos\theta = (2\sigma_1 - \sigma_2 - \sigma_3)/(3\sqrt{2}\,\tau_0); \quad (0^o \leq \theta \leq 60^o) \tag{15}$$

$$\xi = \sqrt{3}\,\sigma_0/f_{cu} \tag{16}$$

$$\rho = \sqrt{3}\,\tau_0/f_{cu} \tag{17}$$

where $\sigma_1 \geq \sigma_2 \geq \sigma_3$ are the principal stresses.

Strain measures

$$\varepsilon_0 = (\varepsilon_1 + \varepsilon_2 + \varepsilon_3)/3 \tag{18}$$

$$\gamma_0 = 2[(\varepsilon_1 - \varepsilon_2)^2 + (\varepsilon_2 - \varepsilon_3)^2 + (\varepsilon_3 - \varepsilon_1)^2]^{1/2}/3 \tag{19}$$

where $\varepsilon_1 \geq \varepsilon_2 \geq \varepsilon_3$ are the principal strains.

Failure surface functions

Tensile meridian

$$\rho_t = a_0 + a_1\xi + a_2\xi^2 \tag{20}$$

Compressive meridian

$$\rho_c = b_0 + b_1\xi + b_2\xi^2 \tag{21}$$

Ellipse connecting the two meridians in the deviatoric plane

$$\rho^f = [\rho_c km + \rho_c \cdot n(km^2 + 5\rho_t^2 - 4\rho_t\rho_c)^{1/2}]/(km^2 + n^2) \tag{22}$$

where $k = (\rho_c^2 - \rho_t^2)$, $m = 2\cos\theta$, $n = (2\rho_t - \theta_c)$.

CHAPTER 4

A CONCRETE MATERIAL MODEL

Klaus-Jürgen Bathe
Massachusetts Institute of Technology
Mechanical Engineering Department
Cambridge, MA 02139

Jan-Anders Sundberg
ADINA Engineering AB
Munkgatan 20D
S-722 12 Västerås, Sweden

ABSTRACT

We present a three-dimensional concrete material model effective for various applications. The model is based on a hypoelastic nonlinear stress-strain description and failure envelopes for tension and compression failures. Although quite general in applicability, the model is relatively simple to use and computationally efficient.

1. INTRODUCTION

The concrete material model published by Bathe and Ramaswamy in 1979 [1] and available in the ADINA program has found considerable use (see, for example, [2-7]). The main attributes of the model are that it is a hypoelastic model based on a nonlinear uniaxial stress-strain relation that is generalized to take biaxial and triaxial stress conditions into account, and that it models tensile failure and compression failure by quite general failure envelopes. The concrete model is defined with a number of input parameters that provide considerable flexibility in its use.

The main aim in the development of the model was to provide a practical analysis tool to describe the major nonlinear constitutive behavior of concrete. The model was to be as simple as possible, but reproduce the important nonlinear strength characteristics consistent with laboratory experimental results, and the model was to be computationally efficient.

The use of the model during the recent years has resulted in most interesting applications. However, as expected, we have also gained experience that has shown us where improvements in the model would be much desirable. Two important improvements that we have implemented are the tension stiffening effect and a gradual stress release in the strain softening region. Also, we refined the computational procedures to allow larger incremental solution steps throughout the complete analysis.

The objective in this paper is to summarize the current status of our concrete material model. To render the paper self-contained we summarize the complete model and not just the improvements. We have implemented the improvements to the model in the ADINA program (version ADINA 86) and give a few applications. These are included in the paper to primarily show the increase in the capabilities reached over the earlier version of the model [1].

2. THE CONCRETE MATERIAL MODEL

The model implemented employs three basic features to describe the material behavior, namely, (i) a nonlinear stress-strain relation including strain-softening to allow for the weakening of the material under increasing compressive stresses, (ii) failure envelopes that define cracking in tension and crushing in compression, and (iii) a strategy to model the post-cracking and crushing behavior of the material. In the solution, the material can be subjected to cyclic loading conditions, i.e., the numerical solution allows for unloading and reloading including deactivation of tensile failures. However, as will become apparent, the cyclic loading conditions are only modeled realistically in essentially proportional loading.

In the following, the material model is described for infinitesimal displacement conditions using the engineering stresses ${}^{t}\sigma_{ij}$ and engineering strains ${}^{t}e_{ij}$. In order to analyze problems with large rotation conditions, the total Lagrangian stress and strain variables must be substituted for the engineering variables [8], and then the model is directly applicable.(*)

2.1 Stress-strain relations

The general multiaxial stress-strain relations are derived from a uniaxial stress-strain relation ${}^{t}\tilde{\sigma}$ versus ${}^{t}\tilde{e}$. In this section, we describe the uniaxial and multiaxial stress-strain relations employed in the model and concentrate on the behavior prior to tensile cracking or compression crushing.

In the following discussion, all uniaxial parameters are identified by a curl (~) placed over them, (i.e., all parameters that have been obtained from fig. 1 carry a curl).

(*) We use the notation of refs. [1,8] in this paper.

2.1.1 Uniaxial conditions

A typical uniaxial stress ${}^t\tilde{\sigma}$ to uniaxial strain ${}^t\tilde{e}$ relation (assuming loading of the material) is shown in fig. 1 [9]. This stress-strain relation shows that there are basically three strain phases; namely, corresponding to ${}^t\tilde{e} \geq 0$, $0 > {}^t\tilde{e} \geq \tilde{e}_c$ and $\tilde{e}_c > {}^t\tilde{e} \geq \tilde{e}_u$ where $\tilde{e}_c$ is the strain corresponding to the minimum (crushing) stress, $\tilde{\sigma}_c$, that can be reached, and $\tilde{e}_u$ is the ultimate compressive strain. If ${}^t\tilde{e} > 0$, i.e. the material is in tension, the stress-strain relation is linear until tensile failure at the stress $\tilde{\sigma}_t$, and a constant Young's modulus $\tilde{E}_0$ is employed,

$$ {}^t\tilde{\sigma} = \tilde{E}_0 \; {}^t\tilde{e} \qquad (1)$$

$$ \frac{d\,{}^t\tilde{\sigma}}{d\,{}^t\tilde{e}} = \tilde{E}_0 \qquad (2)$$

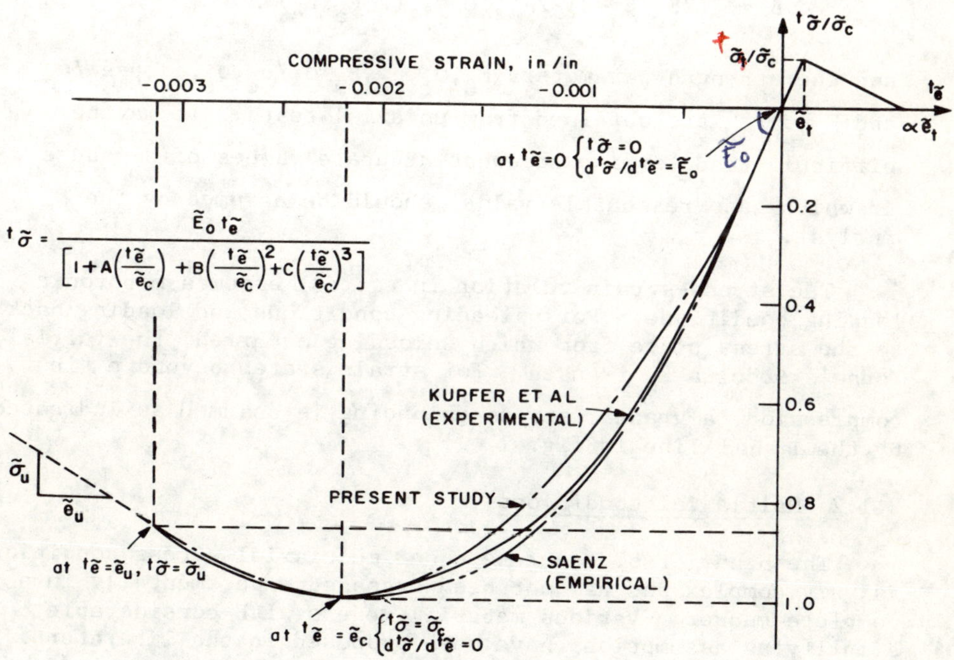

Fig. 1 Concrete model uniaxial stress-strain law

For ${}^{t}\tilde{e} \leq 0$, we assume the following relation,

$$ {}^{t}\tilde{\sigma}/\tilde{\sigma}_c = \frac{(\tilde{E}_0/\tilde{E}_s)({}^{t}\tilde{e}/\tilde{e}_c)}{1 + A({}^{t}\tilde{e}/\tilde{e}_c) + B({}^{t}\tilde{e}/\tilde{e}_c)^2 + C({}^{t}\tilde{e}/\tilde{e}_c)^3} \tag{3} $$

and hence,

$$ {}^{t}\tilde{E} = \frac{\tilde{E}_0[1-B({}^{t}\tilde{e}/\tilde{e}_c)^2 - 2C({}^{t}\tilde{e}/\tilde{e}_c)^3]}{[1+A({}^{t}\tilde{e}/\tilde{e}_c) + B({}^{t}\tilde{e}/\tilde{e}_c)^2 + C({}^{t}\tilde{e}/\tilde{e}_c)^3]^2} \tag{4} $$

where

$$ A = \frac{\tilde{E}_0/\tilde{E}_u + (p^3-2p^2)\tilde{E}_0/\tilde{E}_s - (2p^3-3p^2+1)}{(p^2 - 2p + 1)p} $$

$$ B = (2\tilde{E}_0/\tilde{E}_s-3)-2A \ , \ C = (2-\tilde{E}_o/\tilde{E}_s)+A $$

and the strength parameters $\tilde{E}_0, \tilde{\sigma}_c, \tilde{e}_c, \tilde{E}_s = \tilde{\sigma}_c/\tilde{e}_c, \tilde{\sigma}_u, \tilde{e}_u, p = \tilde{e}_u/\tilde{e}_c$ and $\tilde{E}_u = \tilde{\sigma}_u/\tilde{e}_u$ are obtained from uniaxial tests. It may be difficult to determine somewhat accurate values of $\tilde{\sigma}_u$ and $\tilde{e}_u$, in which case reasonable values should be assumed by the analyst.

The stress-strain relation in eq. (3) assumes monotonic loading conditions. For unloading conditions and loading back to the stress state from which unloading occurred, the initial Young's modulus $\tilde{E}_0$ is used. For strain states beyond $\tilde{e}_u$ in compression, a gradual stress unloading is assumed as indicated by the dashed line in fig. 1.

2.1.2 Multiaxial conditions

The behavior of concrete under multiaxial stress conditions is very complex and has not been assessed experimentally in a complete manner. Various material models with considerable simplifying assumptions have been proposed in the literature. Considering the variability of concrete materials that need be described in practice, the objective in this work was to develop an effective but simple model that provides sufficient flexibility to the analyst to fit the basic material behavior.

The stress-strain relations are evaluated differently depen ing on whether the material is loading or unloading. The Poisso ratio is assumed to be constant under all stress conditions.

To characterize loading and unloading conditions we define a loading function ${}^{t}f$ and a loading scalar ${}^{t}g$,

$$ {}^{t}f = {}^{t}\bar{s}; \quad {}^{t}g = \frac{{}^{t}\sigma_{p3}}{\tilde{\sigma}_{c}} \qquad (5)$$

where ${}^{t}\bar{s}=(\frac{1}{2}\,{}^{t}s_{ij}\,{}^{t}s_{ij})^{1/2}$, ${}^{t}s_{ij}={}^{t}\sigma_{ij}-\delta_{ij}\,{}^{t}\sigma_{m}$, ${}^{t}\sigma_{m}=\frac{1}{3}\,{}^{t}\sigma_{ii}$, δ_{ij} is the Kronecker delta and ${}^{t}\sigma_{p3}$ is the smallest principal stress. The material is loading except when the unloading conditions are determined,

$$ {}^{t}f < f_{max} \quad \text{and} \quad {}^{t}g < g_{max} \qquad (6)$$

where f_{max} and g_{max} are the maximum values of the loading function and loading scalar that have been reached during the complete solution. In unloading, the initial Young's modulus, $\tilde{E}_{0}$, is used to form the incremental stress-strain matrix, both for stiffness and stress calculations.

To obtain the stress-strain relations in loading conditions, the principal stresses are calculated and for each principal stress direction a uniaxial tangent Young's modulus, ${}^{t}\tilde{E}_{pi}$, corresponding to the strain in the principal stress direction, ${}^{t}e_{pi}$, is evaluated using eqs. (2) and (4). When using eq. (4), the current strain ${}^{t}e_{pi}$ is employed and to account for multiaxial stress conditions the material variables $\tilde{\sigma}_{c}, \tilde{\sigma}_{u}, \tilde{e}_{c}$ and $\tilde{e}_{u}$ are replaced by the variables $\tilde{\sigma}'_{c}, \tilde{\sigma}'_{u}, \tilde{e}'_{c}$ and $\tilde{e}'_{u}$ defined in eq. (11). Let ${}^{t}\sigma_{p1}$, ${}^{t}\sigma_{p2}$ and ${}^{t}\sigma_{p3}$ be the principal stresses at time t, with ${}^{t}\sigma_{p3} \leq {}^{t}\sigma_{p2} \leq {}^{t}\sigma_{p1}$ and ${}^{t}\tilde{E}_{p1}$, ${}^{t}\tilde{E}_{p2}$ and ${}^{t}\tilde{E}_{p3}$ the corresponding uniaxial Young's moduli. Hence, the material is considered as orthotropic with the directions of orthotropy defined by the principal stresses.

The stress-strain matrix corresponding to these directions is, considering three-dimensional stress conditions,

$$\underline{C}=\frac{1}{(1+\nu)(1-2\nu)}\begin{bmatrix} (1-\nu)\,{}^t\tilde{E}_{p1} & \nu\,{}^tE_{12} & \nu\,{}^tE_{13} & 0 & 0 & 0 \\ & (1-\nu)\,{}^t\tilde{E}_{p2} & \nu\,{}^tE_{23} & 0 & 0 & 0 \\ & & (1-\nu)\,{}^t\tilde{E}_{p3} & 0 & 0 & 0 \\ & & & \frac{1}{2}(1-2\nu)\,{}^tE_{12} & 0 & 0 \\ & \text{sym.} & & & \frac{1}{2}(1-2\nu)\,{}^tE_{13} & 0 \\ & & & & & \frac{1}{2}(1-2\nu)\,{}^tE_{23} \end{bmatrix} \tag{7}$$

where ν is the constant Poisson ratio, and the shear modulus in a coordinate plane is calculated from the weighted Young's modulus corresponding to that plane,

$${}^tG_{ij} = \frac{{}^tE_{ij}}{2(1+\nu)} = \frac{1}{2(1+\nu)}\,\frac{\left|{}^t\sigma_{pi}\right|{}^t\tilde{E}_{pi} + \left|{}^t\sigma_{pj}\right|{}^t\tilde{E}_{pj}}{\left|{}^t\sigma_{pi}\right| + \left|{}^t\sigma_{pj}\right|} \tag{8}$$

The above stress-strain relations for material loading conditions are only employed in the calculation of the stiffness matrix at time t. Considering the evaluation of the stress increment, $\underline{\sigma}$, from time t to time t+Δt, the stress integration is approximated in the following manner,

$$\underline{\sigma} = \hat{\underline{C}}\,\underline{e} \tag{9}$$

where $\underline{e}$ is the strain increment. The stress-strain matrix $\hat{\underline{C}}$ is as $\underline{C}$ in eq. (7) but the uniaxial Young's moduli ${}^t\tilde{E}_{pi}$ are replaced by the moduli ${}^\tau\tilde{E}_{pi}$. These are evaluated using the uniaxial stress-strain relationship in fig. 1,

$${}^\tau\tilde{E}_{pi} = \left[\tilde{\sigma}\Big|_{\text{at } {}^te_{pi}+e_{pi}} - \tilde{\sigma}\Big|_{\text{at } {}^te_{pi}} \right] \Big/ e_{pi} \tag{10}$$

where ${}^te_{pi}$ and e_{pi} are the strain components and incremental strain components at time t measured in the directions of the principal stresses ${}^t\sigma_{pi}$.

2.2 Material failure envelopes

To model the failure of the material in tension and compression in two- and three-dimensional analysis and to account for multiaxial conditions in the uniaxial stress-strain law of fig. 1, failure envelopes are employed. Fig. 2 shows an example of the failure envelopes in two principal stress directions. Regarding the tensile failure envelope it is noted that considering one principal stress direction the tensile strength of the material in this direction does not change with the introduction of a tensile stress in the other principal stress direction, but a compressive stress decreases this tensile strength.

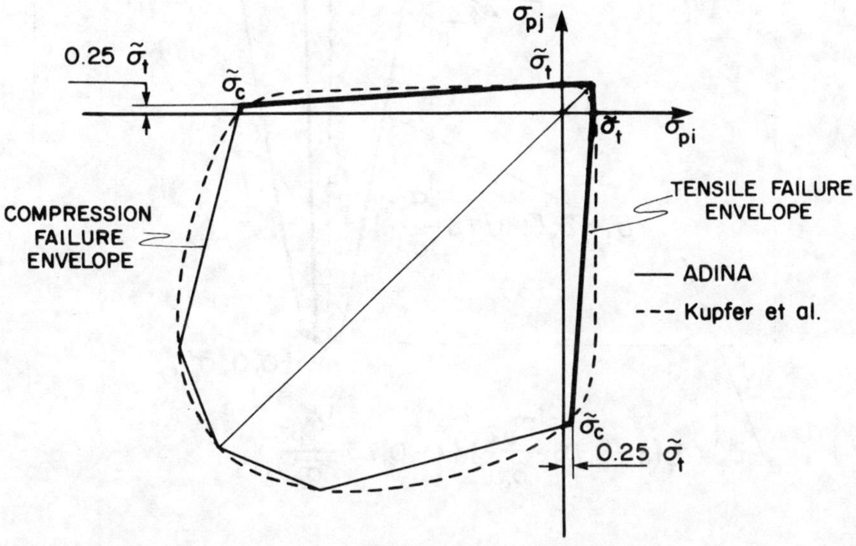

Fig. 2 Triaxial failure envelopes reduced to two-dimensional conditions

The failure envelopes shown in fig. 2 are a special case of the triaxial failure envelopes for tensile and compressive failures shown in figs. 3 and 4.

The triaxial tensile failure envelope is a straight-forward generalization of the envelope for two-dimensional analysis. However, the compressive failure envelope is more complex and requires more data input. The envelope can be used to represent a large number of different envelopes like the biaxial

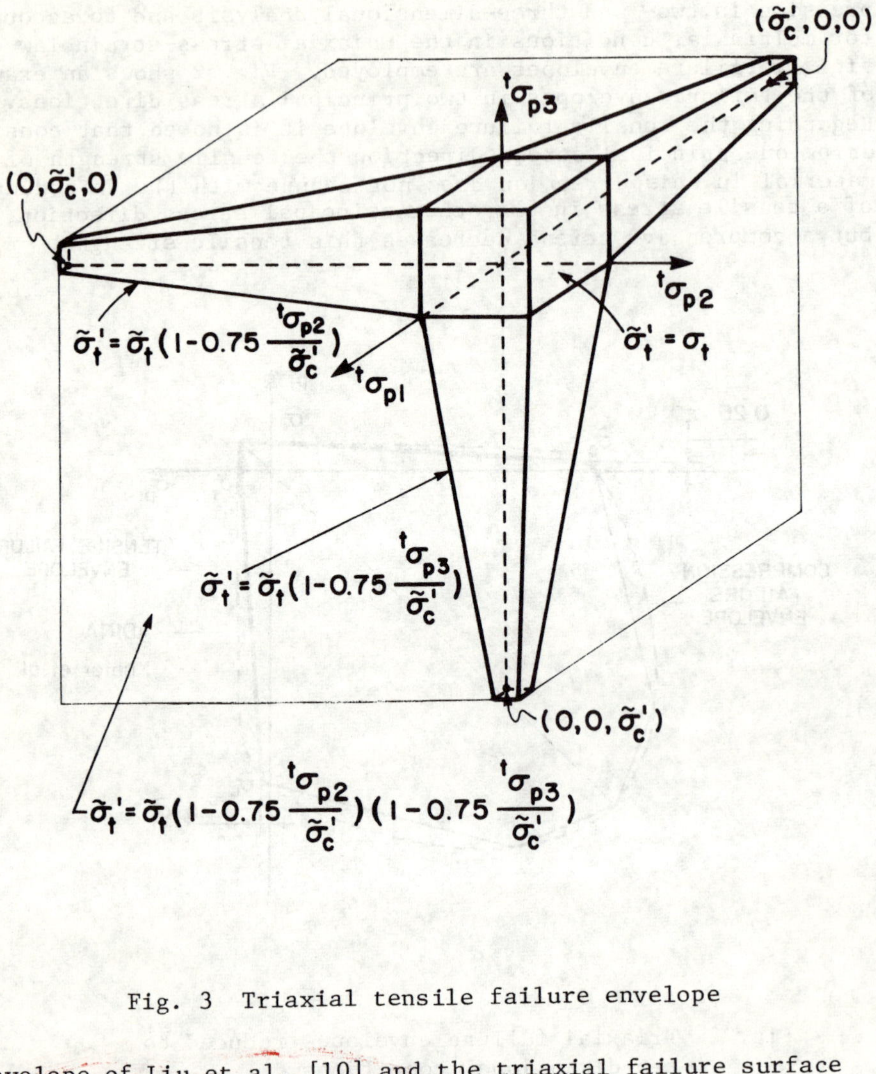

Fig. 3 Triaxial tensile failure envelope

envelope of Liu et al. [10] and the triaxial failure surface of Khan and Saugy [11]. The shape of the compressive failure surface used is largely based on the experimental results reported by Kupfer et al. [9] and Launay and Gachon [12].

The compressive failure envelope is input using 24 discrete stress values. Firstly, the values $\sigma^{i}_{p1}/\tilde{\sigma}_{c}$ are input. These values define at what stress magnitudes ${}^{t}\sigma_{p1}$ the discrete two-dimensional failure envelopes for additional stresses ${}^{t}\sigma_{p2}$ and ${}^{t}\sigma_{p3}$ are input. These failure envelopes are defined by the

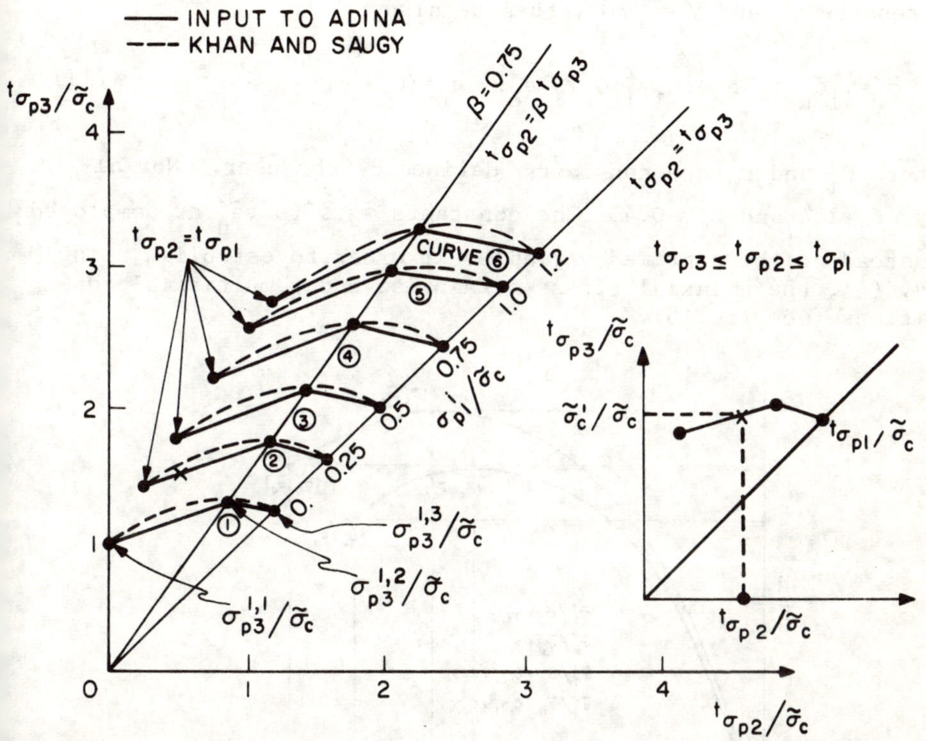

Fig. 4 Triaxial compression failure envelope

failure stress values $\sigma_{p3}^{i,j}/\tilde{\sigma}_c$ (i=1,...,6; j=1,2,3) that correspond to the stress magnitudes ${}^t\sigma_{p2}={}^t\sigma_{p1}$, ${}^t\sigma_{p2}=\beta\,{}^t\sigma_{p3}$ (β is a constant) and ${}^t\sigma_{p2}={}^t\sigma_{p3}$.

The failure envelopes are employed to establish the uniaxial stress-strain law accounting for multiaxial stress conditions, and to identify whether tensile or crushing failure of the material has occurred. Having calculated the current principal stresses, to establish the uniaxial stress-strain law it is assumed that ${}^t\sigma_{p1}$ and ${}^t\sigma_{p2}$ are held constant and the minimum stress that would have to be reached in the third principal stress direction to cause crushing of the material is calculated using the failure envelopes, see fig. 4. Let this stress be σ_c' and $\gamma_1=\tilde{\sigma}_c'/\tilde{\sigma}_c$, then we also use

$$\tilde{\sigma}_u'=\gamma_1\tilde{\sigma}_u \; ; \quad \tilde{e}_c'=(C_1\gamma_1^2+C_2\gamma_1)\tilde{e}_c ; \quad \tilde{e}_u'=(C_1\gamma_1^2+C_2\gamma_1)\tilde{e}_u \tag{11}$$

where C_1 and C_2 are constants defined by the user. Normally, $C_1=1.4$ and $C_2=-0.4$. The constants $\tilde{\sigma}_c',\tilde{\sigma}_u',\tilde{e}_c',\tilde{e}_u'$ are employed instead of the unprimed variables in order to establish using eq. (4), the uniaxial stress-strain law under multiaxial conditions (see fig. 5).

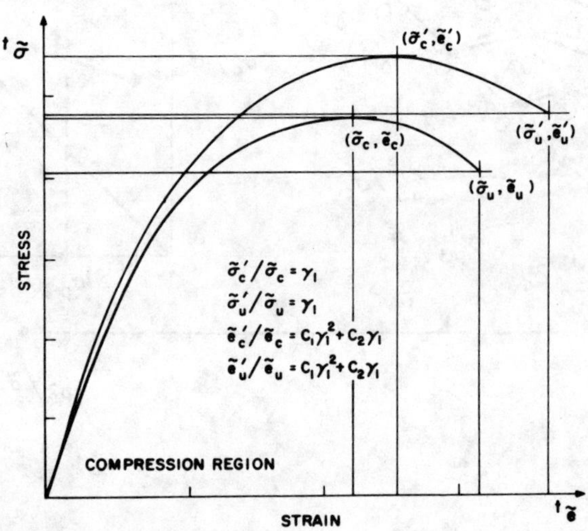

Fig. 5 Increase of strength parameters for model under multiaxial conditions

To identify whether the material has failed, the principal stresses are used to locate the current stress state in the failure envelopes. In the following we consider how one single plane of tensile failure develops and how the material fails in compression crushing.

Tensile failure occurs if the tensile stress in a principal stress direction exceeds the tensile failure stress. In this case it is assumed that a plane of failure develops perpendicular to the principal stress direction. The effect of this material failure is that the normal and shear stiffnesses across the plane of failure, and the corresponding normal stress and shear stresses are reduced.

Prior to tensile failure the stress-strain material law is given by eqs. (7)-(10). Assuming that ${}^{t}\sigma_{p1}$ is larger than the tensile failure stress, the new material stress-strain matrix is, in the stiffness matrix calculation,

$$\underline{C} = \frac{1}{1-\nu^2}\begin{bmatrix} {}^{t}\tilde{E}_{p1}\eta_n & \nu\,{}^{t}E_{12}\eta_n & \nu\,{}^{t}E_{13}\eta_n & 0 & 0 & 0 \\ & {}^{t}\tilde{E}_{p2} & \nu\,{}^{t}E_{23} & 0 & 0 & 0 \\ & & {}^{t}\tilde{E}_{p3} & 0 & 0 & 0 \\ & & & \frac{1}{2}(1-\nu)\,{}^{t}E_{12}\eta_s & 0 & 0 \\ & \text{sym.} & & & \frac{1}{2}(1-\nu)\,{}^{t}E_{13}\eta_s & 0 \\ & & & & & \frac{1}{2}(1-\nu)\,{}^{t}E_{23} \end{bmatrix} \qquad (12)$$

where the ${}^{t}\tilde{E}_{pi}$ are the uniaxial Young's moduli evaluated in the principal stress directions using eqs. (2) and (4) and the ${}^{t}E_{ij}$ are evaluated using eq. (8). The constant η_n is a small value, normally 10^{-4}, to reduce the stiffness normal to the tensile failure plane, and the constant η_s, normally 0.5, reduces the shear stiffness in the tensile failure plane.

For the stress calculation, the following stress-strain matrices are used.

For the tensile stress normal to the tensile failure plane and the shear stresses in this plane, we use the total strains to calculate the total stresses with

$$\hat{\underline{C}} = \begin{bmatrix} E_f & 0 & 0 \\ & G_{12}^{f} & 0 \\ \text{sym.} & & G_{13}^{f} \end{bmatrix} \qquad (13a)$$

where E_f and G^f_{12}, G^f_{13} are evaluated as shown in fig. 6. In this figure α is a variable, input by the user that defines the amount of tension stiffening. Fig. 6 shows that G^f_{ij} is evaluated from ${}^{\tau}G_{ij}$, which is the secant shear modulus of the material corresponding to the total strains. Also, fig. 6 assumes loading from zero stress directly into the tensile region. If the tensile stress is reached by unloading from a compressive stress, the strain normal to the tensile failure plane is measured from the strain where the stress is zero (see the example of fig. 7).

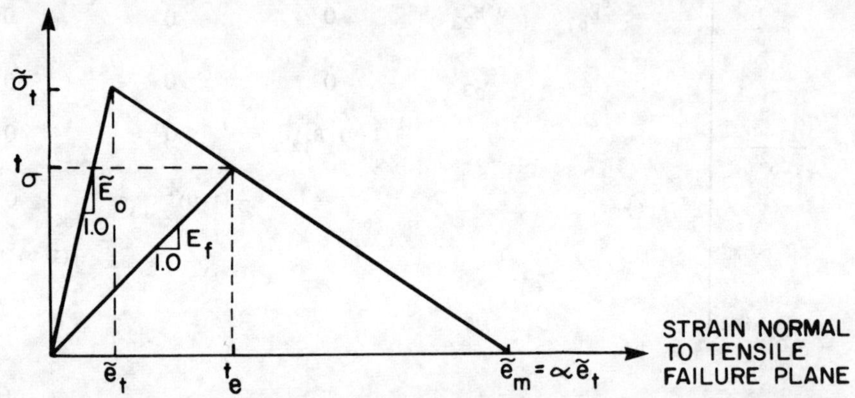

a) Calculation of Young's modulus, E_f, normal to tensile failure plane

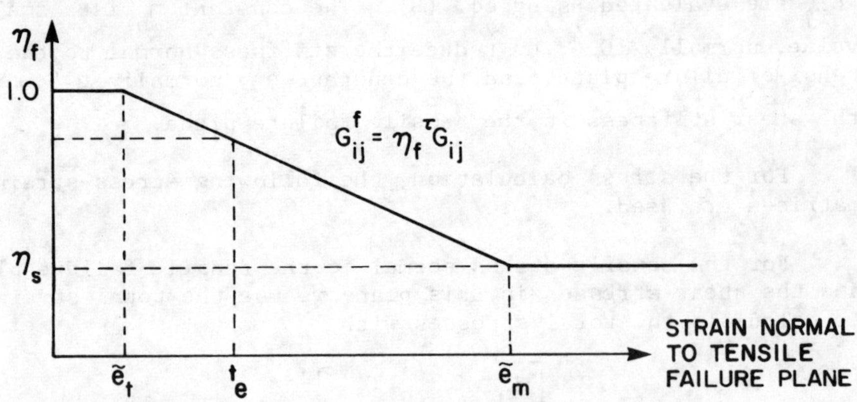

b) Calculation of shear modulus in tensile failure plane

Fig. 6 Material moduli for stress calculation after tensile failure

For the remaining stress components, the increment in the stresses is evaluated from the incremental strains using

$$\hat{\underline{C}} = \frac{1}{1-\nu^2}\begin{bmatrix} {}^{\tau}\tilde{E}_{p2} & \nu\,{}^{\tau}E_{23} & 0 \\ & {}^{\tau}\tilde{E}_{p3} & 0 \\ \text{sym.} & & (1-\nu^2)\,{}^{\tau}G_{23} \end{bmatrix} \tag{13b}$$

where the ${}^{\tau}\tilde{E}_{pi}$ are the uniaxial Young's moduli evaluated using eq. (10) and ${}^{\tau}E_{23}$ and ${}^{\tau}G_{23}$ are evaluated using eq. (8) but with ${}^{\tau}\tilde{E}_{pi}$ instead of ${}^{t}\tilde{E}_{pi}$.

Considering eqs. (12) and (13) we note that by use of eq. (13a) the tensile stress normal to the tensile failure plane and the shear stresses in this plane are gradually released. Also, the equations show that plane stress conditions are assumed to exist at the plane of tensile failure. The factor η_n is not set exactly equal to zero in order to avoid the possibility of a singular stiffness matrix. The value to be employed for η_s must depend on a number of physical factors, and in the numerical solution it is at this point best to leave η_s as a variable that is input by the analyst.

At times, the "plane of tensile failure" is referred to as a "crack", but we should interpret this terminology judiciously because a physical crack does not actually develop at the element integration point. Instead, the material has failed in one principal stress direction.

Eqs. (12) and (13) describe the solution when tensile failure occurs. To identify compression failure, the largest principal stress ${}^{t}\sigma_{p1}$ is employed to establish from fig. 4, by interpolation, the biaxial failure envelope on ${}^{t}\sigma_{p2}$ and ${}^{t}\sigma_{p3}$. The material has crushed if the stress state corresponding to ${}^{t}\sigma_{p2}$ and ${}^{t}\sigma_{p3}$ lies on or outside the biaxial failure envelope.

2.3 Post-tensile cracking behavior

Once a tensile plane of failure has formed, it is checked in each subsequent solution step whether the failure is still active. The failure is considered to be inactive provided the normal strain across the plane becomes negative and less than the strain at which the "last" failure occurred and is active otherwise. Therefore, a tensile failure plane may repeatedly be active and inactive.

If a tensile failure plane has developed, which may or may not be active, the material stress-strain relations are always established as described above but corresponding to the principal stress directions in the failure plane and the direction perpendicular to this plane. Hence, instead of using the principal stresses and corresponding directions as done for the unfailed material, the stress conditions along and normal to the material tensile failure plane are used to evaluate the stress-strain matrix. Also, when a failure plane is or was active, a subsequent failure plane is assumed to form perpendicular to the direction of the one that developed first, once a normal stress along the original failure plane has reached the tensile failure stress. It follows that at any integration point, the direction of the third tensile failure plane is fixed once failure has occurred in two directions.

It may also happen that after tensile failure of the material (in one or two directions) the material fails in compression crushing, which is identified, as usual, by entering the compression crushing envelope in fig. 4 with the principal stress(es) that act(s) along the tensile failure plane(s).

If the material has crushed in compression, it is assumed that the material strain-softens into all directions.

2.4 Post-compression crushing, strain-softening behavior

Consider first uniaxial stress conditions. As shown in fig. 1, for a uniaxial strain smaller than $\tilde{e}_c$, the material has crushed and softens with increasing compressive strain, i.e. ${}^t\tilde{E}$ is negative. The difficulty of including this material behavior lies in that the stiffness matrix can become indefinite if a negative Young's modulus is used. To circumvent this difficulty if ${}^t\tilde{E}$ in eq. (4) is negative, a zero value (actually a small positive value) for ${}^t\tilde{E}$ is employed instead. However, in the calculation of the stress increments the actual negative value of ${}^\tau\tilde{E}$ is used.

A special case corresponds to the use of the automatic load incrementation algorithm available in ADINA [13]. In this solution, the actual negative tangent Young's modulus is used in the stiffness calculation.

Under multiaxial stress conditions the compression crushing is identified using the multiaxial failure envelope, and once the material has crushed isotropic conditions are assumed. As in uniaxial conditions, in the subsequent solution steps the Young's modulus is assumed to be zero (i.e. very small positive) in the stiffness matrix calculations, unless the automatic load stepping algorithm is used, but the stress increments are computed from the uniaxial stress-strain law with the constants $\tilde{\sigma}_c'$, $\tilde{e}_c'$ and so on (see fig. 5) corresponding to the multiaxial conditions at crushing. The Young's modulus ${}^{\tau}\tilde{E}$ corresponding to the current strain increment e_{p3} is evaluated using the uniaxial stress-strain relationship in fig. 1,

$$ {}^{\tau}\tilde{E} = \left[\tilde{\sigma}\big|_{\text{at } {}^{t}e_{p3} + e_{p3}} - \tilde{\sigma}\big|_{\text{at } {}^{t}e_{p3}} \right] \Big/ e_{p3} \qquad (14) $$

where ${}^{t}e_{p3}$ and e_{p3} are the strain component and incremental strain component at time t measured in the direction of the principal stress ${}^{t}\sigma_{p3}$. To obtain the stress increment, eq. (9) is used with the matrix $\underline{\hat{C}}$ corresponding to isotropic material conditions with Young's modulus ${}^{\tau}\tilde{E}$.

Note that, as shown in fig. 1, when ${}^{t}e_{p3}$ becomes equal to or less than $\tilde{e}_u'$, the constant Young's modulus $\tilde{E} = -\tilde{\sigma}_u'/\tilde{e}_u'$ is employed; i.e. ${}^{\tau}\tilde{E} = \tilde{E}$ is then used.

If unloading of the crushed material in the strain-softening region occurs, characterized by $e_{p3} \geq 0$, the initial Young's modulus $\tilde{E}_0$ is used. In all these calculations the principal stresses are checked individually whether a positive value has been reached, and if so, the stress in the corresponding direction is set to zero.

2.5 Temperature effects

In some analyses temperature strains, ${}^{t}\underline{e}^{th}$, need be included. These strains are taken into account by replacing the total incremental strain, $\underline{e}$, in the governing incremental equations by the strain $\underline{e} - \underline{e}^{th}$, where $\underline{e}^{th}$ is the thermal incremental strain and is calculated from the temperature conditions.

3. EXAMPLE SOLUTIONS

Figure 7 summarizes the response obtained when subjecting a single plane stress element to one-dimensional strain-controlled loading.

This solution includes

- initiation, closing and reopening of a crack
- loading, unloading and reloading with large load steps
- crushing of the material with strain-softening behavior.

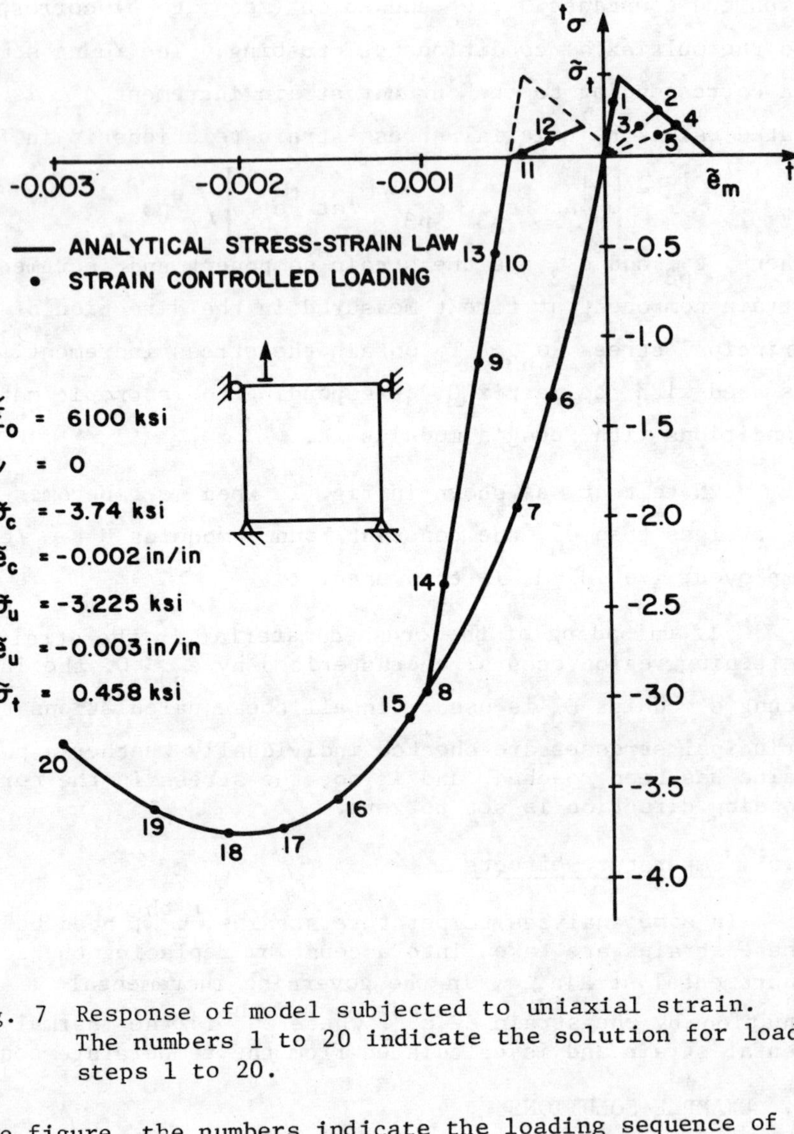

Fig. 7 Response of model subjected to uniaxial strain. The numbers 1 to 20 indicate the solution for load steps 1 to 20.

In the figure, the numbers indicate the loading sequence of the concrete model. Note that the material model can handle relatively large load step increments.

Figure 8 shows the response calculated when a fully constrained plane stress concrete element is subjected to a uniform temperature loading, $^t\theta$. The initial temperature is zero degrees and the temperature is raised to 450°C in 15 equal load steps. Note the increase in the crushing strength of the material due to the biaxial stress conditions.

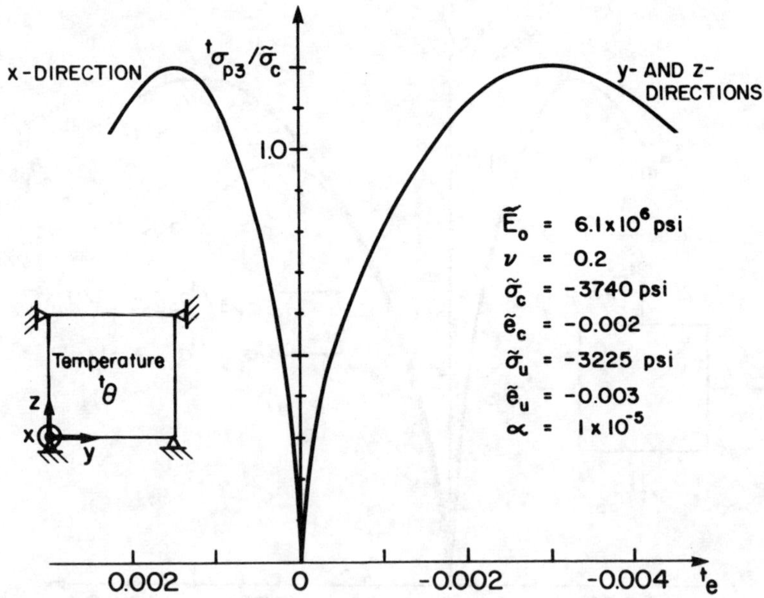

Fig. 8 Response of plane stress element subjected to temperature loading; ${}^t\theta$ varies from 0 to 450°C in 15 equal steps.

A simple plane stress concrete element subjected to biaxial compression loading is considered in fig. 9. The analysis is performed using the automatic load incrementation algorithm in the ADINA program. For equal biaxial compression a strength increase of 20 percent over $\tilde{\sigma}_c$ is achieved. As pointed out above, when using the automatic load incrementation algorithm, ADINA always evaluates for the stiffness matrix calculations the actual tangent stiffness modulus in the strain softening region and calculates directly the post-crushing response.

Figure 10 shows the results obtained in a two-dimensional analysis of a simply-supported concrete beam. The objective with this example is to show the tension stiffening effect on the beam response. Ten six-node two-dimensional elements with 2x2 Gauss integration are used to model the concrete, and the steel reinforcement is modeled using 3-node truss elements with an area of 0.62 in^2. Figure 10 compares the results obtained when α=1 (no tension stiffening [1]) with those of α=8, and also gives the results of Suidan and Schnobrich [14]. The ultimate load including tension stiffening is not significantly different from the response when α=1.0, but the numerical solution is considerably more stable.

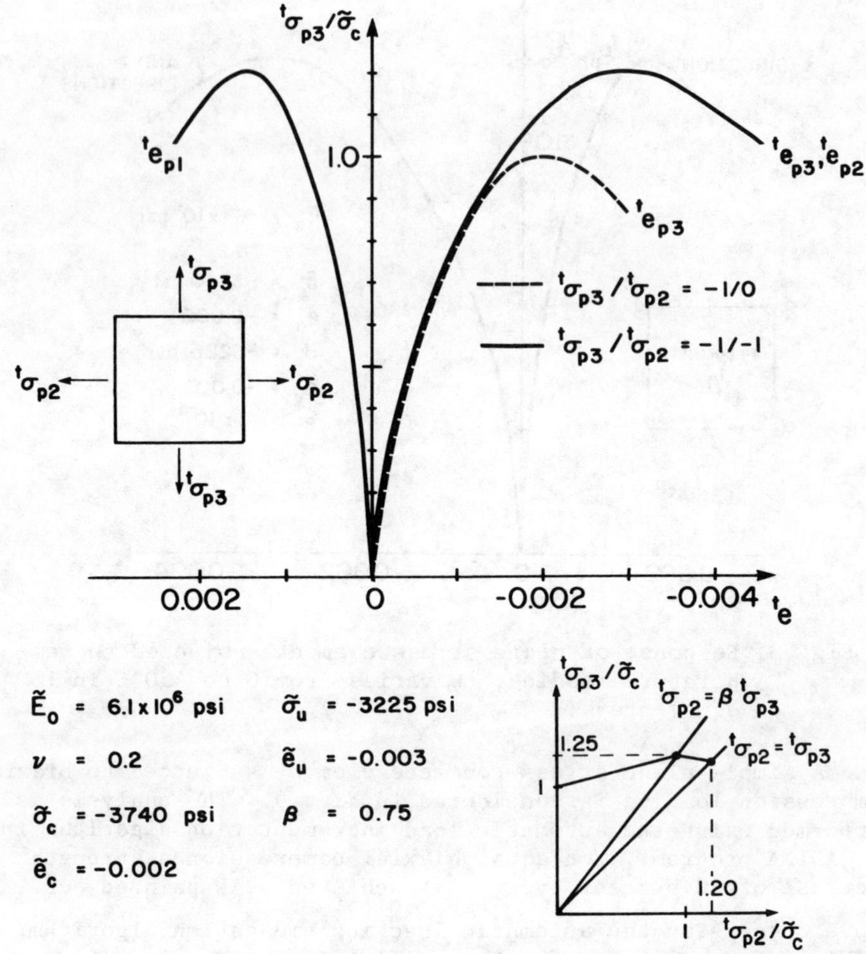

Fig. 9 Response of plane stress element subjected to biaxial loading

4. CONCLUDING REMARKS

The objective in this paper was to summarize the current status of our concrete model. The model can be employed in a number of important concrete analyses, notably the collapse analyses of two- and three-dimensional concrete structures. One important aspect is the generality of the model with respect to the stress-strain law and failure surfaces used. Also, the model is deemed computationally efficient because relatively large incremental load steps can be employed.

Although a powerful capability already, further improve-ments to the concrete model will surely be desirable. These will best be identified in the further study of the model against laboratory and field test results.

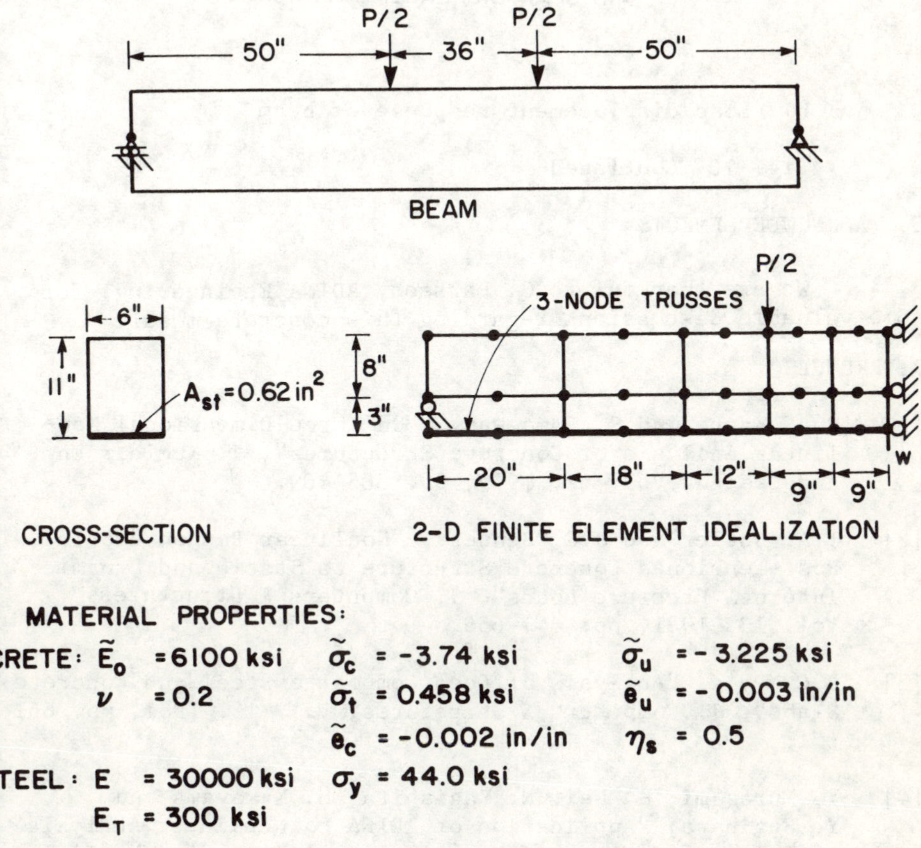

a) Model of beam

Fig. 10 Analysis of a reinforced concrete beam

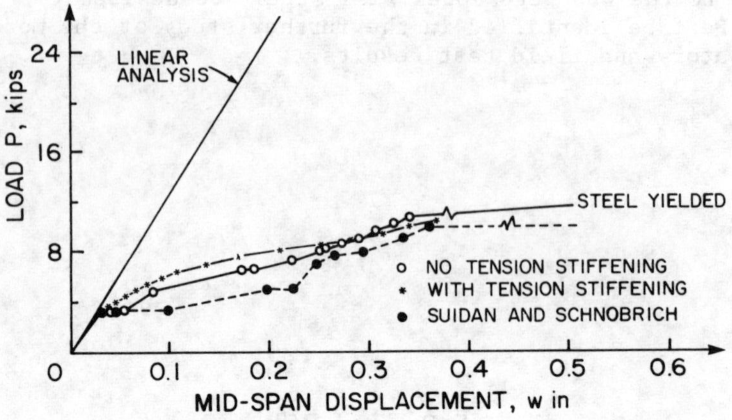

b) Load displacement response of beam

Fig. 10 Continued

5. ACKNOWLEDGEMENTS

We are thankful to G. Larsson, ADINA Engineering, for some valuable discussions regarding this concrete model.

REFERENCES

[1] K.J. Bathe and S. Ramaswamy, "On Three-Dimensional Nonlinear Analysis of Concrete Structures", J. Nuclear Eng. and Design, Vol. 52, 1979, pp. 385-409.

[2] T.A. Butler and J.G. Bennett, "Nonlinear Response of a Post-Tensioned Concrete Structure to Static and Dynamic Internal Pressure Loads", J. Computers & Structures, Vol. 13, 1981, pp. 647-659.

[3] J.C. Pyle, "Analysis of Deep Composite Steel and Concrete Slabs", J. Computers & Structures, Vol. 13, 1981, pp. 661-671.

[4] Y. Murazumi, H. Nei, N. Yagishita, M. Nakayama and Y. Sugihara, "Application of ADINA to Nonlinear Analysis of Reinforced Concrete Shear Walls with Openings", J. Computers & Structures, Vol. 13, 1981, pp. 727-736.

[5] M. Keuser, G. Mehlhorn and V. Cornelius, "Bond Between Prestressed Steel and Concrete—Computer Analysis Using ADINA", J. Computers & Structures, Vol. 17, 1983, pp. 669-676.

REFERENCES

[6] M. Keuser, B. Kepp, G. Mehlhorn and R. Rostásy, "Nonlinear Static Analysis of End-Fittings for GFRP-Prestressing Rods", J. Computers & Structures, Vol. 17, 1983, pp. 719-730.

[7] J.W. Tedesco and W.G. McDougal, "Nonlinear Dynamic Analysis of Concrete Armor Units", J. Computers & Structures, Vol. 21, 1985, pp. 189-201.

[8] K.J. Bathe, Finite Element Procedures in Engineering Analysis, Prentice-Hall, Inc., 1982.

[9] H. Kupfer, H.K. Hilsdorf and H. Rusch, "Behavior of Concrete Under Biaxial Stresses", J. of ACI, Vol. 66, 1969, pp. 656-666.

[10] T.C. Liu, A.H. Nilson and F.O. Slate, "Biaxial Stress-Strain Relations for Concrete", ASCE, J. Struct. Div., Vol. 98, 1972, pp. 1025-1034.

[11] M.H. Khan and B. Saugy, "Concrete for Nuclear Reactors", ACI Special Publication 34, 1972.

[12] P. Launay and H. Gachon, "Concrete for Nuclear Reactors", ACI Special Publication 34, 1972.

[13] K.J. Bathe and E. Dvorkin, "On the Automatic Solution of Nonlinear Finite Element Equations", J. Computers & Structures, Vol. 17, 1983, pp. 871-879.

[14] M. Suidan and W.C. Schnobrich, "Finite Element Analysis of Reinforced Concrete", ASCE, J. Struct. Div., Vol. 99, 1973, pp. 2109-2122.

CHAPTER 5

COMPOSITE FRACTURE MODEL FOR STRAIN SOFTENING AND LOCALISED FAILURE OF CONCRETE

K. Willam
University of Colorado, Boulder, USA

N. Bićanić
University College of Swansea, Wales, UK
University of Zagreb, Yugoslavia

S. Sture
University of Colorado, Boulder, USA

SUMMARY

Localised and distributed concrete fracture are examined within the scope of the composite fracture model of an equivalent homogenised continuum of finite dimension. In one extreme the model reduces to the traditional strain-softening formulation of continuous damage mechanics and on the other it furnishes a non-local softening law which depends on the characteristic fracture dimension as well as the fracture configuration.

These basic concepts for distributed and localised fracture are utilised in a simple triaxial fracture model where tensile debonding and frictional decohesion are considered separately, described by a non-associative Mohr-Coulomb formulation with tension cut-off, in which the tensile strain-softening is controlled by the critical fracture energy release rate.

The performance of the separate composite fracture models in tension and shear is studied on the direct tension and triaxial compression tests and the triaxial composite fracture model is used for the computational prediction of the shear box test of concrete.

The need for the model capabilities to capture the transition from brittle to ductile fracture is discussed as well as the qualitative formulation of frictional hardening and cohesive softening model for the pre- and post-peak responses under triaxial conditions.

1. INTRODUCTION

The postpeak strength behaviour of engineering materials has been of considerable interest in the recent past. In particular the limited ductility of cementitious materials has been investigated with conflicting concepts of the continuous strain-softening models on one side and fracture mechanics concepts on the other. The computational advantages of the smeared approach provide the predominant argument in favour of the continuous damage models in which progressive decohesion due to distributed microdefects is described by an equivalent strain-softening behaviour of a homogenised continuum, but serious questions arise with regard to the validity of computational predictions in the case of fracture localisation when the structural integrity is determined by the severity of a single macrodefect. This type of failure has been traditionally the realm of fracture mechanics in which initiation and propagation of a single discontinuity is considered the determining factor.

Recently developed composite fracture formulation embodies the basic concepts of single and distributed fracture events and reconciles the fundamental differences of strain-softening and fracture mechanics. On one end of the spectrum the composite model enforces a fracture energy release rate which is constant in the case of a highly localised discrete discontinuity, and on the other it reduces to the continuous damage theory for distributed fracture. The constitutive formulation follows strain-softening plasticity theories in which the non-associated flow rule is recast into a non-local fracture rule which includes the characterisitic fracture dimension of the fracture process zone. The underlying composite fracture parameters for tension and shear are calibrated from post-peak experimental data obtained from the direct tension and triaxial compression tests whose principal features are discussed later. These two opposing concepts are implemented within the unified framework of the composite fracture theory which decomposes an elementary volume into a series of intact elastic zones and/or a single fracture zone in which oriented tensile cracking and frictional slip take place in a highly localised manner. A strain softening plasticity model based on the composite fracture concept of homogenising fracture discontinuities into an energy equivalent continuum has recently been extended to mixed mode fracture in the form of splitting compression and shear faulting modes of failure in order to develop a unified approach towards softening in tension and compression [30]. The implications on the computational predictions of the post-peak behaviour are then illustrated with fracture investigations of the direct tension test and the triaxial compression test on cylindrical specimens for which experimental data are available for comparison. The uniqueness of the post-peak response predictions is examined with the aid of systematic mesh sensitivity studies of unnotched test specimens which exhibit pronounced softening behaviour both in tension and shear.

The composite fracture model is also applied to study the post-peak response behaviour of prismatic concrete specimens in the direct shear test, where contained fracture takes place both in the form of tensile cracking as well as frictional slip depending on the ratio between normal and tangential load levels. In this sense the direct shear test exhibits all ingredients of a highly indeterminate structure in which progressive propagation of distributed microdefects leads to an overall degradation of stiffness and strength before coalescence into a single macro-defect takes place which is ultimately responsible for the structural failure. In addition, considerable discussion has recently centred around the well-posedness of dynamic strain softening computations. Some of these arguments are addressed in the context of the elementary model problem for which the composite fracture formulation provides both a unique and mesh-insensitive solution for the fracture energy. In fact, it will be observed that the wave propagation problem is computationally less demanding than the corresponding quasistatic strain softening analysis of continua.

2. PRINCIPAL FEATURES OF CONCRETE SOFTENING BEHAVIOUR

In conjuction with a substantial research effort on finite element prediction of the localised failure in cementitious materials two series of experiments were recently completed which involved stable post-critical loading of cylindrical concrete specimens in direct tension and triaxial compression.

2.1 Direct Tension

In the case of the direct tension the question arises if the fracture energy release rate G_f^I is a proper material property which characterizes the tensile cracking independently of the particular geometry and boundary conditions. This issue was raised previously in the context of fragmentation of rocks by Paul von Rittinger as early as 1867 [21] who stated that the newly created surface area is directly proportional to the energy expended during fracture, i.e. G_f^I = constant. Recently this fundamental fracture mechanics concept was adopted by Hillerborg et al [15] in formulating the fictitious crack band model for concrete, and was extended by Bazant et al [16] to formulate the crack band theory for distributed fracture within a process zone of finite width. Along similar lines the experimental and computational fracture studies by Ingraffea and Saouma [22] were based on the equivalent fracture postulate of a critical stress intensity factor to constitute an invariant fracture property.

To increase the understanding of the tensile fracture mechanisms the following objectives were persued with the direct tension experiments [3]:

(i) Uniqueness of the load-displacement response in the post-peak regime,

(ii) Progressive damage and stiffness degradation during loading and unloading cycles,

(iii) Size effects related to different height specimens.

Fig 1 depicts the axial load-extension response of a typical direct tension experiment conducted at a constant displacement rate with nearly linear ascending branch up to the peak load, when the axial tension suddenly decreased without noticeable formation of a crack band. In fact a single transverse hairline crack only appeared around the periphery of the sample at an overall axial displacement corresponding to the residual tensile strength diminished to near zero load level. At this stage the specimen had separated into two parts which remained entirely intact. Upon continuous loading the increasing displacements reflected simply further crack opening accompanied by rigid body movement of the two separated portions of the specimen.

The displacement-controlled formation of a single crack band in the direct tension specimen can be used to determine the critical fracture energy release rate in tension G_f^I. Following Petersson [23] the integral of nominal normal stress-displacement curve shown in Fig 1 yields a first order approximation of the critical fracture energy release rate in tension for the particular concrete under consideration.

$$G_f^I = \int_0^{u_r} \sigma_z \, du_z \qquad (1)$$

To study the reproducibility of the post-peak response, loading-unloading experiments were performed in order to explore the residual tensile strength and the associated tangent stiffness. Fig 2 depicts the resulting load-displacement response when the stroke is reversed to zero displacement and subsequently increased again. The plot clearly illustrates tensile bonds still remaining intact in the softening regime and the residual tensile strength is fully recovered during reloading. In fact, the post-peak load-displacement response curve approaches that of the monotonic experiment very much along the line of

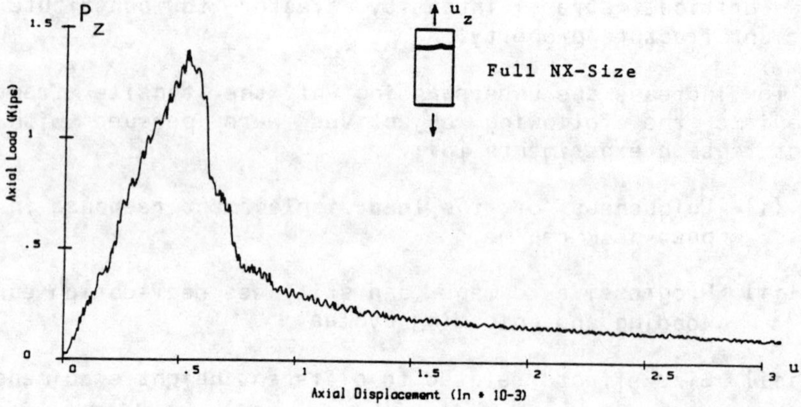

Fig 1 - Monotonic test - full size specimen

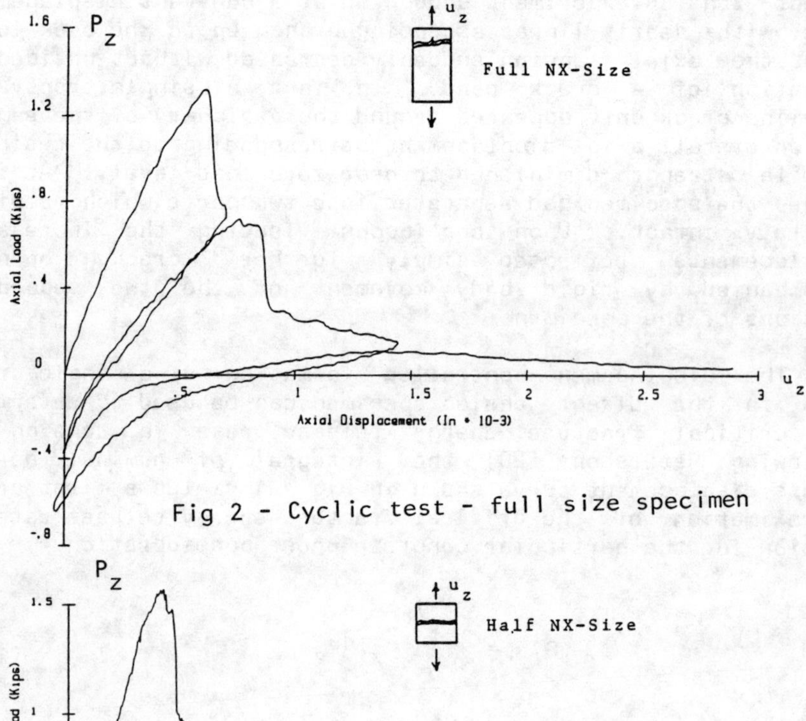

Fig 2 - Cyclic test - full size specimen

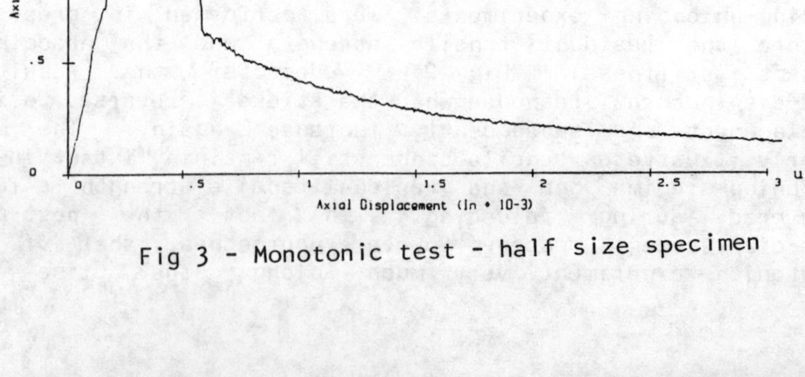

Fig 3 - Monotonic test - half size specimen

strain-softening elastic-plastic behaviour with substantial permanent deformations. A slight decrease of elastic stiffness in the first cycle and a substantial reduction when the tensile crack has fully developed are observed, similarly to the results on notched tension specimens reported recently by Reinhardt [1]. However, the level of elastic-inelastic coupling is not nearly so pronounced as to fully support the progressive fracturing theory of Dougill [8] or the continuous damage theory of Mazars, Hult and Krajčinović [6,9,10].

In order to verify the fracture mechanics hypothesis of a constant fracture energy release rate additional monotonic tension tests were carried out with a specimen of half the height and the same cross-section. For a hypothesis of a constant fracture energy release rate the area under the P-u diagram (actually the nominal axial stress-extension plot) should remain constant, assuming that the tensile strength is not affected by the two different volumes and their different probability of failure. On the other hand, if the concept of continuously distributed cracking should be adopted along the line of the smeared cracking approach, then the area under the P-u diagram, i.e., the strain energy dissipation during the transverse crack process should be reduced by half. Fig 3 illustrates the results of the half-size specimen test. The plot clearly indicates that the area under the P-u diagram is hardly affected by halving the height of the tension specimen, hence the results support the postulate of a constant fracture energy which controls the formation of a single transverse crack.

2.2 Triaxial Compression Experiment

While the fracture mechanics concepts have been widely accepted for tensile macro-cracking, there is considerable debate with regard to shear fracture of concrete and geomaterials. In fact, most geotechnical failure computations are based on the traditional notion of hardening/softening plasticity theories [24,25], while prominent applied mechanicians advocate fracture mechanics models for the development of shear bands in geophysical applications [26, 27, 28].

In cementitious materials such as concrete the situation is aggravated by the heterogeneous composition of the mesostructure which leads, for example, to splitting tensile failure under uniaxial compression. Moreover, the traditional shear tests normally result in a tensile cracking mode of fracture if brittle materials are considered, and a frictional slip mode of failure can only be observed under sufficiently high confining stresses.

Axisymmetric triaxial compression experiments were therefore pursued to study

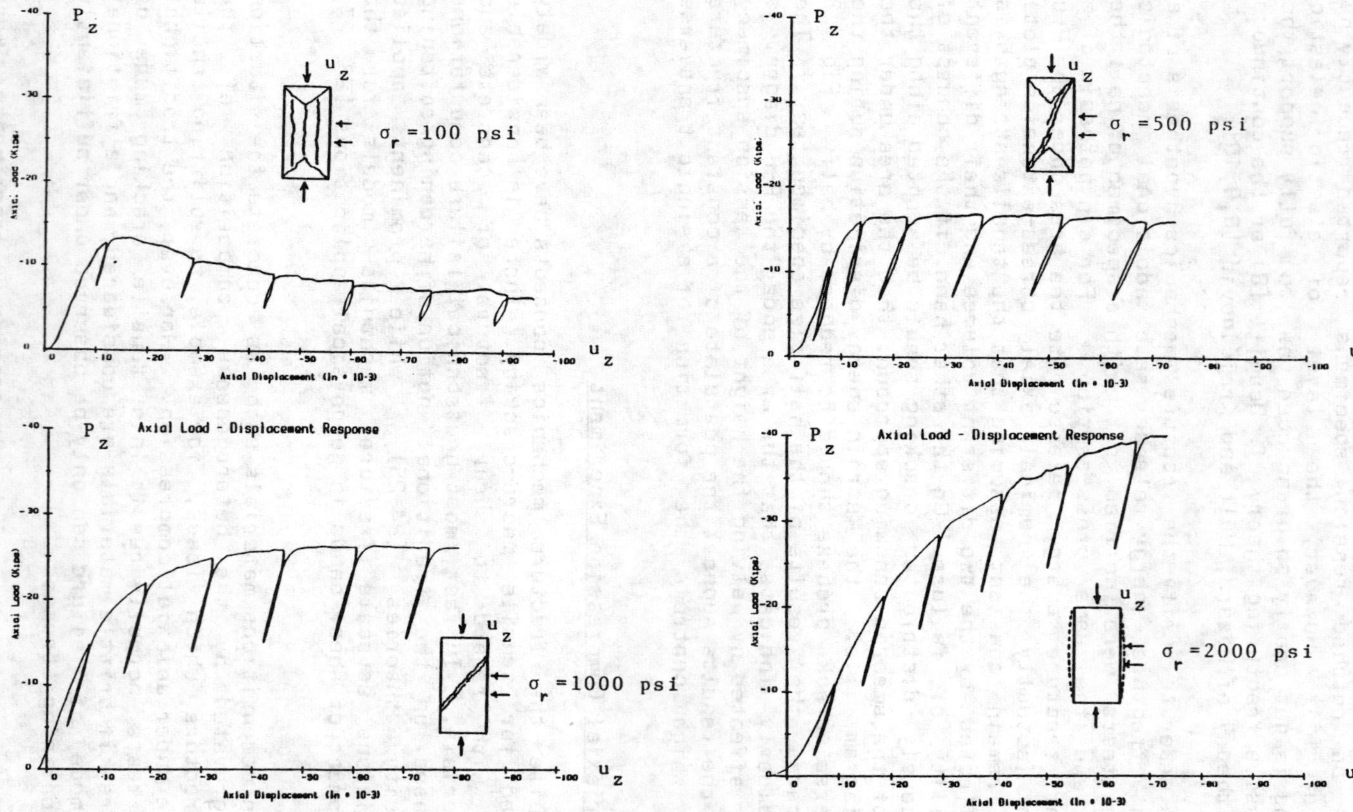

Fig 4 - Load displacement diagrams for the partial unloading experiments in compression

(i) Transition from brittle to ductile fracture at various levels of confinement,

(ii) Shear dilatancy during progressive shear fracture,

(iii) Stiffness degradation during loading - unloading cycles.

To this end conventional triaxial compression experiments were conducted in a Hoek cell device developed previously at the University of Colorado, Boulder. Experimental setup is described in detail elsewhere [29,30], and Fig 5 depicts the axial load-displacement behaviour at four different confining levels σ_r = 100, 500, 1000 and 2000 psi. The experimental results are plotted in the form of global load-displacement results rather than in the form of stress and strain measures since localization of fracture takes place in the softening regime which requires a proper homogenisation concept. The four plots clearly demonstrate the increase of ductility with increasing confinement which suppresses the softening tendencies. The partial unloading cycles indicate very little hysteretic energy dissipation. The linear elastic unloading-reloading behaviour hardly exhibits any degradation of stiffness. This suggests that the elastic-plastic coupling remains negligible and conjectures little damage accumulation of the elastic material texture not only in the pre- but also in the post-peak regime. The resulting failure modes showed a distinct transition from splitting tension under low confinement, over shear faulting under intermediate confinement, to distributed plastic flow at high confinement when no softening was observed altogether.

The dilatation characteristics are shown in Fig 5 in the form of lateral versus axial displacement response curves. Again the lateral expansion diminishes rapidly with increasing confinement pressure. This indicates that volumetric dilatation accompanies strain-softening and that the transition from brittle to ductile fracture is related to the transition from volume dilation to volume compaction.

The triaxial strength results are illustrated in Fig 6 in the Rendulic plane where the data are shown on the compressive meridian. The diagram depicts the maximum strength envelope in conjunction with the residual strength data. The residual strength diminishes to zero at zero confinement and approaches rapidly the maximum strength envelope with increasing confinement when softening is almost completely suppressed.

The direction of inelastic fracture strains are superimposed in Fig 7 onto the stress path in the Rendulic plane. To this end the axial and radial displacements in Fig 5 were converted into strains since their ratio remained fairly constant in the softening regime. It was therefore felt that

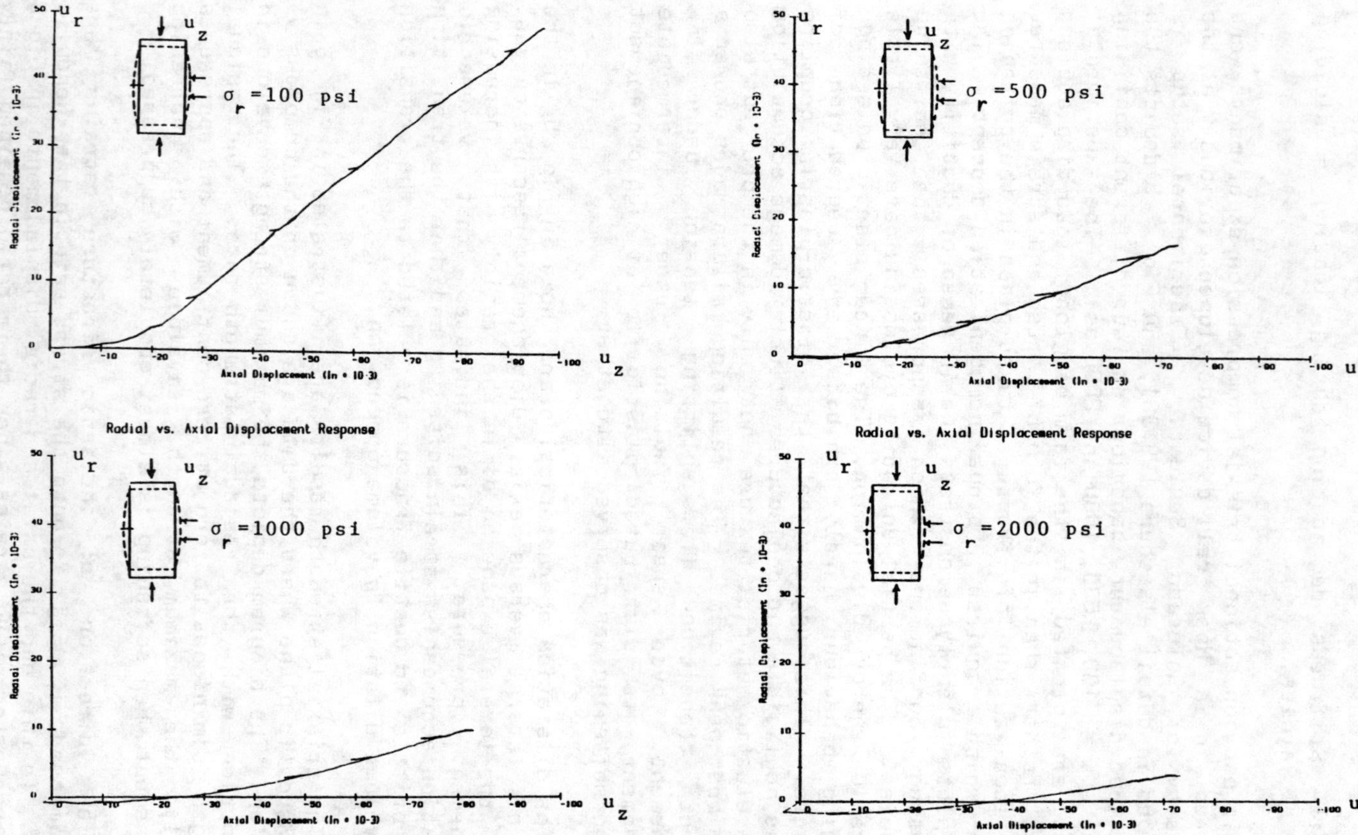

Fig 5 - Dilatation curves for the partial unloading experiments in compression

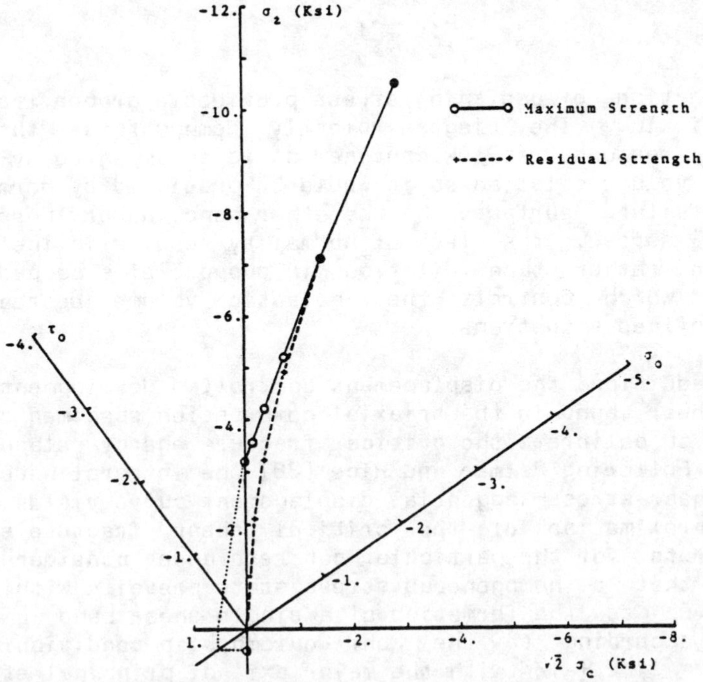

Fig 6 - Triaxial strength envelopes in Rendulic plane

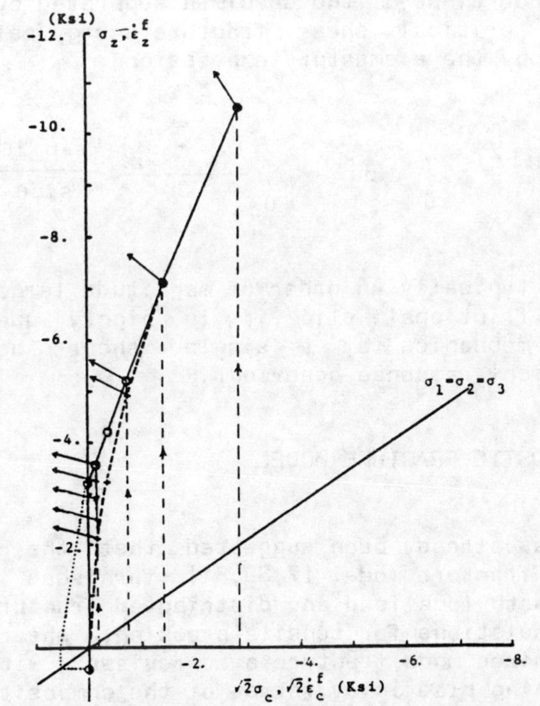

Fig 7 - Post-peak fracture strains in Rendulic plane

the inelastic dilatancy ratio

$$\alpha = \dot{\varepsilon}_r^f / \dot{\varepsilon}_z^f$$

as a function of confining stress provides a proper fracture characteristic. The diagram clearly demonstrates that the post-peak response at low confinement is accompanied by large inelastic volume dilation as it would be predicted by normality to the failure surface. On the other hand, under increasing confining stress, the lack of normality results in inelastic compaction rather than dilation in support of a capped flow potential which controls the inelastic volume decrease in highly confined situations.

In addition, the displacement controlled development of a single shear band in the triaxial compression specimen can be utilised to calibrate the critical fracture energy relase rate in shear. Following Palmer and Rice [38] the integral under the nominal shear stress-tangential displacement curve yields first order approximation of the critical shear fracture energy release rate for the particular concrete under consideration. Assuming that a homogeneous stress state prevails within the specimen up to the formation of a single shear band which is inclined according to the Mohr-Coulomb slip condition at an angle $\theta = \psi/2 + \pi/4$ with the major axis of principal stress, and neglecting the effect of elastic shear deformations within the two portions of the specimen separated by the shear band, then the critical shear fracture energy release rate can be evaluated by the elementary expression

$$G_f^{II} = \int_0^{u_r} \tau \, du = \int_{u_o}^{u_r} \frac{(\sigma_r - \sigma_z)}{2} \frac{\sin 2\theta}{\sin\theta} du_z \tag{2}$$

which is typically an order of magnitude larger than G_f^I in the case of frictional slip it is highly questionable if the fracture mechanics of a single shear band governs the post-critical response behaviour.

3. COMPOSITE FRACTURE MODEL

It has already been suggested that the recently proposed composite fracture model [7,30,31] furnishes formulation that includes both localised and distributed fracture concepts. The basic formulations for tensile cracking, shear slip and the combined mixed mode fracture are reviewed with an emphasis on the softening plasticity format of the composite damage model.

3.1 Composite Fracture Model for Localised Tensile Fracture

In the case of tensile cracking the elementary volume V is composed of the undamaged elastic zone V_e and the localised crack band V_t of width d_t and length b_t in which oriented microcracking is confined. Resorting to the maximum stress criterion for cracking, the orientation of the tensile crack band is defined by the angle between the reference axis and the minor principal stress direction (Fig 8).

From the equivalence of the energy release during cracking it can be shown [7] that the softening modulus E_c^{ts} in tension (loss of tensile strength related to the total strain) depends on the fracture energy release rate G_f^I as well as the crack geometry

$$E_c^{ts} = \frac{1}{\dfrac{2\, G_f^I\, b_t}{\sigma_t^2\, a\, b} - \dfrac{1}{E}} \tag{3}$$

This expression clearly demonstrates that the strain-softening modulus E_c^{ts} of the equivalent continuum element depends also on the geometry of the elementary volume and thus exhibits a pronounced size effect. In order words, E_c^{ts} varies with the mesh size of the particular finite element layout if the fracture energy release rate G_f^I constitutes an invariant fracture property.

Contrary to some earlier proposals, the equivalent softening modulus is completely defined by two parameters, the initial tensile strength σ_t and the fracture energy release rate G_f^I and there is no need to speculate on the width of the blunt crack band d_t or the material behaviour in the localised fracture zone.

If the angle of localisation band is ignored, i.e. $b_t = b$, $h_t = a$, it holds that

$$E_c^{ts} = \frac{1}{\dfrac{2\, G_f^I}{\sigma_t^2}\dfrac{1}{h_t} - \dfrac{1}{E}} \tag{4}$$

Alternatively, the gradual softening can be described in terms of softening modulus E_c^{fs}, where the loss of tensile strength is related to the fracture strain of the homogenised continuum (Fig 8).

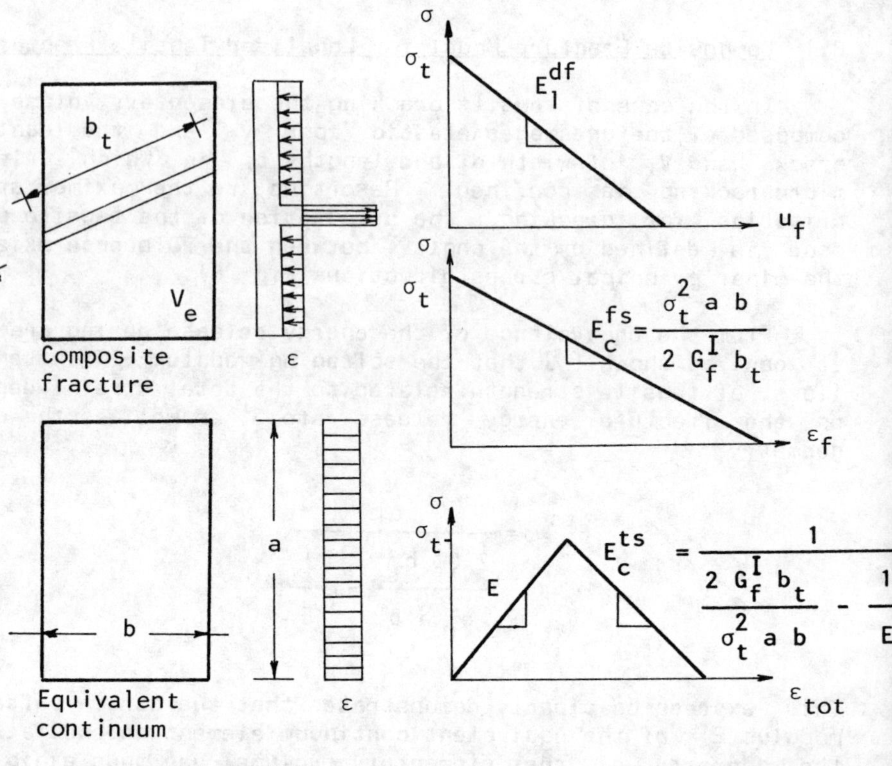

Fig 8 - Composite fracture model for tensile cracking

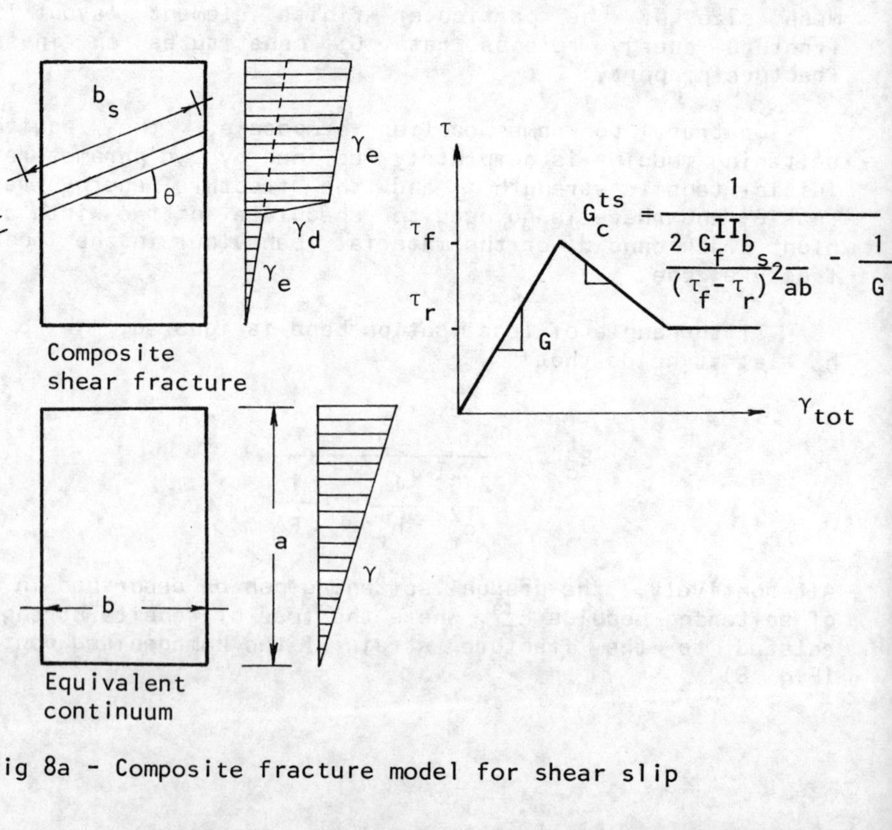

Fig 8a - Composite fracture model for shear slip

$$E_c^{fs} = \frac{E\,E_c^{ts}}{E + E_c^{ts}} = \frac{\sigma_t^2\, a\, b}{2\, G_f^I\, b_t} \qquad (5)$$

The above composite fracture formulation can also be viewed as an attempt to incorporate the concept of a constant fracture energy release rate in the strain-softening formulation of an equivalent elastic-plastic continuum.

For simplicity the two-dimensional case of an elementary volume V=abt subject to a tensile state of stress is considered again (Fig 8). Adopting the maximum stress criterion results in orientation of the localised crack band $A = 2b_t t$ normal to the direction of major principal stress "n". Incipient cracking is controlled by the initially isotropic fracture condition

$$F\,(\boldsymbol{\sigma},\kappa) = \sigma_n - \sigma_t \qquad \sigma_n = \sigma_1\,,\ \sigma_t = \sigma_t\,(\kappa) \qquad (6)$$

where the normal tensile stress in the crack band coincides initially with the major principal stress $\sigma_n = \sigma_1$. The tensile strength σ_t describes the degradation of bond stresses across the crack band as a function of the tensile crack opening u_f in analogy to the fictious crack band model of Hillerborg et at [15]. The associated fracture rule with $Q(\sigma) = F(\sigma,k)$

$$\dot{\boldsymbol{\varepsilon}}_f = \dot{\lambda}\,\mathbf{n} \quad \text{where} \quad \mathbf{n} = \frac{\partial Q}{\partial \boldsymbol{\sigma}} \quad \text{and} \quad \begin{bmatrix} \dfrac{\partial Q}{\partial \sigma_1} \\ \dfrac{\partial Q}{\partial \sigma_2} \\ \dfrac{\partial Q}{\partial \sigma_3} \end{bmatrix} = \begin{bmatrix} 1 \\ 0 \\ 0 \end{bmatrix} \qquad (7)$$

defines direction and magnitude of the tensile strain-softening resonse in accordance to the flow rule of traditional flow theory of plasticity. Restricting the attention to Mode I type cracking the fracture potential $Q(\sigma)$ coincides with the fracture condition $F(\sigma,k)$ along the line of an associated flow rule and degenerates in this case to the unit vector in the principal stress space. In view of the strain-space setting of the computational algorithms the consistency condition

$$\dot{F} = \frac{\partial F}{\partial \boldsymbol{\sigma}}^T \mathbf{E} \, (\dot{\boldsymbol{\varepsilon}} - \dot{\lambda} \mathbf{n}) - (\frac{\partial \sigma_t}{\partial \kappa} \frac{\partial \kappa}{\partial \varepsilon_f})^T \dot{\lambda} \, \mathbf{n} = 0 \tag{8}$$

leads to the tensile fracture parameter $\dot{\lambda}$. With $\frac{\partial F}{\partial \sigma} = n$ one obtaines

$$\dot{\lambda} = \frac{\mathbf{n}^T \mathbf{E} \dot{\boldsymbol{\varepsilon}}}{E_c^{fs} + \mathbf{n}^T \mathbf{E} \, \mathbf{n}} \quad \text{for } \dot{\lambda} > 0 \quad \text{progressive cracking} \tag{9}$$

$$\text{and} \quad \dot{\lambda} = 0 \quad \text{for} \quad \mathbf{n}^T \mathbf{E} \dot{\boldsymbol{\varepsilon}} \leqq 0 \quad \text{elastic behaviour}$$

where the additive decomposition of strain rate $\dot{\boldsymbol{\varepsilon}} = \dot{\boldsymbol{\varepsilon}}_e + \dot{\boldsymbol{\varepsilon}}_f$ has been inferred and where $\mathbf{E}$ denotes the elastic material law. Considering the hardening or rather the softening modulus E_c^{fs} which provides the key to incorporate the fracture energy concept within the framework of strain-softening plasticity, the softening modulus is formally defined by the chainrule of differentation used in the consistency condition as

$$E_c^{fs} = (\frac{\partial \boldsymbol{\sigma}_k}{\partial \kappa} \frac{\partial \kappa}{\partial \boldsymbol{\varepsilon}_f})^T \mathbf{n} \tag{10}$$

There are several possibilities to evaluate the softening rule - however, in order to control the fracture energy release rate immediately via E_c^{fs} the following approach will be pursued.

Considering the invariance of the softening branch in the $\sigma - u_f$ space and the inherent definition of fracture energy in terms of

$$G_f^I = \int_0^{u_r} \sigma \, du_f \tag{11}$$

it is only natural to monitor softening directly in terms of the softening slope E_1^{df} depicted in Figure 8. In this case the mapping between the crack opening displacement du_f and the equivalent tensile fracture strain $dk = d\varepsilon_f$ leads in the case of strain-softening to

$$E_c^{fs} = \left(\frac{\partial \sigma_t}{\partial u_f} \frac{\partial u_f}{\partial \varepsilon_f} \frac{\partial \varepsilon_f}{\partial \boldsymbol{\varepsilon}_f} \right)^T \mathbf{n} \quad \text{where} \quad \dot{\varepsilon}_f = < \dot{\boldsymbol{\varepsilon}}_f^T \, \dot{\boldsymbol{\varepsilon}}_f >^{\frac{1}{2}} \tag{12}$$

This provides the key for the homogenisation of the localised crack discontinuity over the elementary volume of size V = abt. Denoting by h_t the height of the elementary volume normal to the crack band A = $b_t t$, the projection along the normal vector "n" yields the geometric "gauge length" h_t for the evaluation of the softening modulus for cracking

$$E_c^{fs} = E_l^{fd} \, h_t \quad \text{where} \quad E_l^{fd} = \frac{\partial \sigma_t}{\partial u_f}, \quad h_t = \frac{\partial u_f}{\partial \varepsilon_f}, \quad \left(\frac{\partial \varepsilon_f}{\partial \boldsymbol{\varepsilon}_f}\right)^T \mathbf{n} = 1 \tag{13}$$

As a consequence the resulting tangential material law

$$\dot{\boldsymbol{\sigma}} = \mathbf{E}_T \, \dot{\boldsymbol{\varepsilon}} \quad \text{where} \quad \mathbf{E}_T = \mathbf{E} - \frac{\mathbf{E}\,\mathbf{n}\,\mathbf{n}^T\,\mathbf{E}}{E_c^{fs} + \mathbf{n}^T\,\mathbf{E}\,\mathbf{n}} \tag{14}$$

will depend on the mesh size through the geometry of the elementary volume similar to non-local continuum theories. In fact composite fracture theory in which the surface area A_t of the crack band is related to the intact volume V leads to a generalisation of the softening modulus in Eq.13 in the form of

$$E_c^{fs} = E_l^{fd} \frac{V}{A_t} = E_l^{fd} \, h_t \tag{15}$$

where the volume fraction $A_t/V = 1/h_t$ should be viewed as microcrack density per unit volume.

Assuming again the constant slope E_l^{fd}, the fracture energy release rate can be expressed as

$$G_f^I = \frac{\sigma_t^2}{2 E_l^{fd}} \quad \text{or} \quad E_l^{fd} = \frac{\sigma_t^2}{2 G_f^I} \tag{16}$$

leading to the expression for the softening modulus of the homogenised continuum

$$E_c^{fs} = \frac{\sigma_t^2 h_t}{2 G_f^I} = E_1^{fd} h_t \tag{17}$$

which is the same as obtained before (Eq. 15)

Finally, it can be stated again that the softening modulus E_c^{fs} has to be adjusted to the mesh size of the particular finite element idealisation, if the fracture energy release rate G should remain constant. As a result the fracture mechanics concept leads to a non-local format of the governing softening properties which is fundamentally different from the local (geometry independent) format of strain-softening plasticity and progressive fracture theories.

3.2 Composite Fracture model and Localised Shear Slip

For the analysis of shear slip localization within the unified approach of composite fracture model, the elementary volume V is again composed of the undamaged elastic zone V_e and the localized shear band V_s of width d_s and length b_s in which oriented slip is confined. Resorting to the Mohr-Coulomb condition of frictional slip, the orientation of the frictional shear band is defined by the angle $\sigma = \psi/2 + \pi/4$ with the major axis of principal stress (Fig 8a).

From the energy equivalence the softening modulus shear G_c^{ts} depends on the fracture energy release rate G_f^{II}, and the crack geometry according to

$$G_c^{ts} = \frac{1}{\dfrac{2 G_f^{II} b_s}{(\tau_f - \tau_r)^2 a b} - \dfrac{1}{G}} \tag{18}$$

The shear softening modulus G_c^{ts} of the equivalent continuum element depends on the geometry of the elementary volume and thus exhibits a size effect. Again G_c^{ts} depends on the particular finite element mesh layout if the fracture energy release rate G_f^{II} constitutes invariant fracture property. As a result, the underlying fracture mechanics concept leads also in the case of shear slip to a non-local format of the equivalent softening relation which is again fundamentally different from the local constitutive theories of strain-softening plasticity and continuous damage mechanics.

It can be concluded that the composite fracture models for localised tension and shear have essentially the same structure. In both cases the fracture energy release concept results in a non-local format of the equivalent softening continuum.

3.3 Composite Fracture Model for Tension and Shear

The concept of a composite fracture model for Mode I type tensile cracking has been extended to the case of Mode II type shear fracture in order to capture the pronounced softening response in triaxial compression under low confinement. The simplest approach employs the straight combination of the two criteria described in sections 3.1 and 3.2, i.e. Mohr- Coulomb condition augmented by an independent tension cut-off. In that case tensile debonding was controlled by the fracture energy release rate G_f^I for Mode I type cracking, and the decohesion due to frictional slip was monitored by the fracture energy release rate G_f^{II} for Mode II type fracture which governs the formation of shear bands. As already indicated because of the increase of uniaxial compressive versus tensile strength, G_f^{II} is an order of magnitude larger than G_f^I when faulting along a single shear plane is considered along the line of the Mohr-Coulomb falure hypothesis. The transition from tensile cracking to shear faulting was defined by the inter- section of the tension cut-off with the Mohr-Coulomb criterion in the tension-compression regime [20]. However it should be noted that mixed mode fracture is actually initiated by tensile splitting and the formation of a single shear band is thus the result of coalescing crack feathers along an inclined slip plane. Therefore, the postulate of an independent faulting mechanism is really the consequence of the limited resolution of current computational tools to trace each current individual tensile crack preceding the formation of a single shear discontinuity, see also [17].

Since tensile cracking and shear faulting are clearly interrelated, an improved approach based on a continuous failure condition which comprises both tensile as well as shear fracture within a single criterion has been developed [30]. The criterion follows the line of empirical criterion proposed by Hoek and Brown, which can be viewed as a parabolic variation of the Mohr-Coulomb criterion with the tension cut-off condition

$$\sigma_3 = \sigma_1 - (c\,\sigma_c^2 - m\,\sigma_c\,\sigma_1)^{\frac{1}{2}} \qquad (19)$$

where σ_c denotes the uniaxial compressive strength and m and c define the frictional and cohesive strength material parameters. By rewriting this failure condition strain-hardening/softening is accommodated by the variation of the two parameters m and c

$$F(\boldsymbol{\sigma},\kappa) = \tau - (c - m\,\sigma)^{\frac{1}{2}} = 0 \qquad [20]$$

$$\text{where } \sigma = \sigma_1/\sigma_c \text{ and } \tau = |\sigma_1 - \sigma_3|/\sigma_c$$

and where cohesive and frictional strength parameters [c and m] vary during softening between the limiting values [initial c_o, m_o and residual c_r and m_r]. Introducing the fracture potential

$$Q(\boldsymbol{\sigma}) = \tau - (c_o - m_o\,\sigma)^{\frac{1}{2}} \qquad [21]$$

a tangential material law for non-associated decohesion has been proposed

$$\mathbf{E}_T = \mathbf{E} - \frac{\mathbf{E\,n\,m}^T\mathbf{E}}{E_f + \mathbf{m}^T\mathbf{E\,n}} \qquad [22]$$

where the principal issue of the non-associated softening formulation revolves around the proper definition of E_f for the mixed mode fracture. To this end it was assumed that tensile splitting in compression as well as shear faulting is preceded by diffuse cracking and that the actual crack surface is in this case a multiple of that in the direct tension experiment. Utilising the fact that the inelastic dilatancy angle forms a fairly constant ratio of lateral versus axial displacement in the softening regime, the softening modulus of equivalent continuum was obtained as

$$E_f = \frac{(m_o - m_r)}{2(c_o - c_r)}\,\frac{\sigma}{(c - m\sigma)^{\frac{1}{2}}}\,E_c\,\frac{1}{\alpha}\,h_c \qquad [23]$$

where E_c is the decohesion modulus and h_c defines the gauge length or the crack spacing under compressive splitting in analogy to h_t in the case of tensile cracking. A fracture energy argument similar to that developed for tensile cracking leads to the following definition of the crack spacing under compression

$$h_c = \frac{V}{A_c} = \frac{G_f^{I}}{G_f^{II}}\,\frac{V}{A_t} \qquad [24]$$

which represent the reciprocal value of the microcrack density in compression.

This microcrack density V/A_c plays the fundamental role of the geometric "gauge length" (which had the simple physical interpretation of the crack spacing h in the case of the direct tension experiment) and may be evaluated in terms of the ratio between the fracture energy release rate in tension G_f^I and in shear G_f^{II}.

4. COMPUTATIONAL RESPONSE PREDICTIONS

In the following, the composite fracture model is applied to four computational model problems - the direct tension test, the triaxial compression test, the direct shear test and the uniaxial wave problem involving strain softening and localisation.

4.1 Post-peak Response Computations of Direct Tension Test

The post-peak response predictions of the direct tension test using the composite fracture model are illustrated, and the upper half of the cylindrical NX specimen is idealised with three different meshes of biquadratic elements. The initial axial load-displacement relation corresponds to the bilinear idealisation of the uniaxial tension test, whereby the specimen is loaded under displacement control with radial restraints at the loading platens.

Figure 9 depicts the computational response predictions for the case of constant fracture energy release rate in tension i.e. G_f^I - const. The reference solution agrees fairly closely with the results of the square 4 x 8 mesh as well as the rectangular 2 x 4 and 4 x 2 idealisations in which the softening modulus is adjusted according to Eq. 3 in order to enforce a constant fracture energy release rate. The constant fracture energy release rate formulation captures the strain localisation in the single crack band fairly well. Figures 10 and 11 clearly illustrate the overall motion and the redistribution of strains and stresses within the tensile test specimen during the formation of the single macrodefect. In fact, the comparison between the results at peak strength and the solution at low residual strength demonstrates the consequences of strain localisation which initiates at the right top corner of the tension specimen because of the stress concentration due to radial restraint at the loading platen.

In contrast, the computational post-critical predictions for the case of mesh independent softening modulus exhibit the almost perfectly brittle response shown in Fig 12. This excessive decrease of ductility is a direct consequence of the mesh refinement which introduces multiple cracking into the idealisation instead of a single macrocrack. As a result the energy dissipation capacity of the specimen diminishes to the

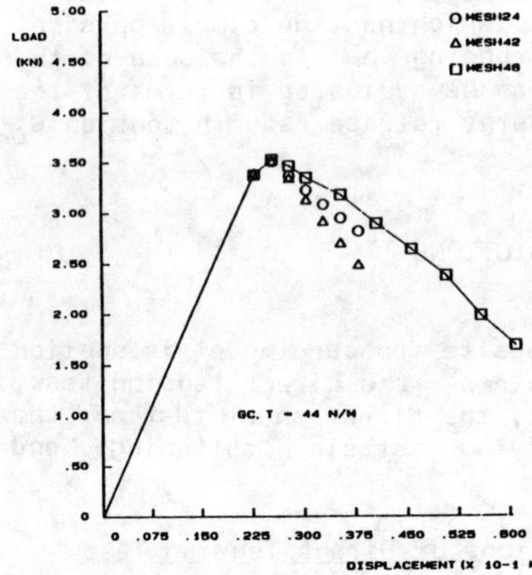

Fig 9 - $G_f^I = 44$ N/m
Load/displacement curves

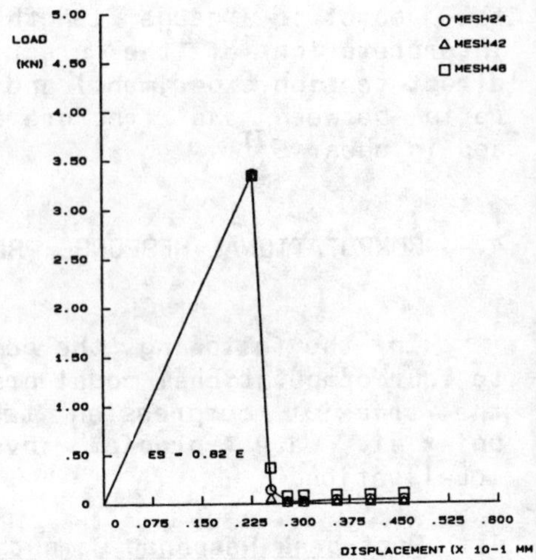

Fig 12 - $E_c^{ts} = 0.82$ E
Load/displacement curves

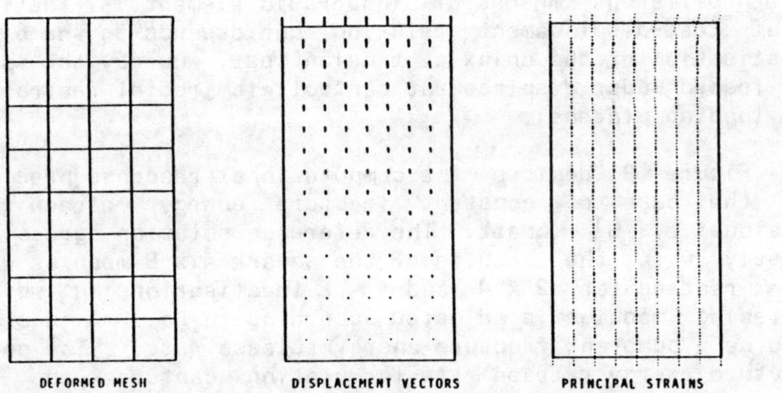

Fig 10 - Direct tension test - peak response

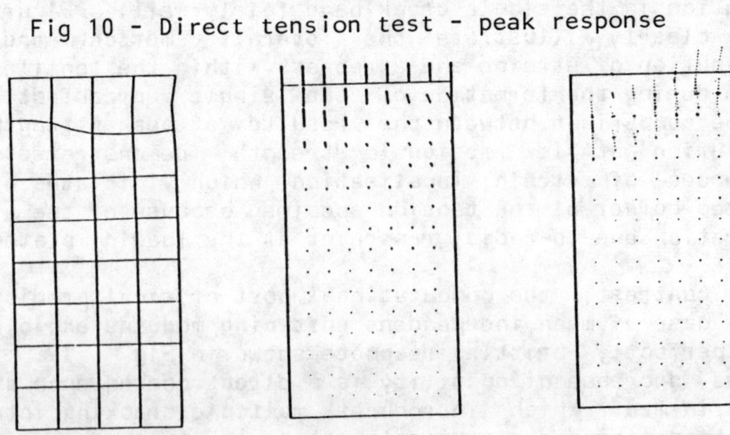

Fig 11 - Direct tension test - post-peak response

limiting condition of elastic perfectly brittle response behaviour if the height of the specimen is subdivided into a refined mesh.

4.2 Post-peak Response Computations of Triaxial Compression Test

In view of the different interpretation of the experimental data in the post-peak regime, the computational response predictions for the compression test using the composite fracture model are examined. The upper half of the cylindrical NX specimen is idealised with three different mesh layouts of biquadratic elements.

Figure 13 illustrates the computational response predictions for the case where the softening modulus is mesh independent. The numerical results indicate that the post-critical response is too brittle but fairly independent of the particular mesh-size and underestimates the ductility. This conclusion coincides with previous observations [32,33] for distributed shear fracture which can be summarized by stating that the constant softening formulation yields post-critical response predictions which are quite consistent and independent of mesh size. The constant softening modulus formulation captures fracture in the form of a shear cone and radial splitting within the compression specimen. Figures 14 and 15 clearly illustrate the overall motion and the strain localisation along the cone, whereby the axisymmetric idealisation imposes some restriction on the failure mode. However, the comparison between the results at peak strength and the solution at the residual strength level clearly demonstrates the formation of the compressive shear cone at the loading platen and the subsequent radial splitting at mid-height.

On the other hand, the basic fracture mechanics postulate of an invariant energy release rate leads to the results depicted in Fig 16. The excessive ductility during the formation of the shear band leads to a post-critical response prediction which exhibits very little softening. In fact the entire specimen constitutes the fracture process zone which is mobilised in the energy exchange during shear failure. As a result the constant softening modulus formulation for uniformly distributed microdefects is in this case more appropriate.

Frictional slip type fracture takes place in cementitious materials in a far more distributed fashion than tensile cracking. In fact, a single shear band develops only under sufficiently confined conditions when premature cracking is suppressed by compressive stress states. Therefore the fracture mechanics hypothesis of a constant fracture energy

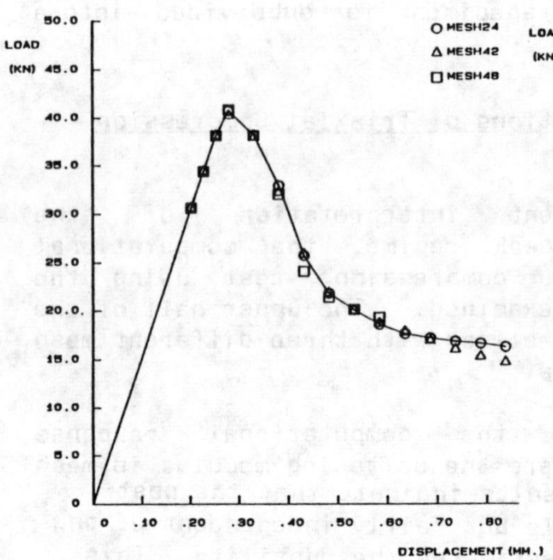

Fig 13 - G_c^{ts} = 0.05 G
Load/displacement curves

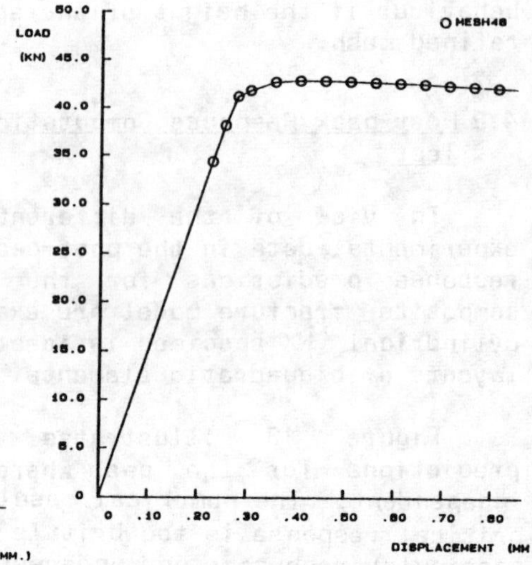

Fig 16 - G_f^{II} = 2452 N/m
Load/displacement curves

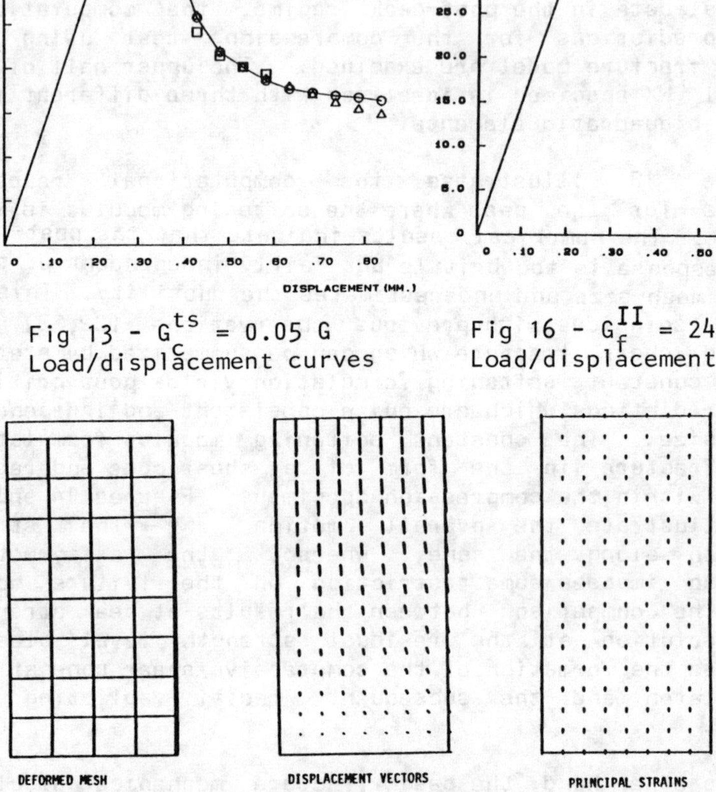

Fig 14 - Triaxial compression test - peak response

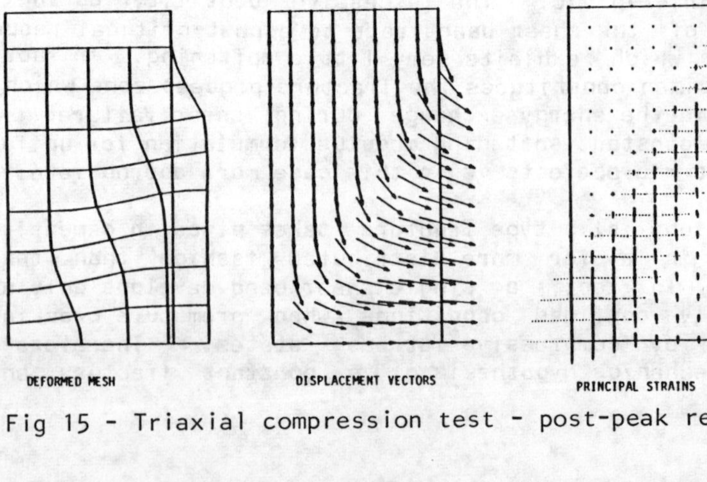

Fig 15 - Triaxial compression test - post-peak response

release rate in shear is not of immediate concern for traditional structures but is of paramount interest in geotechnical and geophysical applications.

4.3 Post-peak Response Calculations of the Direct Shear Test

In the final example the post-peak behaviour of a direct shear experiment in which both tensile cracking and frictional slip take place simultaneously is examined. In this case distributed fracture is contained by the two steel box compartments, whose motions are monitored by two servo-controlled actuators in the stiff shear test frame built at the University of Colorado, Boulder [34]. In this sense the direct shear test specimen constitutes a miniature structure in which contained fracture propagates in a manner that is similar to the failure behaviour in confined structures of high indeterminacy. The normal pressure is applied to a prescribed load level and held constant during the subsequent tangential loading monitored under displacement control in order to capture the full post-critical response branch down to the residual load level. Depending on the ratio of normal to tangential load level, the full range of predominantly tensile cracking to highly confined frictional slip is mobilized within the "shear test" specimen.

Because of the planar mode of fracture a two-dimensional plane strain idealisation is used to model the specimen and contiguous shear box. The finite element mesh layout includes the actual test specimen as well as the upper and lower shear box compartments in order to ensure that the tangential displacement control lies in the prevalent shear plane at mid-height of the specimen.

The composite fracture model for tensile cracking and frictional slip are combined in the simplest manner indicated in section 3.3 for describing constant softening modulus or alternatively constant fracture energy release in tension and shear. The actual properties of the concrete specimen were calibrated before in the context of the direct tension and triaxial compression tests. Therefore the direct shear test constitutes an entirely independent example of different size which will assert the previous findings on distributed and localised fracture in tension and shear. The post-critical behaviour is in one case described by the constant softening model both in tension and shear and in the other case by the constant fracture energy release rates G_f^I and G_f^{II}. In addition, a third post-peak strategy is pursued in which the fracture energy release rate in tension is kept constant, while the shear softening modulus remains invariant. For frictional slip a non-associated flow rule is used throughout in order to capture the splitting tension mode of fracture during shearing with a flow potential in the form of the maximum stress

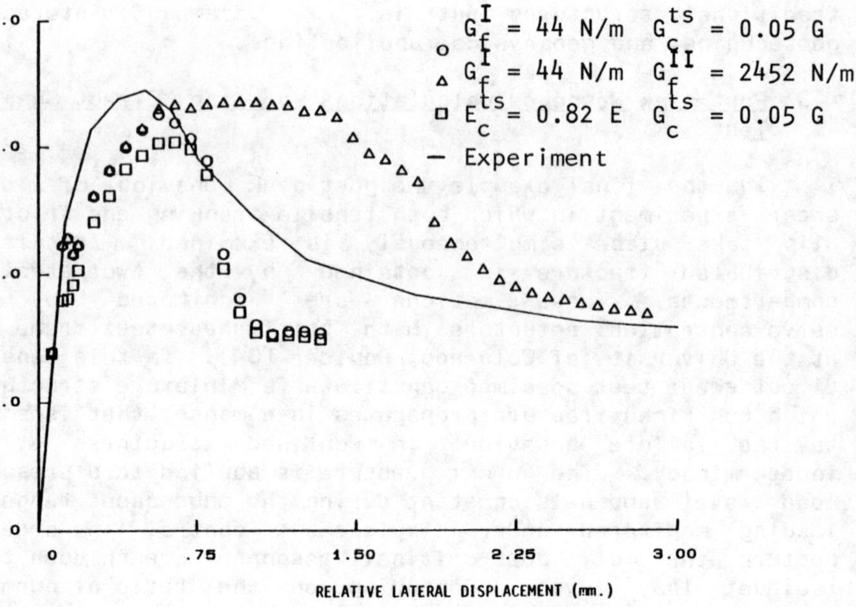

Fig 17 - Direct shear test - numerical predictions for various softening strategies

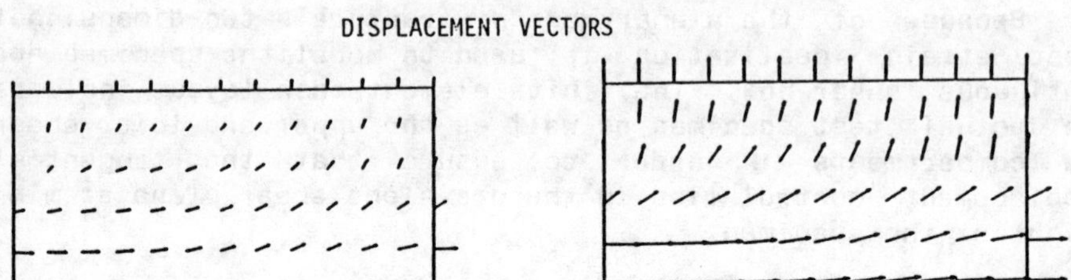

Fig 18 - Displacement vectors for the direct shear specimen

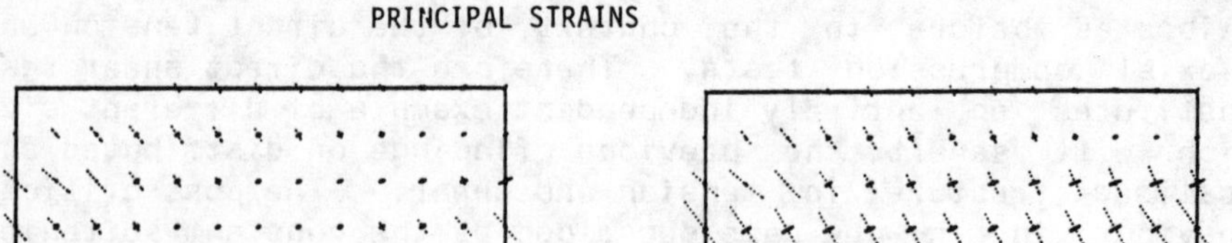

Fig 19 - Strain distribution for the direct shear specimen

criterion. In the case of tensile cracking this flow rule reduces to the traditional associated flow rule used to monitor the growth of tensile cracking strains.

Figure 17 depicts the overall load-deformation behaviour in terms of the nominal shear stress in the prevalent shear plane versus the relative tangential displacement between the upper and lower shear box compartments. The three computational predictions exhibit considerable variation as compared to the experimental results. The constant shear softening modulus approach shows the best agreement with regard to both strength and overall ductility while the constant fracture energy formulation indicates far too ductile behaviour. Clearly, in view of the confined fracture propagation within the shear specimen one could argue that a single slip band only develops in the final stage at the residual strength level. In addition, this localised shear band involves extensive tensile cracking in a direction which differs substantially from the prevalent shear plane. Therefore, one could conclude that contained fracture is beyond the realm of fracture mechanics of single macrodefects. However, in engineering applications this final mode of fracture in cementitious materials is always accompanied by extensive zones of distributed damage in which the constant softening modulus concept prevails.

Figure 18 illustrates the shearing motion in terms of the total displacement vectors for load steps corresponding to the peak and post-critical regimes. Observe the localisation into a wide shear band which is accompanied by considerable vertical dilation due to tensile splitting. The total strain distribution is depicted in Fig 19 at the same loading steps. The plots emphasize the localisation of shear strains in the prevalent shear plane while the upper and lower portions of the specimen unload progressively.

The results clearly indicate that the direct shear test is not a simple material test when cementitious materials are considered. In fact, the contained mode of fracture results in a complex combination of tensile cracking and frictional slip which is highly distributed before final localisation of a macrodefect takes place.

4.4 Dynamic Study of the model problem

The model problem (Fig 20) was analysed with the aid of the explicit finite element code DEGDYN developed for the dynamic investigation of concrete dams [37]. In this case the uniaxial bar was subdivided into 9, 21 and 97 elements of equal length and subjected to the constant velocity c = 0.50 at both ends, see Fig 20. In contrast to the quasistatic problem, no special provision was necessary to capture localisation of the strain softening zone in the stress wave problem. The analytic solution was recently reviewed by Bažant and Belytschko [35]

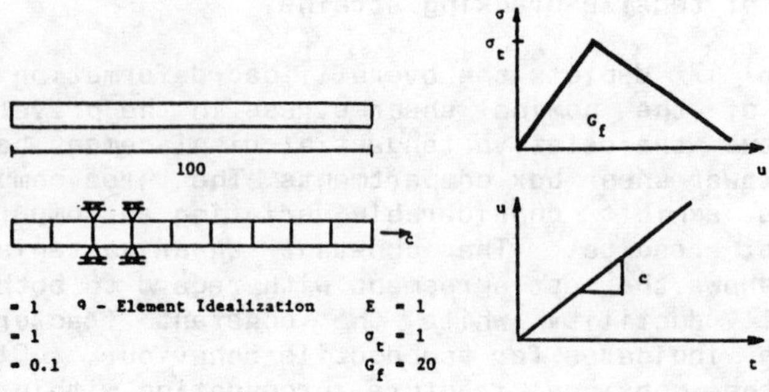

Fig 20 - Dynamic model problem

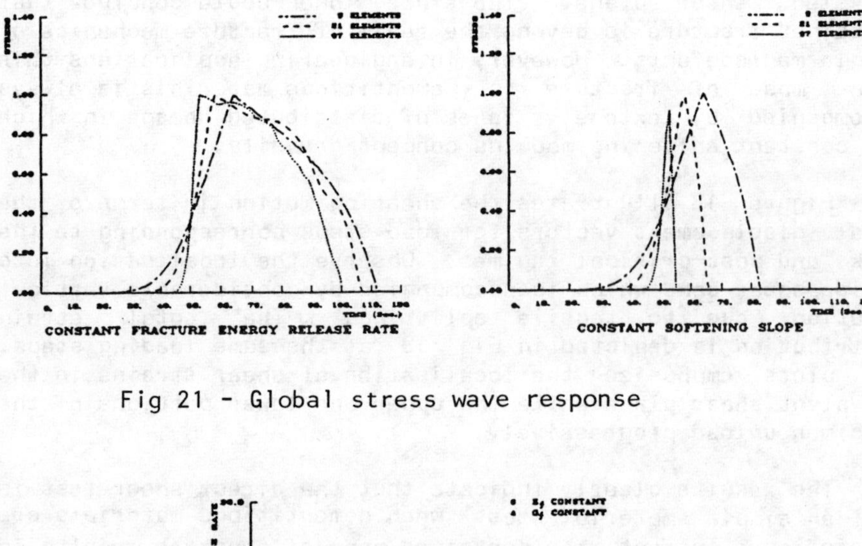

Fig 21 - Global stress wave response

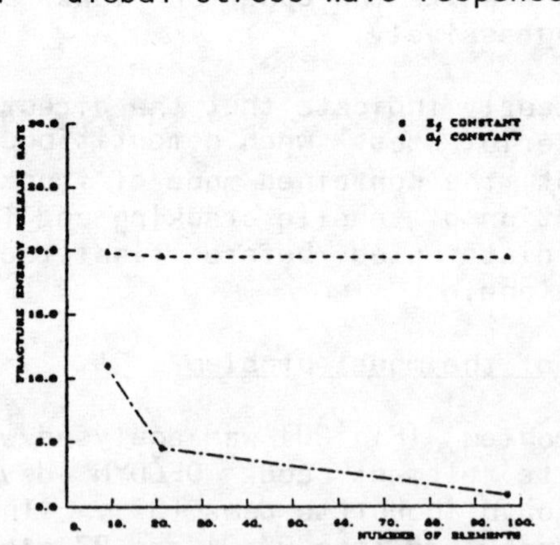

Fig 22 - Fracture energy mesh sensitivity

for which a well posed solution was found irrespective of the loss of hyperbolicity of the underlying characteristic wave equation in the softening regime.

Fig 21 and 22 compare the numerical results of the composite fracture model with those of the traditional strain softening approach in which the softening modulus is kept constant irrespective of the mesh size. The resulting stress histories and the associated values of the fractue energy release rates clearly demonstrate the extreme mesh sensitivity of those quantities when E_f = const. On the other hand, the composite fracture model maintains constant fracture energy release rates without resorting to non-local continum theories along the line of the recent imbricate continuum proposal [36] for resolving this basic deficiency of the traditional strain softening approach.

5. CONCLUDING REMARKS

The principal feature of the composite fracture model is a unified approach towards localised and distributed fracture in the form of tensile cracking, frictional slip and mixed mode fracture. In the case of localised fracture the composite model provides a non-local softening law of the equivalent continuum which depends on the characteristic fracture dimension as well as the geometric configuration of the fracture band within the elementary volume of finite dimension. In fact, the composite fracture model recovers in this case the constant fracture energy release concept without resorting to the fracture mechanics of discrete macrodefects. On the other end of the spectrum the composite damage model degenerates to the continuous damage theory of distributed fracture when the volume fraction of localised fracture approaches unity. In this case the softening modulus of the equivalent continuum constitutes a local material property which holds pointwise and is independent of the particular fracture configuration.

The composite model was applied to the post-critical response behaviour of the direct tension and triaxial compression tests on cylindrical specimens for which test data were available for comparison. While the constant fracture energy release concept captured the localisation of a single crack band in tension, the constant softening modulus formulation reproduced distributed fracture in shear.

In the example of the direct shear test both tensile cracking as well as frictional slip modes of contained fracture were mobilized, similar to the failure behaviour in highly confined structures. In this case the combination of the constant fracture release concept for tensile cracking and the constant softening formulation for shear slip furnished the best agreement with the experimental post-critical response behaviour.

The exposition was based on simplification in order to arrive at elementary algebraic relations which clearly exhibit the fundamental differences of distributed and highly localised fracture. In reality fracture is an evolutionary process in which the localised damage zone increases gradually and where the critical fracture energy release rate varies in accordance with the degradation of strength in the post- critical response regime. In addition inelastic dissipation mechanisms take place which are mobilised particularly in shear. As such the composite fracture model is a first order approximation which captures the limiting conditions of the local size-independent softening relation as well as the non- local fracture energy release rate concept. For the continuous transition from distributed to localised fracture an evolutionary damage process should provide an immediate solution to combine these two extreme cases.

To that end the combined mechanism of cohesive softening and frictional hardening within the scope of strain softening plasticity format of the composite fracture model covers the entire failure spectrum. It seems that an overall constitutive formulation for triaxial concrete behaviour should also include a compressive cap in addition to the tensile softening cap and the frictional hardening/softening mechanism associated with the shear strength.

REFERENCES

1. Reinhardt, H.W., "Fracture Mechanics of an Elastic Softening Material Like Concrete", Heron Publ. Vol. 29, No.2, p. 42, Delft, 1984.

2. Gopalaratnam, V.S. and Shah, S.P., "Softening Response of Concrete in Direct Tension", Technological Institute, Report, Northwestern University, Evanston, 1984.

3. Hurlbut, B., "Experimental and Computational Strain-Softening Investigation of Concrete", M.Sc. Thesis, CEAE Department, University of Colorado, Boulder, Colorado, 1985.

4. van Mier, J.G.M., "Strain-Softening of Concrete under Multiaxial Loading Conditions", Ph.D. Dissertation, University of Technology, Eindhoven, 1984.

5. Ortiz, M., "A Constitutive Theory for the Inelastic Behaviour of Concrete", Division of Engineering Research Report, Brown University, Providence, Rhode Island, 1984.

6. Mazars, J., "Application de la Mecanique de l'Endommagement au Comportment Non Lineaire et a la Rupture du Beton de Structure", Ph.D. Dissertation, Universite Pierre et Marie Curie, Paris, 1984.

7. Willam, K.J., Bićanić, N. and Sture, S., "Constitutive and Computational Aspects of Strain-Softening and Localization in Solids", ASME/WAM '84 meeting, New Orleans, Symposium on Constitutive Equations, Macro and Computational Aspects, ASME Vol. G00274, ed. K. William, 233-252, New York, 1984.

8. Dougill, J.W., "On Stable Progressively Fracturing Solids", J. Appl. Math. Physics, ZAMP, Vol. 27, 423-436, 1976.

9. Janson, J. and Hult, J., "Fracture Mechanics and Damage Mechanics, a Combined Approach", J. de Mechanique Appliquee, Vol.1, 69-84, 1977.

10. Krajčinović D. and Fonseka, G.U., "The Continuous Damage Theory of Brittle Materials, - Part I: General Theory", ASME J. Appl. Mech. Vol.48 809-815, 1981.

11. Stankowski, T. "Concrete under Multiaxial Load Histories", M.Sc. Thesis, CEAE Department, Universty of Colorado, Boulder, 1983.

12. Stankowski, T., and Gerstle, K.H., "Simple Formulation of Concrete Behaviour Under Multiaxial Load Histories", acceped for publ. J. ACI, 1984.

13. Bažant, Z.P. and Belytschko, T.B., "Continuum Theory for Strain- Softening", Trans. American Society of Civil Engineers, EMD Vol. 110, No.12, 1666-1692, 1984.

14. Belytschko, T. and Bažant, Z., "Strain-Softening Materials and Finite Element Solutions", ASME/WAM'84, New Orleans, Symposium on Constitutive Equations, Macro and Computational Aspects, ASME Vol. G00274 ed. K. Willam, 253-277, New York, 1984.

15. Hillerborg, A., Modeer and Petersson, "Analysis of Crack Formation and Crack Growth in Concrete by Means of Fracture Mechanics and Finite Element", Cement and Concrete Research, Vol. 6, 773-781, 1976.

16. Bažant, Z.P. and Cedolin, L., "Blunt Crack Bound Propagation in Finite Element Analysis", Trans. American Society of Civil Engineers, EMD, Vol.105, No. EM2, 297-315, 1979.

17. Willam, K. and Sture, S., "A composite Fracture Model for Localized Failure in Cementitious Materials", Second Sumps. on Interaction of Non- Nuclear Munitions with Structures, Panama City Beach, Florida, April 15- 19, 1985.

18. Hoek, E. and Brown, E.T., "Underground Excavation in Rock", The Institution of Mining and Metallurgy, London, p.527, 1980.

19. Vermeer, P.A., and De Borst, R. "Non-Associated Plasticity for Soils, Concrete and Rock", Heron Publ. Vol.29, No.3, p.64, Delft, 1984.

20. Day, S.E., "A Softening Plasticity Model for Concrete", M.Sc. Thesis, CEAE Department, University of Colorado, Boulder, Colorado, 1985.

21. Von Rittinger, P., "Lehrbuch der Aufbereitungskunde", Ernst und Korn, Berlin, 1967.

22. Saouma, V.E. and Ingraffea, A.R., "Fracture Mechanics Analysis of Discrete Cracking", Proc. IABSE Conf. on Advanced Mechanics of Reinforced Concrete, 393-416, Delft, June 2-4, 1981.

23. Petersson, P.E., "Fracture Energy of Concrete: Method of Determination", Cement and Concrete Research, Vol.10, 78-79, 1980.

24. Desai, C.S. and Siriwardane, H.J., "Constitute Laws for Engineering Materials with Emphasis on Geologic Materials", Prentice-Hall, Inc., Englewood Cliffs, N.J., 1984.

25. Resende, L., and Martin, J.B., "A Progressive Damage Continuum Model for Granular Materials", Comp. Meth. Appl. Mech. Eng., Vol.42, 1-18, 1984.

26. Rudnicki, J.W., and Rice, J.R., "Conditions for the Localization of Deformation in Pressure-Sensitive Dilatant Materials", J. Mechanics and Physics of Solids, Vol.23, 371-394, 1975.

27. Costin, L.S., "A Microcrack Damage Model for Brittle Rock", Sandia Report SAND, 83-1590, 1983.

28. Pietruszczak, St. and Mroz, Z., "Finite Element Analysis of Deformation of Strain-Softening Materials", Int. J. Num. Meth. Eng., Vol.17, 327-334, 1981.

29. Willam, K.J., Sture, S. Bićanić, N., Christensen, J. and Hurlbut, B., "Identification of Strain-Softening Properties and Computational Predictions of Localized Fracture", Structural Research Series No. 8404, University of Colorado, Boulder, 1984.

30. Willam, K.J., Hurlbut, B. and Sture, S., "Experimental, Constitutive and Computational Aspects of Concrete Failure", US-Japan Joint Seminar on Finite Element Analysis of Reinforced Concrete Structures, University of Tokyo, May 21-24, 1985, Proc. Vol.1, 149-171, 1985.

31. Willam, K., Bićanić, N., Pramono, E. and Sture, S., "Composite Fracture Model for Strain Computations of Concrete", Int. Conf. on Fracture Mechanics of Concrete, Lausanne, Oct. 1-3, 1985.

32. Willam, K.J., "Experimental and Computational Aspects of Concrete Fracture", Invited paper at Int. Conf. Computer-Aided Analysis and Design of Concrete Structures, Split, Yugoslavia, Sept. 17-20, 1984.

33. Bićanić, N., Sture, S., Hurlbut, B., and Day, S. "On the Prediction of the Peak and Post-Peak Behaviour of Concrete Structures", paper at Int. Conf. Computer-Aided Analysis and Design of Concrete Structures, Split, Yugoslavia, Sept. 17-21, 1984.

34. Christensen, J., Ickert, K., Stankowsi, T., Sture, S., and Willam, K.J., "Numerical Modeling of Strength and Deformation Behaviour in the Direct Shear Test", Proc. Int. Conf. on Constitutive Laws for Engineering Materials, eds. C.S. Desai & R.H. Gallagher, 537-544, Tucson, AZ, 1983.

35. Bažant, Z.P. and Belytschko, T.B., "Continuum Theory for Strain- Softening", ASCE J. Eng. Mech. Vol.110, No.12, 1666-1692, 1984.

36. Bažant, Z.P. and Belytschko, T.B., "Wave Propagation in a Strain- Softening Bar: Exact Solution", ASCE J. Eng. Mech., Vol.111, No.3, 381- 389, 1985.

37. Bićanić, N., "Nonlinear Finite Element Transient Response of Concrete Structures", Ph.D. Dissertation, C/Ph/50/78, University College of Swansea, Wales, 1978.

38. Palmer, A.C. and Rice, J.R., "The Growth of Slip Surfaces in the Progressive Failure of Overconsolidated Clay", Proc. Royal Soc., A, 332, 527-548, London, 1973.

CHAPTER 6

ENDOCHRONIC CONSTITUTIVE MODELING OF MARINE FIBER REINFORCED CONCRETE

by

D.V. Reddy
Department of Ocean Engineering
Florida Atlantic University
Boca Raton, Florida,
U.S.A.

K.R. Gopal
Department of Civil Engineering
Carleton University
Ottawa, Ontario, Canada

SUMMARY

Endochronic theory (a unified viscoplastic theory) is applied to the constitutive modeling of marine fiber reinforced concrete. This theory has the unique capability of predicting both loading and unloading behavior without the definition of a yield surface.

Endochronic constants are determined by curve fitting the experimental stress-strain curve determined from the tests on concrete specimens with steel fibers subjected to varying periods of circulating seawater exposure. The flexural behavior of beams is predicted based on these endochronic constants, and the tensile contribution of fibers intersecting a unit cross-sectional area. Comparison between the experimental and analytical results indicates reasonable agreement.

The chapter includes a brief discussion of certain criticisms of earlier formulations of the theory, and the recently proposed modified versions to overcome the indicated deficiencies.

1. INTRODUCTION

1.1 General

The main reason for the deterioration of concrete in the coastal and offshore environments is its inability to resist the internal tensile stresses created by the corrosion products of steel reinforcement. One way to overcome this problem is to include fibers in the concrete, which will bridge the cracks restraining their further propagation.

Addition of fibers to concrete was found to significantly increase its tensile and flexural strengths and transform its behavior from brittle to that of a pseudo-ductile material.

1.2 Endochronic Modeling

In plasticity theories, plastic or irrecoverable strains are assumed to form after the yielding takes place. But in reality they may occur right from the beginning due to internal grain rearrangement like in soil or concrete, which is not accounted for in the plasticity theories. Another limitation of plasticity theories is that there is a certain ambiguity associated with the choice of a proper yield surface. Loading and unloading criteria, needed to determine the plastic strains, cause unnecessary complications in the numerical computations.

The endochronic theory proposed by Valanis [20] overcomes some of the limitations of plasticity formulations. The theory, derived on the basis of irreversible thermodynamics, has been applied to predict the mechanical response of metals under complex strain histories. Jain [10], provided an excellent summary of the thermodynamic derivation of the endochronic constitutive equations.

The basic concept underlying the endochronic formulations is the characterization of the material strain state in terms of an intrinsic time, which is a non-decreasing scalar variable that depends on the increments of strain as well as time, and geometrically represents the length of the path traced by the states of the material in a strain-time space of suitable metric, i.e., the memory of the material is expressed in terms of an intrinsic parameter. The early versions of the theory were rate-independent; time was not involved in the intrinsic time terms. The distance along a strain path between two strain states P and P' in Fig. 1 denoted as $d\zeta$, termed the intrinsic time measure, is defined as follows:

$$d\zeta^2 = P_{ijkl} \, d\varepsilon_{ij} \, d\varepsilon_{kl} \tag{1.1}$$

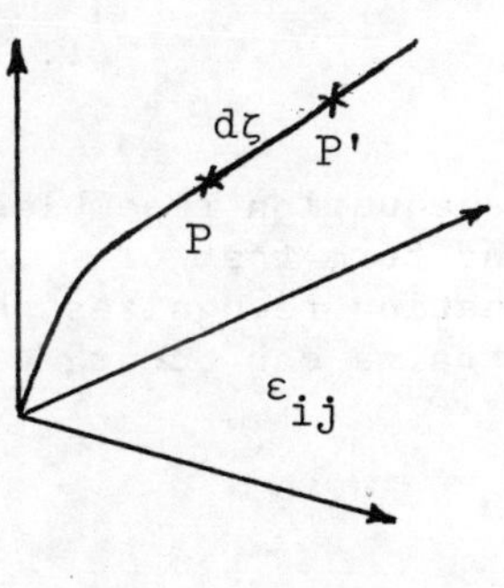

Fig. 1 Path in strain space

The tensor P_{ijkl} is a material property tensor as it varies from material to material. The stress at point 'P' is not determined simply by the strain at 'P', but by the history of the strain along the path OP. Valanis termed the materials for which the stress is a functon of the strain history with respect to an intrinsic time scale, as endochronic (Greek: endo=inner, chronos=time), and the corresponding theory as endochronic theory.

To account for the experimental observation that the materials strain harden, the intrinsic time scale Z is defined as follows:

$$dZ = \frac{d\zeta}{f(\zeta)} \; ; \; \frac{dZ}{d\zeta} > 0; \; Z > 0 \tag{1.2}$$

in which $f(\zeta)$ is a monotonically increasing material property-dependent hardening function. The intrinsic time scale Z is representative of the deformation hisotry of the material. The basic stress-strain relations of Valanis are presented below:

If the temperature effects are neglected, Eqs. 3.24 and 3.25 of Valanis [20] can be simplified to a form:

$$\sigma_{ij} = 2 \int_0^Z \mu(Z - Z') \frac{\partial \varepsilon_{ij}}{\partial Z'} \, dZ' \tag{1.3}$$

Considering a single exponential term for the modulus,

$$\mu(Z) = \mu_0 e^{-\alpha Z} \tag{1.4}$$

then

$$\sigma_{ij} = 2 \mu_0 \int_0^Z e^{-\alpha (Z-Z')} d\varepsilon_{ij} (Z'), \tag{1.5}$$

$$\text{in which} \quad d\varepsilon_{ij} = \frac{\partial \varepsilon_{ij}}{\partial Z'} \, dZ \tag{1.6}$$

Integrating Eq. 1.5 following the Leibnitz rule,results in

$$d\varepsilon_{ij} = \frac{d\sigma_{ij}}{2\mu_0} + \frac{\alpha}{2\mu_0} dZ \, \sigma_{ij} \tag{1.7}$$

The first term on the right in the above equation resembles the classical elastic term and the second term that of plastic term. The form of the above equation recognizes the experimental observation that plastic strains can occur immediately following the loading. The plastic strain term in Eq. 1.7.

$$d\varepsilon_{ij}^{p} = \left(\frac{\alpha}{2\mu_0}\, dZ\right) \sigma_{ij}\,, \tag{1.8}$$

is similar to the Prandtl-Reuss relation of classical plasticity.

$$d\varepsilon_{ij}^{p} = (d\lambda)\sigma_{ij} \tag{1.9}$$

Endochronic theory contains these relations of classical plasticity as a special case. But this formulation is continuous and 'dZ' does not depend on any postulations or inequalities. On the contrary, $d\lambda$ needs the postulation of a yield surface and the loading and unloading criteria.

1.2.1 Applications to concrete

Endochronic theory was first applied to concrete by Bazant and Bhat [3] by extending the formulation for metals, developed by Valanis [20], to include many nonlinear features of concrete like hydrostatic pressure sensitivity, inelastic dilaltancy due to deviatoric strains, strain hardening, and softening. The model was able to predict stress-strain diagrams for concrete of different strengths, failure envelopes of uniaxial, biaxial, and triaxial stress-strain diagrams, lateral strains and volume expansion, hysteresis loops for repeated compression, strain rate effects, and creep behavior. This is the only constitutive theory which can model all the above nonlinear features of concrete at the present time. Various endochronic constants were determined by fitting large sets of experimental data. It was indicated that those constants would be the same for any normal weight concrete with compressive strengths in the range of 3500 to 6000 psi. The hysteretic behavior of concrete beams was predicted by Bazant and Bhat [4] using the above model. The same formulation was applied to predict the finite element behavior of cubes, beams, and other concrete structural elements by Sorensen, Arnesen, and Bergan [18].

1.2.2 Recent Modifications

Although capable of modeling many complex phenomena, early endochronic formulations were unable to properly characterize the response to small cyclic stress oscillations superimposed on large static stresses. Those formulations do not predict a positive energy dissipation or closed hysteresis loops, and yield a reloading path that is less steep than the unloading slope for small load-unload oscillations. Sandler [17] and Rivlin [16] pointed out that these theories implied a physically impossible condition of positive energy creation violating the Drucker stability postulate. It was indicated that this feature my lead to unstable solutions in wave propagation problems. Valanis [22] has addressed some of these criticisms. Details of those criticisms are not discussed here as both of them were directed on the early formulations, which overcome the deficiencies of earlier models, are discussed.

1.2.3 Plastic strain formulations

Valanis [21] redefined intrinsic time in terms of plastic strains, instead of the total strain, which enables the formation of closed hysteresis loops. With this new definition, $d\zeta$ in Eq. 1.1 is rewritten as

$$d\zeta^2 = P_{ijkl}\, d\varepsilon^p_{ij}\, d\varepsilon^p_{kl} \qquad (1.10)$$

in terms of the plastic strain increments. It was established for the one-dimensional case that the rate of dissipation at the onset of loading, unloading, and reloading is zero, leading to closed hysteresis loops. The new model implies the existence of a yield surface in an 'a priori' fashion . Various versions of classical plasticity theories were shown to be particular cases of the endochronic formulation. This model was applied by Valanis and Read [25] to describe the soil response to cyclic shear loading, hydrostatic compression, and densification under cyclic shear. Lin and Wu [11] applied the new endochronic formulation to obtain the cyclic behavior and plastic strain wave propagation in titanium alloys. Some of the recent developments of the endochronic theory and their applications were discussed by Valanis and Lee [23]. More recently, Valanis and Lee [24] proposed an endochronic theory of cyclic plasticity which predicts very closely the experimental behavior of mild steel and grade 60 steel under various loadings.

1.2.4 'Jump-kinematic hardening' formulation

Bazant [1] introduced the concept of 'jump-kinematic hardening' in endochronic theory to obtain closed hysteresis loops that ensure positive energy dissipation. The concept of a yield surface with a moving center was used. The basic definitions of inelastic deviatoric and volumetric strain increments were modified to include the hardening coefficients. In the original paper by Bazant and Bhat [3] they were defined as follows:

$$de''_{ij} = \frac{s_{ij}}{2G} dZ \qquad (1.11)$$

and

$$de'' = d\lambda + \frac{\sigma}{3K} dZ' \qquad (1.12)$$

where

de''_{ij}, $d\varepsilon''$ = deviatoric and volumetric strain increments,

dZ, dZ' = intrinsic time increments,

$d\lambda$ = dilatancy parameter,

s_{ij} = deviatoric stress tensor,

and

σ = volumetric stress.

In the kinematic hardening model Eqs. 1.11 and 1.12 are modified as follows:

$$de''_{ij} = (s_{ij} - \alpha_{ij})\, \rho \frac{dZ}{2G} \qquad (1.13)$$

and

$$de''\, d\varepsilon = d\lambda + (\sigma - \alpha)\psi \frac{dZ'}{3K} \qquad (1.14)$$

in which

α_{ij}, α = current centers of the deviatoric and volumetric loading surfaces in stress space,

and

ρ, ψ = coefficients (less than unity) which reduce the rate of accumulation of inelastic strains in cases of unloading and reloading

ρ and ψ satisfy a certain inequality so as to assure closed hysteresis loops. For non-cyclic loading α_{ij} and α can be set to zero, and ρ and ψ to unity. For the case of cyclic loading, the centers of the loading surface must jump into the extreme stress points whenever loading changes to unloading, or unloading to reloading. This can be explained as follows with reference to Fig. 2.

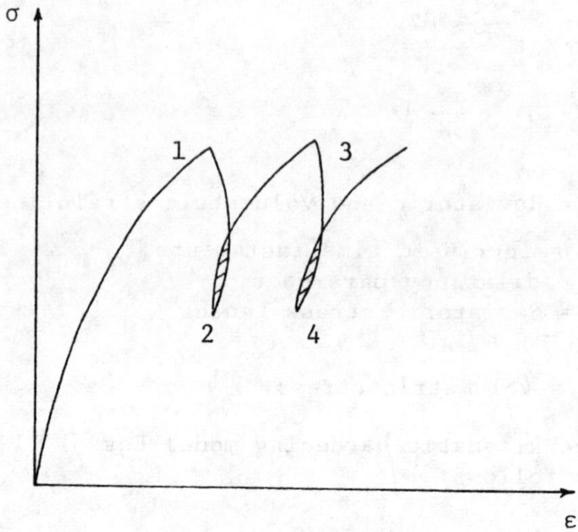

Fig. 2 Closed hysterisis loops in the 'jump-kinematic' hardening model

When the loading changes to unloading, the centers of the loading surfaces, α_{ij} and α, are moved to the extreme stress point in the loading path, i.e., to point 1. When the unloading is changed to reloading, the centers are moved to the extreme stress point in the unloading path, i.e., point 2. Between the loading reversals they are kept constant. This process was found to give closed hysteresis loops even for an infinitesimal load-unload cases. More details on this formulation can be obtained from Bazant, Ansal, and Krizek [2], and Bazant, Krizek, and Shieh [5].

1.3 Scope

i) Review of basic endochronic theory including recent modifications ii) Extension to model various nonlinear features of concrete, iii) Development of an analytical model to predict the load-deflection behavior fiber reinforced concrete beams, iv) Experimental determination of stress-shear and flexural behavior, v) Determination of endochronic constants by curve fitting techniques, vi) Comparison of the experimental and predicted behavior, discussion and conclusions.

2. ENDOCHRONIC STRESS-STRAIN MODEL FOR CONCRETE

The endochronic formulation proposed by Bazant and Shieh [6], used in this investigation, belongs to the family of theories with jump-kinematic hardening criteria, as described in Section 1.2.1. As the details of this formulation are discussed at considerable length by Bazant and Bhat [3] and Bazant and Shieh [6], only the salient points are presented here.

The deformations are assumed to be so rapid that the material is modeled as a rate-insensitive material by dropping the time-dependent intrinsic terms. The elastic properties are assumed to be incrementally isotropic and the stress-strain relations are written separately in terms of deviatoric and volumetric components as follows:

$$de_{ij} = de^e_{ij} + de^p_{ij} = \frac{ds_{ij}}{2G} + \frac{s_{ij}}{2G}\, dZ \qquad (2.1a)$$

$$d\varepsilon = d\varepsilon^e + d\varepsilon^p = \frac{d\sigma}{3K} + d\lambda + \frac{\sigma}{3K} dZ' + d\lambda' \qquad (2.1b)$$

where

$e_{ij} = \varepsilon_{ij} - \delta_{ij}\,\varepsilon$ = deviatoric components of the strain tensor,

$\varepsilon = \frac{1}{3}\varepsilon_{kk}$ = volumetric strain,

δ_{ij} = Kronecker delta,

$s_{ij} = \sigma_{ij} - \delta_{ij}\,\sigma$ = deviatoric components of the stress tensor, σ

$\sigma = \frac{1}{3}\sigma_{kk}$ = volumetric stress,

K, G = bulk and shear moduli,

λ = inelastic dilatancy,

λ' = shear compaction,

Z, Z' = intrinsic times for distortion and compaction respectively.

and

The increments of intrinsic times are defined as follows:

$$dZ = \frac{d\zeta}{Z_1}, \qquad (2.2)$$

and

$$dZ' = \frac{d\zeta'}{Z_2}, \qquad (2.3)$$

where

$$d\zeta = \frac{d\eta}{f(\eta,\varepsilon,\sigma)};\ d\eta = F(\varepsilon,\sigma)d\xi$$

$$d\zeta' = \frac{d\eta'}{h(\eta')};\quad d\eta' = H(\sigma)d\xi'$$

$$d\xi = \sqrt{J_2\ (d\varepsilon)} = \sqrt{1/2\ de_{ij}\ de_{ij}}\ ,$$

and

$$d\xi' = \sqrt{[I_1(d\varepsilon)]^2} = |d\varepsilon_{11} + d\varepsilon_{22} + d\varepsilon_{33}|$$

The increments of the inelastic dilatancy and shear compaction are defined as follows:

$$d\lambda = \ell\ (\lambda\ )\ L(\ \lambda,\varepsilon,\sigma)d\xi, \tag{2.4}$$

and

$$d\lambda' = \ell'\ (\lambda\ ')\ L'(\lambda',\varepsilon,\sigma)d\xi, \tag{2.5}$$

where

J_2 = second invariant of the deviator of the tensor which follows in parentheses,

I_1 = first invariant of the tensor which follows in parentheses,

Z_1, Z_2 = constants,

$f(\eta,\varepsilon,\sigma)$, $F(\varepsilon,\sigma)$ = distortion hardening and softening functions respectively,

$h(\eta\ ')$, $H(\sigma)$ = compaction hardening and softening functions respectively,

$\ell(\lambda)$, $L(\lambda,\varepsilon,\sigma)$ = dilatancy hardening and softening functions respectively,

and

ξ, ξ' = non-decreasing scalar variables termed distortion and compaction measures respectively.

2.1 Strain-hardening functions

Inelastic strain increments, whether due to plastic slip or due to microcracking, decrease as the residual strain increases. This is due to the fact that the potential locations of stress peaks gradually decrease with the accumulation of strain, which is evident more in the cyclic loading. As ζ and ζ' are adopted as the measures of the accumulated inelastic strain (or damage), functions 'f' and 'h' must increase as ζ and ζ' increase. By analysis of test data, the strain hardening functions were defined as follows by Bazant and Shieh [6]:

$$f(\eta,\varepsilon,\sigma) = \left(1 + \frac{\beta_1\eta + \beta_2\eta^2}{1 + F_2/a_7} \right) F_3 , \tag{2.6}$$

$$F_3 = 1 + \frac{a_{10}}{J_2(\varepsilon)\,(1+a_9/\eta^2)} ,$$

and

$$h(\eta') = 1 + \frac{\eta'}{\beta_3} + \left(\frac{\eta'}{\beta_4}\right)^2 , \tag{2.7}$$

In the above equations, F_3 is a function which makes the cyclic stress loops to contract more strongly at low cyclic deviatoric strains. Function F_2 governs the strain-softening behavior after the peak.

2.2 Strain-softening and hydrostatic pressure sensitivity functions

If the functions $F(\varepsilon,\sigma)$ and $H(\sigma)$ remain constant, the uniaxial stress-strain diagram would approach an asymptote of positive slope in which case the failure of the material will not be modeled. To obtain a gradual decrease of slope on approach to peak stress, the inelastic strain increments must be increased, obtained by an increase in F and H. Inelastic volume change due to hydrostatic pressure is taken into consideration through the intrinsic time variable, ζ', called compaction measure. By analysis of test data the following softening functions have been found:

$$F(\varepsilon,\sigma) = F_1 + F_2 \tag{2.8}$$

where

$$F_1 = \frac{a_o\,(1-g_1)}{1-a_5(I_3^\sigma)^{1/3}\,(1+g_2)}$$

and

$$F_2 = \frac{a_2\,\sqrt{J_2(\varepsilon)}\,\left(1+|a_6 I_2^\sigma|^{1/4} + F_5\right)}{\{1 - a_1 I_1^\sigma + |a_8 I_2^\sigma|^{1/4} F_4 - a_3 I_3^\sigma\,[J_2(\sigma)]^{1/8}\,(1+g_2)\}}$$

in which

$$g_1 = g_{11}\, g_{12},$$

$$g_2 = g_{21}\, g_{22}\, g_{23},$$

$$g_{12} = 1 - \left[\, 1 + \left(\frac{\sigma_{min}}{a_{17}\,(\sigma_{max} - a_{23})}\right)^4 \right]^{-1},$$

$$g_{11} = a_{14}\,[\,J_2\,(\varepsilon)]^{\frac{\sigma_{med}-\sigma_{min}}{\sigma_{max}-a_{23}}}\left\{a_{15}\left(\frac{\sigma_{med}-\sigma_{min}}{\sigma_{max}-a_{23}}\right)^{4/3} - a_{16}\right\},$$

$$g_{21} = \left\{\frac{a_{18}\left(\dfrac{\sigma_{med}-\sigma_{min}}{\sigma_{max}-a_{23}}\right) - 1}{a_{19}\left(1 - a_{20}\dfrac{|\sigma_{min}|}{\sigma_{max}-a_{23}}\right)(\sigma_{min} - a_{23})}\right\}^{5/4},$$

$$g_{22} = \left[\, 1 + a_{21}\left(\frac{\sigma_{min}}{\sigma_{max} - a_{23}}\right)^4\right]^{-1},$$

$$g_{23} = \left(\frac{\{J_2\,(\varepsilon)\}^{1/4}}{a_{22} + \{J_2\,(\varepsilon)\}^{1/2}}\right)^3,$$

$$F_4 = \left(\frac{\{J_2\,(\varepsilon)\}^{1/4}}{a_4 + \{J_2(\varepsilon)\}^{1/2}}\right)^3,$$

and

$$F_5 = a_{11}\sigma_{min}\,(1+a_{12}\sigma_{min})\left\{\frac{J_2\,(\varepsilon)^{1/4}}{|a_{13}\sigma_{min}|^{1/4} + \{J_2(\varepsilon)\}^{1/2}}\right\}^3$$

where

σ_{min}, σ_{med}, and σ_{max} are the minimum, medium, and maximum principal stresses. a_0, a_1, ... a_{22} are endochronic constants.

2.3 Dilatancy and compaction functions

The dilatancy functions govern the microcracking in the concrete. Since microcracking is primarily due to deviatoric strains, it is expressed as a function of $J_2(\varepsilon)$. The increments of the dilatancy parameter $d\lambda$, are given as follows:

$$d\lambda = \ell\,(\lambda)\; L(\lambda,\varepsilon,\sigma)\,d\xi \qquad (2.9)$$

where

$$\ell(\lambda) = 1 - \frac{\lambda}{\lambda_o}$$

and

$$L(\lambda,\varepsilon,\sigma) = \frac{c_3}{1 - c_1 I_1(\sigma)} \{ (\frac{\lambda}{\lambda_o})^2 + (\frac{c_4 J_2(\varepsilon)}{c_2^2 + J_2(\varepsilon)})^2 \}$$

Since the inelastic strains are due to the formation of microcracks, the incremental elastic moduli should decrease as the microcracking increases. The shear and bulk moduli, G and K, are expressed as follows:

$$G = \frac{1}{1 + c_5\lambda} \quad \frac{E_o}{2(1 + \nu)}, \tag{2.10a}$$

and

$$K = \frac{1}{1 + c_5\lambda} \quad \frac{E_o}{3(1 - 2\nu)}. \tag{2.10b}$$

In the above equations c_1, c_2 ... c_8 are endochronic constants and ν is the Poisson ratio.

Inelastic strains due to hydrostatic pressure are modeled by a separate intrinsic term called shear compaction. It was assumed that the shear compaction will not have any effect on the moduli. The following functions are constructed:

$$\ell'(\lambda') = c_6 \; (1 - \frac{|\lambda'|}{\lambda'_o})$$

$$L'(\lambda',\varepsilon,\sigma) = \frac{\sigma_{min} \; g_3^{1/3}}{1 + (g_3/c_8)^3}$$

and

$$g_3 = |c_7 \; \sigma_{min}|^{0.93} - \sqrt{J_2(\varepsilon)}$$

in which c_6, c_7, and c_8 are the endochronic constants.

3. FLEXURAL BEHAVIOR OF FIBER REINFORCED CONCRETE BEAMS

3.1 Material Models

3.1.1 Endochronic theory for concrete

The endochronic formulation described in Section 2 is used, with the constants for steel fiber reinforced concrete given in Section 5.

3.1.2 Model for fiber action

Fibers are assumed to participate only in the tensile region. Their effect in compression is accounted for by curve-fitting the stress-strain behavior under compressive loading. A crack is assumed to form normal to the principal stress when it exceeds the tensile strength of concrete. Experimental data on the fiber pull-out force and the crack opening displacement COD for steel fibers were reported by Naaman and Shah [14], and Matsuishi and Iwata [12]. Before the full interfacial bond strength, τ, is developed, the fiber force is linearly proportional to the crack opening and then it gradually decreases with a further increase in the crack opening. When the bond stress exceeds the limiting value τ, fiber force is found from the following expression proposed by Wecharatana and Shah [26]:

$$\sigma = \sigma_{max} \left(1 - \frac{\eta}{\eta_{max}}\right)^2 \qquad (3.1)$$

where

σ = fiber stress,

σ_{max} = fiber stress when the full interfacial bond strength is developed,

$= \tau \dfrac{\ell}{d}$

η = crack mouth displacement,

η_{max} = maximum pull-out displacement of fiber,

$= \ell/4$

and

ℓ, d = length and diameter of fiber respectively.

The total tensile resistance provided by the fibers is computed by considering the probabilistic number of fibers which intersect a unit cross-sectional area as reported by Matsuishi and Iwata [12] and Naaman, Moavenzadeh, and McGarry [13] in the following form:

$$N_s = \frac{2V_f}{\pi d^2} \qquad (3.2)$$

where

N_s = number of fibers intersecting a unit cross-sectional area,

and

V_f = volume content of fibers.

3.1.3 Model for reinforcing bars

The reinforcing bars are assumed to behave elasto-plastically until the initiation of strain hardening. The strain hardening portion is modeled by the following method as proposed by Burns and Siess [7]:

$$f_s = f_y \left\{ \frac{112\,(\varepsilon_s - \varepsilon_{sh}) + 2}{60\,(\varepsilon_s - \varepsilon_{sh}) + 2} + \frac{\varepsilon_s - \varepsilon_{sh}}{\varepsilon_u - \varepsilon_{sh}} \left(\frac{f_u}{f_y} - 1.7 \right) \right\} \quad (3.3)$$

where

f_s = stress in the steel corresponding to a strain of ε_s

f_y = yield stress of steel,

ε_{sh} = strain at which the hardening will commence,

and

ε_u = strain at ultimate stress of steel.

If the strain exceeds the yield value in one direction of loading, the stress-strain curve would no longer be elasto-plastic due to the Bauschinger effect; it will be a function of the residual strain for the load reversal. Upon load reversal, the following expression is used for stress between zero and yield:

$$f_s = f_y \left[1 - \exp\left(\frac{-2.05\ \varepsilon_s}{\varepsilon'_{sh}} \right) + \frac{0.129\ \varepsilon_s}{\varepsilon'_{sh}} \right] \quad (3.4)$$

where

ε'_{sh} = effective strain hardening strain

It is a function of the load history and the initial plastic strain ε_{ip}, defined as the difference in strain between two sequential points of zero stress immediately preceding a given load cycle, expressed as follows:

$$\varepsilon'_{sh} = \frac{\varepsilon_{sh}}{1.38} \ln \frac{\varepsilon_{ip}}{\varepsilon_y} \tag{3.5}$$

When the stresses are greater than the yield stress on reversal, the virgin stress-strain curve is used by substituting ε'_{sh} for ε_{sh} in Eq. 3.3. A perfect bond is assumed between the reinforcing bars and concrete.

3.2 Load-deflection Relations of Beam

A layered beam element approach is used, which is similar to those used in earlier investigations, e.g. Park, Kent, and Sampson [15]. The internal moment for any given curvature increment is determined by satisfying the equilibrium of the cross section. Each section is divided into layers within which the strain is assumed to be constant. The moment developed at the center of the beam for any curvature increment is prorated to different sections along the length to determine the corresponding curvatures. Central deflection is found on the basis of the curvatures at different sections. The description in this section closely follows that of Bazant and Bhat [4], except for the inclusion of fiber effects and the deflection calculations.

3.2.1 Incremental equilibrium equations

When the concrete is uncracked it is modeled as an isotropic material. The stress and strain are related as follows:

$$\{ d\sigma + d\sigma'' \} = [D] \{d\varepsilon\} \tag{3.6}$$

where

$d\sigma$, $d\sigma''$ = elastic and inelastic stress increments,

D = material property matrix,

and

$d\varepsilon$ = strain increment.

For uncracked concrete, D is written as follows:

$$[D] = \begin{bmatrix} D_{11} & D_{12} & D_{13} & 0 & 0 & 0 \\ & D_{22} & D_{23} & 0 & 0 & 0 \\ & & D_{33} & 0 & 0 & 0 \\ & \text{sym} & & D_{44} & 0 & 0 \\ & & & & D_{55} & 0 \\ & & & & & D_{66} \end{bmatrix} \tag{3.7}$$

where

$$D_{11} = D_{22} = D_{33} = K + \frac{4}{3} G,$$

$$D_{12} = D_{13} = D_{23} = K - \frac{2}{3} G,$$

$$D_{44} = D_{55} = D_{66} = 2G,$$

G, K = shear and bulk moduli of concrete determined from the endochronic theory.

Inelastic stresses are determined from Eqs. 2.1a and 2.1b as follows:

$$d\sigma''_{ij} = 2G\, de^{p}_{ij} + 3K\, d\varepsilon^{p}\, \delta_{ij} \tag{3.8}$$

When the principal stress exceeds the tensile strength of concrete, cracks are assumed to form normal to the maximum principal stress direction. After cracking, concrete is modeled as an orthotropic material, with zero stress across the crack. The concrete modulus matrix in the local coordinate system of cracks shown in Fig. 3 as given as follows:

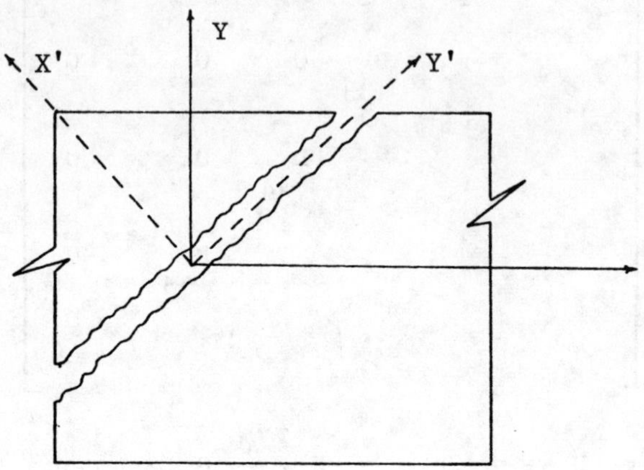

Fig. 3 Local coordinate system for crack

$$
[D'] = \begin{bmatrix}
0 & 0 & 0 & 0 & 0 & 0 \\
 & D'_{22} & D'_{23} & 0 & 0 & 0 \\
 & & D'_{33} & 0 & 0 & 0 \\
 & \text{sym} & & \alpha D_{44} & 0 & 0 \\
 & & & & \alpha D_{55} & 0 \\
 & & & & & D_{66}
\end{bmatrix}
$$

where

$$D'_{22} = D_{22} - \frac{D_{12}^2}{D_{11}},$$

$$D'_{23} = D_{23} - \frac{D_{13} D_{21}}{D_{11}},$$

$$D'_{33} = D_{33} - \frac{D_{31}^2}{D_{11}},$$

and

α = shear transfer factor.

[D'] matrix is obtained by a transformation of the matrix [D] to account for the loss of stiffness normal to the crack plane. A shear transfer factor is needed to account for the frictional force along the crack plane. A value of 0.5 is used for α, reported to be a reasonable value by Suidan and Schnobrich [19]. The cracked concrete modulus matrix [D'] is transformed back to the global coordinates by the matrix [L] of the direction cosines of the principal stresses as follows:

$$[D''] = [L] \; [D'] \; [L] \qquad (3.10)$$

[D"] is used in Eq. 3.6 in place of [D] when a crack occurs.

Fig. 4 shows the coordinate system for the beam. The cross sections of the beam are assumed to remain plane after bending. The shear strains, ε_{xy}, ε_{yz}, and ε_{zx}, and the transverse normal strains, ε_{yy} and ε_{zz}, are taken into consideration in the analysis. They are assumed to be constant throughout each cross section. Both curvature and normal shear ε_{xz}, are assumed to contribute to deflections.

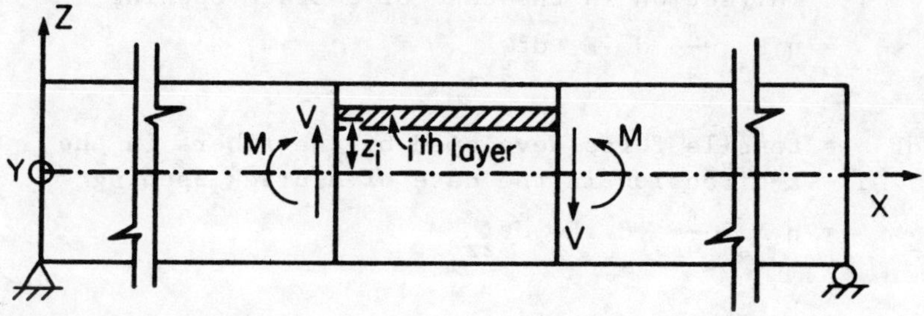

Fig. 4 Coordinate system for beam [4]

The longitudinal normal strain at any level of the beam, z_i, can be written as follows:

$$d\varepsilon_{xx_i} = de - z_i\, dk \tag{3.11}$$

where

$d\varepsilon_{xx_i}$ = incremental longitudinal strain at a distance z_i from the reference axis,

de = incremental longitudinal normal strain at the reference axis,

and

dk = incremental curvature of the beam.

The shear strains, ε_{xy} and ε_{yz}, and the corresponding shear stresses can be neglected for in-plane loading of the beam in xz-plance, Fig. 3.2. Then the cross-sectional equilibrium equations can be written in the following form:

$$dM = \sum_i z_i A_{xi}\, d\sigma_{xxi} + \sum_i z_i\, dF_{x_i}, \tag{3.12}$$

$$dN_x = \sum_i A_{x_i}\, d\sigma_{xx_i} + \sum_i dF_{x_i} \tag{3.13}$$

$$dN_y = \sum_i A_{y_i}\, d\sigma_{yy_i} + \sum_i dF_{y_i} \tag{3.14}$$

$$dN_z = \sum_i A_{z_i}\, d\sigma_{zz_i} + \sum_i dF_{z_i} \tag{3.15}$$

and

$$dV = \sum_i A_{x_i}\, d\sigma_{xz_i} \tag{3.16}$$

The summation is carried out on all the layers. A_{x_i}, A_{y_i}, and A_{z_i} and the areas of the layers in the yz, xz and xy planes respectively.

dF_{x_i} = longitudinal tensile force developed by the fibers in the case of a crack opening

$= \eta N_s \frac{\pi}{4} d^2 E_f\, d\varepsilon_{xx_i}$

dF_{y_i} = tensile force developed by the fibers in the direction in the case of a crack opening

$= \eta N_s \frac{\pi}{4} d^2 E_f\, d\varepsilon_{yy_i}$

and

dF_{z_i} = tensile force developed by the fibers in the z-direction in the case of a crack opening

$= \eta N_s \frac{\pi}{4} d^2 E_f\, d\varepsilon_{zz_i}$

where

η = bond efficiency factor for the fibers, taken as 0.8,

and

E_f = Young's modulus of the fiber material.

By substituting the expressions for $d\sigma_{xx}$, $d\sigma_{yy}$, $d\sigma_{zz}$, and $d\sigma_{xz}$ in terms of de, dk, $d\varepsilon_{yy}$, $d\varepsilon_{zz}$, and $d\varepsilon_{xz}$ into the cross-sectional equilibrium equations, 3.12 - 3.16, five simultaneous force equilibrium equations are obtained, represented in a matrix form as follows:

$$\{df^c\} = [\ R^c\ ]\ \{d\varepsilon\} - \{df''\}, \qquad (3.17)$$

$\{df^c\}^T$ = internal force vector of concrete
= $\{dM, dN_x, dN_y, dN_z, dV\}$

$\{d\varepsilon\}$ = $\{de, dk, d\varepsilon_{yy}, d\varepsilon_{zz}, d\varepsilon_{xz}\}$

$\{df''\}$ = vector of inelastic forces,

and

$[R^c]$ = stiffness matrix due to concrete alone.

The force equilibrium equations for the reinforcing steel can also be written as:

$$\{df^s\} = [R^s]\ \{d\varepsilon\} \qquad (3.18)$$

The total forces in the cross section are obtained as a sum of the contributions of concrete and steel, Eqs. 3.17 and 3.18, which yields

$$\{df\} = [\ R\ ]\ \{d\varepsilon\} - \{df''\} \qquad (3.19)$$

where

$$[\ R\ ] = [\ R^c] + [R^s]$$

Solution techniques for Eqs. 3.19 have been discussed at length by Bazant and Bhat (1977).

3.2.2 Determination of deflections

The bending moment at the center of the beam is prorated to two other sections, and the corresponding curvatures are determined. Then the central deflection is found by the moment-area method. The load acting on the beam is found from the central moment. The computer program discussed by Bazant and Bhat [4] has been modified to include the fiber effects and to predict the central deflections from the curvatures.

4. EXPERIMENTAL INVESTIGATION

4.1 Stress-Strain Behavior

4.1.1. Techniques for for obtaining complete stress-strain curves

A complete stress-strain curve consists of an ascending part during which the stress in the specimen increases, and a descending part during which the stress decreases; in both regions the strain continues to increase. A complete stress-strain curve is needed to predict nonlinear behavior accurately for extreme loading conditions, e.g., impact and earthquake loadings. The improvement in the concrete performance due to the fiber addition is reflected by the larger area under the stress-strain curve. After the concrete cracks, the fibers bridging the cracks provide resistance to the further crack growth and separation of concrete into fragments. Fiber reinforced concrete, therefore, behaves like a pseudo-ductile material, with higher stresses after the peak in the unloading portion compared to plain concrete.

When a concrete specimen is loaded in a conventional testing machine, both the specimen and the machine deform. When the stress in the specimen reaches a peak value, there is a reduction in the load which results in a sudden release of stored strain energy by the machine. This released strain energy causes uncontrolled failure of the specimen. As a result, a stable descending part cannot be observed. In order to avoid uncontrolled failure, the load on the specimen must be reduced at a sufficiently fast rate to control the rate of strain energy release of the testing system.

There are basically two methods of obtaining a complete stress-strain curve. They are as follows:

1) use of a closed-loop testing system, and
2) in parallel-testing with a coaxial special alloy steel tube.

In this investigation the second method of testing is adopted. A heat treated special alloy tube was used to encase the concrete cylinder for the stress-strain testing. The jacket is of SAE 4140 grade steel, heat treated to a Brinnel hardness of 445. The chemical composition of the alloy is given as follows.

Table 1
Chemical Composition of Alloy

Item	% volume	Item	% volume
carbon	0.38 - 0.43	manganese	0.75 - 1.0
phosphorus	0.04	sulphur	0.04
silicon	0.20 - 0.25	nickel	0.01
chromium	0.80 - 1.10	molybdenum	0.15 - 0.25

4.1.2 Test Procedure

Complete stress-strain curves were obtained by testing the concrete specimens in parallel with a coaxial steel tube. One end of concrete cylinder was capped with sulphur capping material, according to ASTM C-617 standards. The concrete cylinder was then placed inside the steel tube and capped with hydrostone at the other end, to make it flush with the top of the steel tube. Fig. 5 shows the schematic diagram of the load sharing mechanism. A foil type strain gage of 350 ohms resistance, 0.5 inch length was fixed at each end of the diameter at mid-section of the steel tube. The strains were measured using a quarter bridge circuit in a Vishay digital strain indicator, Model P-350A.

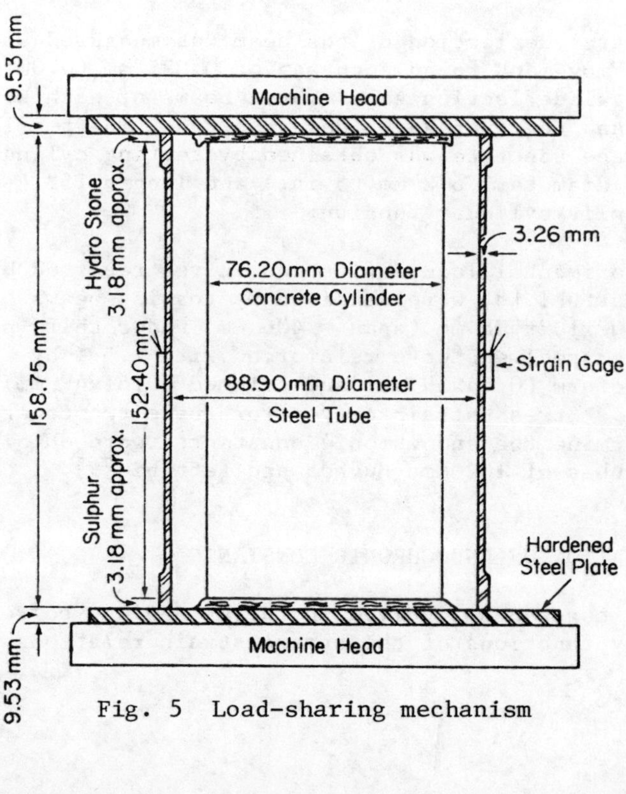

Fig. 5 Load-sharing mechanism

The tests were carried out in a Forney hydraulic testing machine of capacity 1335 kN (300 kips). The rate of straining was 10 micro-strains per second, which corresponds to approximately 30 psi per second on the concrete cylinder. This rate lies within the loading rates of 20 - 50 psi per second, recommended by ASTM C-39 standards, for compressive strength testing of concrete cylinders.

4.2 Flexural Behavior

The experimental data used was that from the author's tests on fiber reinforced concrete beams exposed to circulating sea water and that reported by other investigators. The experimental results of Hughes and Fattuhi [8] were used to check the computer program.

As part of this investigation, steel fiber reinforced beams made of the same mix described in Section 3.4, exposed to circulating sea water were considered. The beams, were of dimensions 102x102x357 mm (4x4x12 in.), reinforced on both the compression and tension sides with steel bars of diameter 6.4 mm (#2 bars) and provided with stirrups of diameter 3.2 mm (#1 bars) at a spacing of 45 mm (1.8 in.). They were tested in 3-point loading over a span of 305 mm (12 in.) at a constant deflection rate of 1.25 mm (0.05 in.) per minute. The tests were carried out in a Tinius Olsen fixed head hydraulic testing machine of capacity 267 kN (60,000 lb.).

The central deflection of the beam was measured as the machine head movement to an accuracy of 0.025 mm (0.001 in.) by a mechanical deflection gage. Three beams of each kind were tested and the average values reported. The tensile strength of the concrete was obtained by testing cylindrical specimens of diameter 76.2 mm (3 in.) and length 152.4 mm (6 in.) under split cylinder tension.

The experimental load-deflection curves reported by Hughes and Fattuhi [8] were obtained by testing beams of dimensions 102x102x508 mm (span = 406 mm) under third point loading. The specimens were reinforced with 1.5 % by volume of Duoform (0.64x59 mm) and crimped (0.51x49 mm) steel fibers. Stress-strain curves for these two concretes, used to determine the endochronic constants, were obtained by testing cubes of 102 mm, Hughes and Fattuhi [9].

5. DETERMINATION OF ENDOCHRONIC CONSTANTS

Various constants involved in the hardening, softening, and dilatancy functions of the stress-strain relations were found by curve-fitting the experimental behavior. At first, the values of the constants were varied manually to obtain a curve as close to the experimental one as possible. Then, an IMSL (International Mathematical and Statistical Library) subroutine "ZXSSQ", was used to minimize the sum of the squares of deviations between the experimental and predicted curve. The subroutine gives a local minimum of the sum of squares of M real functions in R real variables by using a finite difference Levenberg-Marquardt algorithm. The ordinates at various strain values were used as the real functions. The global minimum was obtained by repeating the optimization process with the new constants found in the previous iteration. Various steps involved are shown in Fig. 6.

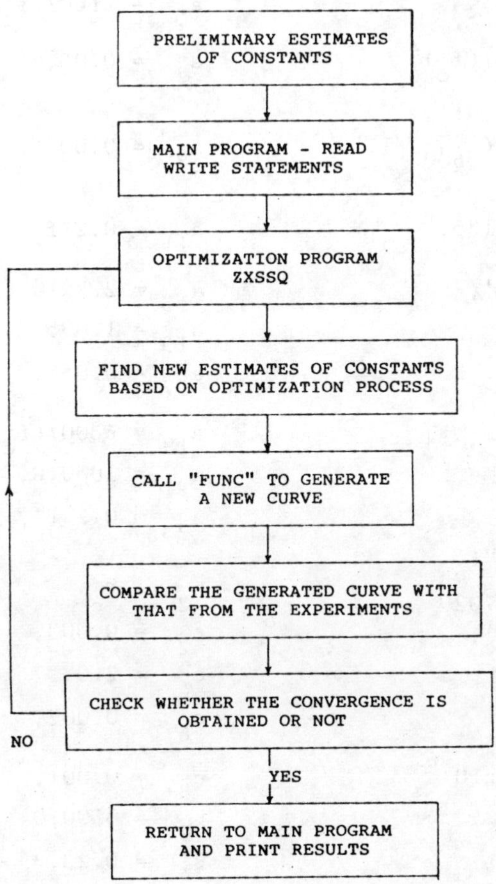

Fig. 6 Flow chart

6. RESULTS AND DISCUSSION

6.1 Endochronic Constants

Fig. 7 shows the endochronic curve fit obtained for the steel fiber reinforced concrete. The constants for the biaxial and triaxial cases could not be determined due to the lack of experimental data in those load cases. Those constants were assumed to be same as those for plain concrete reported by Bazant and Shieh [6]. The constants obtained are as follows:

$a = 0.65,$ $a_1 = 0.68/f'_o$

$a_2 = 1175\left(\frac{f'_c}{4650}\right)^{1/2},$ $a_3 = 324000/(f'_c)^4,$

$a_4 = 0.045,$ $a_5 = 2160/(f'_c)^2,$

$a_6 = 0.15/(f'_c)^2,$ $a_7 = 0.05,$

$a_8 = \frac{15}{f'^2_c}\left(\frac{f'_c}{3600}\right)1.5$ $a_9 = 0.0015,$

$a_{10} = 0.000125,$ $a_{11} = 0.2/f'_c,$

$a_{12} = 0.8/f'_c,$ $a_{13} = 2.2\mathrm{x}10^{-5}/f'_c,$

$a_{14} = 25.0,$ $a_{15} = 1.095,$

$a_{16} = 1.216,$ $a_{17} = 0.055,$

$a_{18} = 0.94,$ $a_{19} = 6300/(f'_c)^2,$

$a_{20} = 14.0,$ $a_{21} = 1000.0,$

$a_{22} = 0.04,$ $a_{23} = 0.2\ f'_c,$

$b_1 = 9.1\left(\frac{f'_c}{7029}\right).$ $b_2 = f'_c$

$c_1 = 2.0/f'_c,$ $c_2 = 0.003,$

$c_3 = 0.50,$ $c_4 = 2.0,$

$c_5 = 250.0,$ $c_6 = 0.002,$

$c_7 = 1.052\mathrm{x}10^{-6}.$ $c_8 = 0.001,$

$\beta_1 = 30.0,$ $\beta_2 = 3500.0$

$\beta_3 = 0.08,$ $\beta_4 = 0.23,$

$Z_1 = 0.0015,$ $Z_2 = 0.0125,$

$\lambda_o = 0.007,$ $\lambda'_o = 0.007,$

and

$\nu = 0.182.$

The initial slope formula used was that given by ACI as,

$$E_o = 57{,}000 \sqrt{f_c'} \qquad (3.11)$$

where

f_c' = compressive strength of concrete (psi).

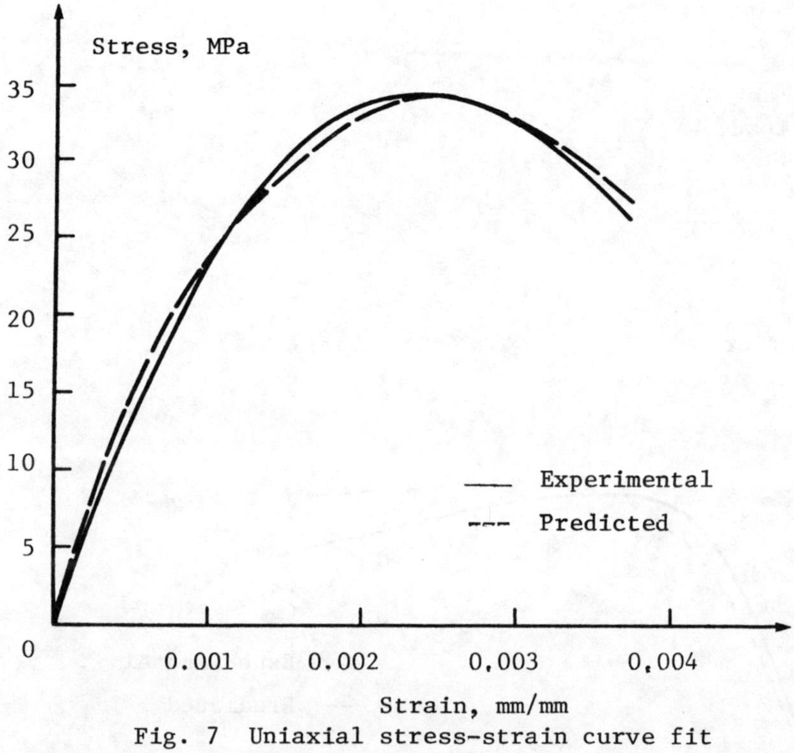

Fig. 7 Uniaxial stress-strain curve fit

6.2 Comparison of Analytical and Experimental Flexural Behavior

The comparison between the experimental and predicted behavior of crimped, Duoform, and hooked-end steel fiber concrete beams is shown in Figs. 8, 9, and 10 respectively. The model over-predicts for crimped fibers, under-predicts for Duoform fibers, and gives almost exact values for hooked-end steel fibers. This seems to be due to the inadequate modeling of fiber action. Duoform fibers are a special type with high bond strength in contradistinction to crimped fibers, in which the stress concentrations cause matrix cracking and disbonding of fibers. The bond properties assumed in the model are more relevant to the hooked-end steel fibers as indicated by the good agreement in Fig. 10. In the ideal case, fiber pull out vs. load-slip distance curves for each kind should be incorporated in the model which would lead to improved predictions.

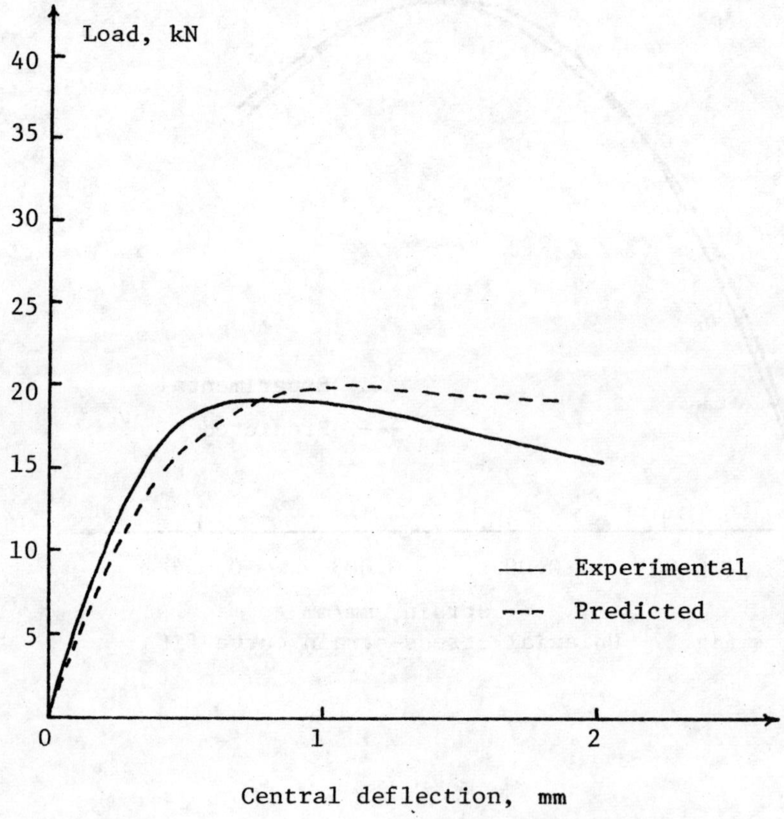

Fig. 8 Predicted and experimental flexural behavior of crimped steel fiber concrete beams - From Hughes and Fattuhi [8]

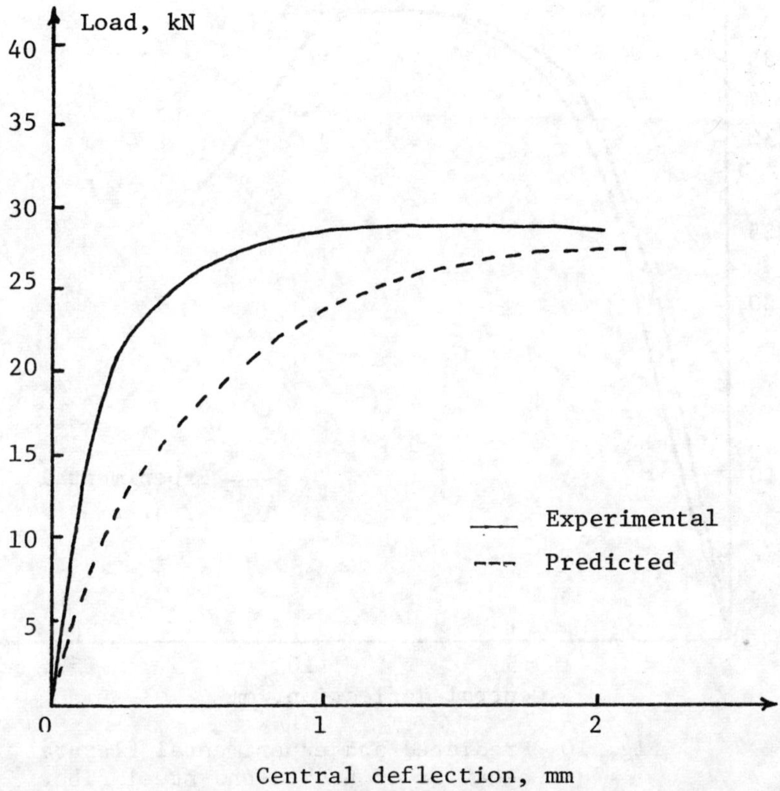

Fig. 9 Predicted and experimental flexural behavior of duoform steel fiber concrete beams - From Hughes and Fattuhi [9]

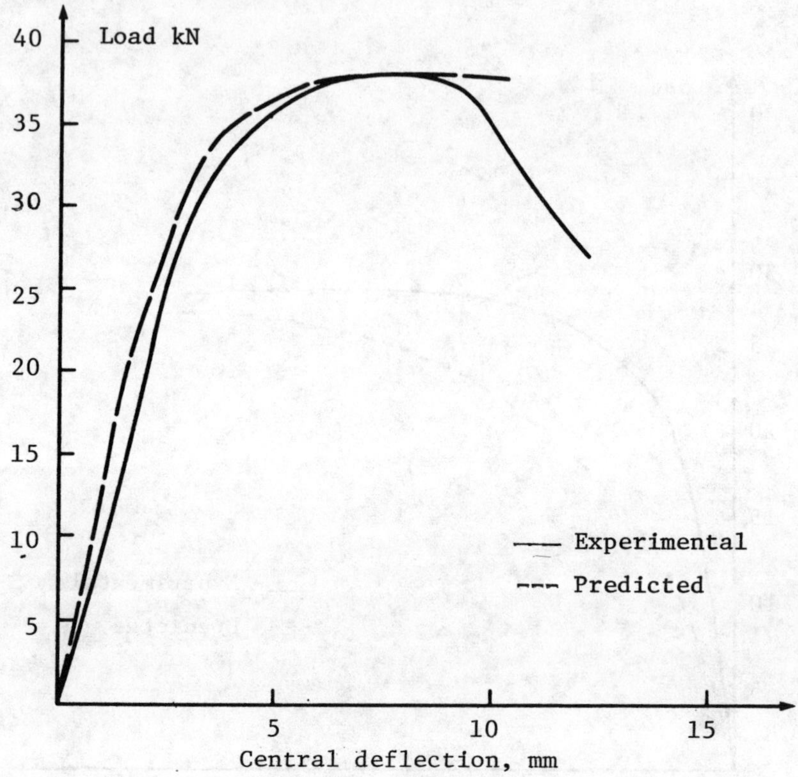

Fig. 10 Predicted and experimental flexural behavior of hooked-end steel fiber concrete beams

6.3 Suggestions for Further Studies

The following suggestions are made for future research efforts in this area:

1) A model should be developed for random fiber concretes after cracking under cyclic loading by adapting similar models for continuous fibers. This kind of a model is needed to study the behavior of fiber reinforced concrete structural elements subjected to dynamic loading.

2) It is desirable to simplify the present endochronic formulation for concrete to enable easy determination of the constants for new sets of data.

3) In view of its capabilities, endochronic theory provides a vast scope for developing accurate fracture analysis of fiber concretes. Most of the previous studies have been based on linear elastic methods which are not realistic for concrete due to its nonlinearities. Ductility due to fibers and the tensile softening in the matrix process zone near the crack tip can be easily accounted for by endochronic formulations.

4) Another interesting field is in the probabilistic characterization of the endochronic parameters by fitting largely scattered data from different sources. Then the statistical mean response of the structure and the associated reliability coefficients can be determined. Since each set of constants represents as particular type of behavior, the probabilistic endochronic approach would provide a good basis for this kind of analysis.

7. ACKNOWLEDGEMENT

The authors would like to thank Dr. S.E. Dunn, Professor and Chairman, Department of Ocean Engineering, Dr. T.E. Shoup, Dean of Engineering, Florida Atlantic University, and Dr. H. Wang, Professor and Chairman, Department of Coastal and Oceanographic Engineering, University of Florida, for their interest and enthusiastic support. The work described forms part of two Florida Sea Grant funded projects: i) No. R/C-7 entitled "Strength and Durability of Fiber and Polymer-Modified Concrete for Marine Structural Applications, and ii) No. R/C-D-10 entitled "Effect of Marine Exposure Upon Service Life of Reinforced Concrete". This support is gratefully acknowledged. Thanks are due to Ms. Fran Arvidson for excellent typing of this manuscript.

8. REFERENCES

1. Bazant, Z.P., "Endochronic Inelasticity and Incremental Plasticity," International Journal of Solids and Structures, Vol. 14, No. 9, 1978, pp. 691-714.

2. Bazant, Z.P., Ansal, A.M., and Krizek, R.J., "Endochronic Models for Soils," Proceedings of the International Symposium on Soil Mechanics-Transient and Cyclic Loads, Ed. Pande, G.N., and Zienkiewicz, O.C., John Wiley & Sons Ltd., New York, 1982, pp. 419-438.

3. Bazant, Z.P., and Bhat, P.D., "Endochronic Theory of Inelasticity and Failure of Concrete," Journal of the Engineering Mechanics Division, ASCE, Vol. 102, No. 4, 1976, pp. 701-722.

4. Bazant, Z.P., and Bhat, P.D., "Prediction of Hysteresis of Reinforced Concrete Members," Journal of the Structural Division, ASCE, Vol. 103, No. 1, 1977, pp. 153-167.

5. Bazant, Z.P., Krizek, R.J., and Shieh, C.L., "Hysteretic Endochronic Theory for Sand," Journal of the Engineering Mechanics Division, ASCE, Vol. 109, No.4, 1983, pp. 1073-1095.

6. Bazant, Z.P., and Shieh, C.L., "Endochronic Model for Nonlinear Triaxial Behavior of Concrete," Nuclear Engineering and Design, Vol. 47, 1978, pp. 305-315.

7. Burns, N.H., and Siess, C.P., "Load-Deformation Characteristics of Beam-Column Connections in Reinforced Concrete," Structural Research Series No. 234, Civil Engineering Studies, University of Illinois, Urbana, January, 1962.

8. Hughes, B.P., and Fattuhi, N.I., "Load-deflection "Curves for Fiber Reinforced Concrete Beams in Flexure," Magazine of Concrete Research, Vol. 29, No. 101, 1977a, pp. 199-206.

9. Hughes, B.P., and Fattuhi, N.I., "Stress-Strain Curves for Fiber Reinforced Concrete in Compression," Cement and Concrete Research, Vol. 7, 1977c, pp. 173-184.

10. Jain, S.K., "On the Thermodynamic Derivation of the Constitutive Equations of the Endochronic Theory of Viscoplasticity," Engineering Publications, Blacksburg, Virginia, 1982.

11. Lin, H.C., and Wu, H.C., "On the Rate-Dependent Endochronic Theory of Viscoplasticity and its Application to Plastic Wave Propagation," International Journal of Solids and Structures, Vol. 19, No. 7, 1983, pp. 587-599.

12. Matsuishi, M., and Iwata, S., "Probabilistic Strength of Steel Fiber Reinforced Concrete," Proceedings of the Third International Conference on the Behavior of Offshore Structures(BOSS), Boston, Massachusets, Vol. 2, 1982, pp. 837-846.

13. Naaman, A.E., Moavenzadeh, F., and McGarry, F.J., "Probabilistic Analysis of Fiber Reinforced Concrete," Journal of the Engineering Mechanics Division, ASCE, Vol. 100, No. 2, 1974, pp. 397-413.

14. Naaman, A.E., and Shah, S.P., "Pull-Out Mechanism in Steel Fiber Reinforced Concrete," Journal of the Structural Division, ASCE, Vol. 98, No. 8, 1976, pp. 1537-1548.

15. Park, R., Kent, D.C., and Sampson, R.A., "Reinforced Concrete Members with Cyclic Loading," Journal of the Structural Division, ASCE, Vol 98, No. 7, 1972, pp. 1341-1359.

16. Rivlin, R.S., "Some Comments on the Endochronic Theory of Plasticity," International Journal of Solids and Structures, Vol. 17, 1981, pp. 231-248.

17. Sandler, I.S., "On the Uniqueness and Stability of Endochronic Theories of Material Behavior," Journal of Applied Mechanics, ASME, Vol. 45, No. 2, 1978, pp. 263-266.

18. Sorensen, S.I., Arnesen, A., and Bergan, P.G., "Nonlinear Finite Element Analysis of Reinforced Concrete Using Endochronic Theory," Proceedings of the International Conference on Finite Elements in Nonlinear Solids and Structural Mechanics, Geilo, Norway, 1977, pp. 167-189.

19. Suidan, M., and Schnobrich, W.C., "Finite Element Analysis of Reinforced Concrete," Journal of the Structural Division, ASCE, Vol. 99, No. 10, 1973, pp. 2109-2122.

20. Valanis, K.C., "A Theory of Viscoplasticity Without a Yield Surface - Theory and Application," Archives of Mechanics, Vol. 23, No. 4, 1971, pp. 517-551.

21. Valanis, K.C., "Fundamental Consequences of a New Intrinsic Time Measure - Plasticity as a Limit of the Endochronic Theory," Archives of Mechanics, Vol. 32, No. 1, 1980, pp. 171-191.

22. Valanis, K.C., "On the Substance of Rivlin's Remarks on the Endochronic Theory," International Journal of Solids and Structures, Vol. 17, 1981, pp. 249-265.

23. Valanis, K.C., and Lee, C.F., "Some Recent Developments of the Endochronic Theory with Applications," Nuclear Engineering and Design, Vol. 69, 1982, pp. 327-344.

24. Valanis, K.C., and Lee, C.F., "Endochronic Theory of Cyclic Plasticity with Applications," Journal of Applied Mechanics, ASME, Vol. 51, No. 2, 1984, pp. 367-374.

25. Valanis, K.C., and Read, H.E., "A New Endochronic Plasticity Model for Soils," Proceedings of the International Symposium on Soil Mechanics-Transient and Cyclic Loads, Ed. Pande, G.N., and Zienkiewicz, O.C., John Wiley and Sons Ltd., New York, 1982, pp. 375-417.

26. Wecharatana, M., and Shah, S.P., "A Model for Predicting Fracture Resistance of Fiber Reinforced Concrete," Cement and Concrete Research, Vol. 13, 1983, pp. 819-829.

PART 2

NUMERICAL MODELS AND THEIR APPLICATION

CHAPTER 7

A FINITE ELEMENT MODEL FOR REINFORCED CONCRETE SLABS WITH CONSIDERATION OF TRANSVERSE SHEAR DEFORMATIONS

D. G. Klein
WÖLFEL Beratende Ingenieure, Höchberg-Würzburg,
formerly Institut für Massivbau, TH Darmstadt, FRG

SUMMARY

A finite element model is developed to investigate the influence of lateral shear on the behaviour of reinforced concrete slabs. A multilayered plate element is used with independent rotations in each layer to model the nonlinear distribution of shear strains across the slab thickness. The nonlinear deformation behaviour of concrete in a threedimensional stress state, cracking of concrete in the tensile range, and crushing under compression are included. The smeared crack approach is used with consideration of stress transfer across the rough crack surfaces. Reinforcement is idealized as uniaxial, smeared element layer with linear elastic - ideal plastic material behaviour. The nonlinear equations are solved by a mixed incremental iterative procedure. The application of the finite element model is examined by analytical verification of two tests from a series of square prestressed concrete slabs, clamped with one edge in square columns.

1. INTRODUCTION

A major problem in all analytical investigations of the load bearing behaviour of reinforced concrete slabs is the consideration of transverse shear forces and their deformations. A common case of significant influence of lateral shear are flat plates with a concentration of bending moments and shear forces at the connection between slab and column. In order to retain realistic results, any calculation method must handle progressive cracking through the slab thickness at inclined angles to the slab mid-plane, nonlinear deformation behaviour of concrete in threedimensional stress states and yielding of reinforcement.

One powerful tool to handle these problems is the finite element method with three-dimensional elements and an incremental, iterative procedure to solve the nonlinear equations. A disadvantage of threedimensional elements, however, is a large number of unknown degrees of freedom associated with a large bandwidth and thus a comprehensive amount of computation time. To model the successive changes and nonlinear distribution of stiffness through the slab thickness, a partition into several, superimposed elements is necessary. On the other hand a reduction of the number of elements over the slab area is limited because threedimensional elements of small thickness relative to its planar dimensions result in ill-conditioned equations and cause numerical problems. A common practice to reduce the amount of computation time is the limitation of threedimensional elements to areas with dominating shear deformations and connection of these elements with slab or flat shell elements in the rest of the area by special transition elements.

In the presented paper an alternate finite element model is introduced. To avoid the just mentioned difficulties with threedimensional elements, a layered flat shell element is used. This element is based on the thick shell element of Ahmad et al. [1]. The nonlinear distribution of shear strains across the slab thickness is considered with additional, independent rotations in each layer at a nodal point. The deformation behaviour of plain concrete is described by a nonlinear elastic, isotropic stress-strain relation. Cracking of concrete in the tensile range and crushing under compression is defined by a three-dimensional, stress dependent failure envelope. The smeared crack approach is used with consideration of stress transfer across the rough crack surfaces. Reinforcement is idealized as uniaxial, smeared element layers that function with perfect bond to the surrounding concrete. The stress-strain relation of reinforcement is linear elastic - ideal plastic.

The application of the program is limited to short time, monotonically increasing loads although nonproportional load histories are possible and locally limited unloading is allowed. The complicate behaviour of a deformed, cracked slab under reversal of the load is not included in the material descriptions.

The developed computer code is an extension of the program system for calculation of thin reinforced concrete slabs with consideration of realistic material behaviour, developed at the Institute of Massivbau, Technische Hochschule Darmstadt, as research project under support of the Deutsche Forschungsgemeinschaft ([2],[3]).

2. FINITE ELEMENT MODEL

A multilayered plate element similar to the formulation of Mawenya and Davis [4] is used. This element is based on the quadratic, isoparametric, curved shell element of Ahmad/Irons/Zienkiewicz [1]. The Ahmad element has been derived as a modification of a threedimensional, isoparametric, solid element by introducing the constraint of straight "normals" across the slab thickness and neglecting the strain energy corresponding to stresses perpendicular to the middle surface. However, the normal to the middle surface does not remain normal after deformation. This permits the shell to experience transverse shear deformations. A typical eight node plate element is shown in Fig. 1. The nodal parameters are the three displacements u, v, w along the global cartesian coordinates x, y, z with z normal to the mid plane and the two rotations α, β about the y- and x-axes, respectively.

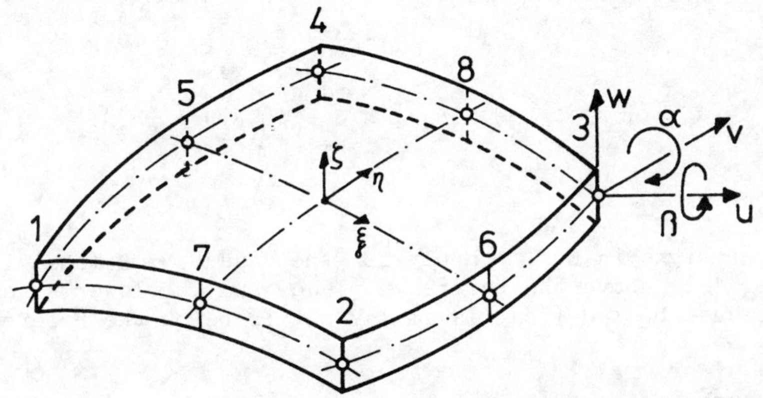

Fig. 1: Plate Element and Definition of Nodal Parameters

The layered element is an assembly of plate elements with the same nodal coordinates in the x- and y-directions, however, at different distances b from the reference plane. Each of these elements may have different material properties and a different thickness h. Fig. 2 represents the compound, layered element.

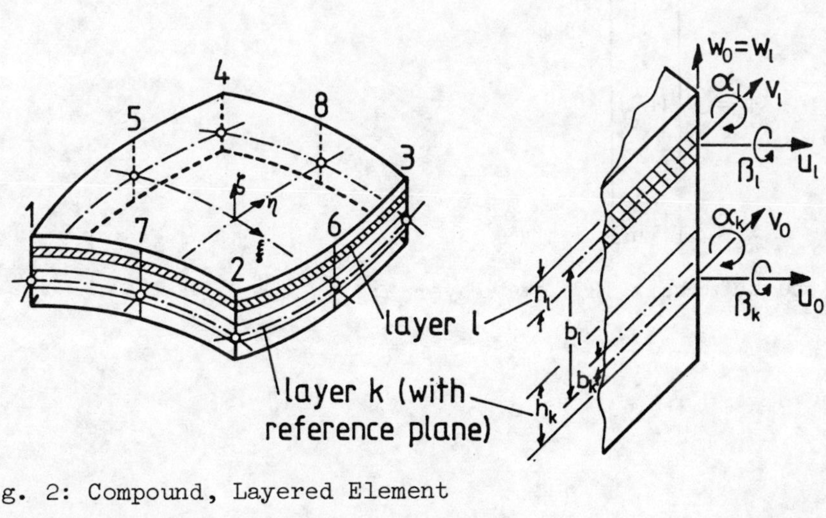

Fig. 2: Compound, Layered Element

The deformations of the layered element are described by the displacements u_o, v_o, w_o of the reference plane and two normal rotations α and β of each constituent layer. The normal rotations vary from layer to layer, and are independent of the lateral displacement. The lateral or transverse displacement w is assumed to be the same for all layers. By considering continuity of the in-plane displacements at the interfaces of the layers the following relations between the deformations of any layer ℓ and the deformations of the reference plane are obtained:

$$u_\ell = u_o \pm (\frac{1}{2}h_k \pm b_k) \cdot \alpha_k \pm \sum_{i=k+1}^{\ell-1} (h_i \alpha_i) \pm \frac{1}{2}h_\ell \alpha_\ell \quad (1a)$$

$$v_\ell = v_o \mp (\frac{1}{2}h_k \pm b_k) \cdot \beta_k \mp \sum_{i=k+1}^{\ell-1} (h_i \beta_i) \mp \frac{1}{2}h_\ell \beta_\ell \quad (1b)$$

$$w_\ell = w_o \quad (1c)$$

The upper signs in Eqs. 1 apply for the case $b_\ell > 0$ (that is if layer ℓ is above the reference plane) and the lower signs for the case $b_\ell < 0$ (that is if layer ℓ is below the reference plane).

Assembling the displacements of all layers at a nodal point "i" in the vector $\{\delta^i\}$ and the nodal point displacements of the compound, layered element in the vector $\{\delta_o^i\}$, Eqs. 1 can be written in matrix notation:

$$\{\delta^i\} = [H^i] \cdot \{\delta_o^i\} \quad (2)$$

with the transformation matrix:

$$[H^i] = \begin{bmatrix} [H_1^i] \\ \vdots \\ [H_\ell^i] \\ \vdots \\ [H_n^i] \end{bmatrix}$$

and the submatrix $[H_\ell^i]$ for layer ℓ:

$$[H_\ell^i] = \left[\begin{array}{ccc|c|cc|c|cc|c|cc|c}
1 & 0 & 0 & & (\frac{1}{2}h_k + b_k) & 0 & & h_i & 0 & & \frac{1}{2}h_\ell & 0 & \\
0 & 1 & 0 & & 0 & -(\frac{1}{2}h_k + b_k) & & 0 & -h_i & & 0 & -\frac{1}{2}h_\ell & \\
0 & 0 & 1 & 0 & 0 & 0 & \cdots & 0 & 0 & \cdots & 0 & 0 & 0 \\
0 & 0 & 0 & & 0 & 0 & & 0 & 0 & & 1 & 0 & \\
0 & 0 & 0 & & 0 & 0 & & 0 & 0 & & 0 & 1 & \\
 & & & \underbrace{}_{2(k-1)} & & & & \underbrace{}_{2(\ell-k-1)} & & & & & \underbrace{}_{2(n-\ell)}
\end{array}\right]$$

(Upper indices refer to the nodal points and lower indices to the layers.)

According to the nodal deformations a corresponding relation for the inherent nodal forces can be derived by the condition that the work done by the nodal forces of the layered element is equal to the summation of the work done by the nodal forces of each particular element layer at that nodal point. That yields the relation:

$$\{F_o^i\} = [H^i]^T \cdot \{F^i\} \tag{3}$$

It remains to combine the nodal stiffness relations $[C_\ell^{ij}]$ of each particular element layer into one supermatix $[C^{ij}]$:

$$\{F^i\} = [C^{ij}] \cdot \{\delta^j\} \tag{4}$$

Introducing Eqs. 2 and 3 into Eq. 4 yields the final nodal stiffness relation for the compound, layered element:

$$\{F_o^i\} = [H^i]^T \cdot [C^{ij}] \cdot [H^j] \cdot \{\delta_o^i\} = [C_o^{ij}] \cdot \{\delta_o^j\} \tag{5}$$

The combination of the different nodal stiffness matrices into the final element stiffness matrix is computed using standard procedures.

Reinforcement is implemented into the stiffness relation as smeared element layer that deforms only in the direction of the steel bars. Flexural stiffness and shear stiffness (inplane shear deformations as well as lateral shear deformations) are neglected in each reinforcement layer, so that nodal rotations of the reinforcement layers can be eliminated. Furthermore perfect bond is assumed for the reinforcement, that is the reinforcement strains and the strains of the surrounding concrete are the same. Using these assumptions, the nodal displacements of the reinforcement can be calculated using Eqs. 1. The stiffness of the reinforcement layers is added to the element stiffness matrix following multiplication by the transformation matrix [H]. The rotations of the reinforcement layers, however, are set to zero and the thickness of the concrete layer that contains the reinforcement is replaced by the proper distance of the reinforcement layer to the contact surface of the next concrete layer.

3. MATERIAL DESCRIPTION

3.1 Constitutive Relation of Plain Concrete

The stress component lateral to the slab midplane is neglected, $\sigma_{zz} = 0$. Nevertheless, a threedimensional state of stress exists in each concrete layer because the presence of lateral shear stresses. Within large regions of the slab, however, and mainly between cracks, the state of stress in the concrete may be close to biaxial or even uniaxial behaviour. This requires a realistic stress-strain relation for concrete which covers the full range of multiaxial stress states. Because of the large scatter of results by the various methods from tests on the behaviour of concrete under multiaxial stress states (Gerstle et al. [5]) and the randomly distributed value of the stiffness of the concrete over the slab area, dependent on the aggregates, the density, the consolidation, etc., a simple material formulation which requires only a few, readily available constants should be preferably used. Furthermore because the solution method uses a stepwise increasing load, an incremental formulation is necessary that includes nonproportional load histories.

All these requirements are fulfilled by the formulation of Gerstle [6], supplemented by the extension for nonproportional load histories, described by Cornelius [7]. In this theory, the incremental strain-stress relation for multiaxial stress states is formulated, for isotropic behaviour of concrete, in octahedral form:

$$d\varepsilon_o = \frac{d\sigma_o}{3K_T(\sigma_o)} + \frac{d\tau_o}{H_T(\sigma_o)} \qquad (6a)$$

$$d\gamma_o = \frac{d\tau_o}{2G_T(\tau_o)} \qquad (6b)$$

with the tangent values of the three moduli defined by

$$K_T = K_o(1 - .67\,\frac{\sigma_o}{\sigma_{o_u}}) \qquad (7a)$$

$$G_T = G_o(1 - \frac{\tau_o}{\tau_{o_u}}) \qquad (7b)$$

$$H_T[\mathrm{N/mm^2}] = (10 - \frac{300}{\sigma_o + 10}) \cdot 10^3 \qquad (7c)$$

The bulk modulus K_T and the shear modulus G_T relate volumetric and deviatoric strains and stresses; the coupling modulus H_T accounts for the experimentally observed fact that stress deviation causes a volume decrease prior to failure. The formulation (7c) for H_T is defined only for $\sigma_o < -10$ N/mm². Because the value of H_T increases radically with decreasing hydrostatic compression, and the portion of the volumetric change due to coupling becomes essentially neglegible for smaller hydrostatic stresses, the coupling modulus is set to $H_T = 10^9$ N/mm² if $\sigma_o > -10.0003$ N/mm².

Eqs. 7 require the initial values K_o and G_o and the limit values σ_{ou} and τ_{ou} of the octahedral stresses. K_o and G_o can be obtained from uniaxial compression tests which allow determination of the modulus of elasticity E_o and Poisson's ratio ν.

$$K_o = E_o/(1-2\nu) \tag{8a}$$

$$G_o = E_o/(1+\nu) \tag{8b}$$

The limit values σ_{ou} and τ_{ou} must be obtained from an appropriate multiaxial strength envelope. The failure criterion used for plain concrete and the computation of σ_{ou} and τ_{ou} are described in subsection 3.2.

Concrete is introduced as a hyperelastic material. On unloading, that is if $d\sigma_o > 0$, it is assumed that the strains and stresses follow the same relations, Eqs. 6, as during the loading path. This is inconsistent with the real material's behaviour. Because the application of the program system is for monotonically increasing load histories, unloading may occur only locally during redistribution of stresses or as fictitious , corrective stresses during the iteration process, so that unloading is of minor influence on the over all behaviour of the structure.

In Eqs. 6 the incremental flexibility is formulated with octahedral stresses. To implement these relations into the program system it is necessary to derive the values of the incremental flexibility matrix in the direction of the cartesian, global coordinate system (x, y, z). Correspondent to the assumptions of Cornelius [7], the transformation of the octahedral to the coordinate stress and strain increments is based on the coincidence of deviator stress and strain increment vectors. In tensor notation this coincidence can be written as:

$$de_{ij} = \beta \cdot dS_{ij} \tag{9}$$

with deviator strain increments

$$de_{ij} = d\varepsilon_{ij} - \delta_{ij} \cdot d\varepsilon_o \; , \tag{10a}$$

deviator stress increments

$$dS_{ij} = d\sigma_{ij} - \delta_{ij} \cdot d\sigma_o \ , \tag{10b}$$

and the octahedral stress and strain increments:

$$d\varepsilon_o = \frac{1}{3}\, d\varepsilon_{ii}\, ,\ d\gamma_o = \frac{1}{3\gamma_o}\, e_{ij} \cdot de_{ij} \tag{11a}$$

$$d\sigma_o = \frac{1}{3}\, d\sigma_{ii}\, ,\ d\tau_o = \frac{1}{3\tau_o}\, S_{ij} \cdot dS_{ij} \tag{11b}$$

(δ_{ij} is the Kronecker delta. In Eqs. 9 to 11 and in the following derivations of this subsection the summation convention on repeated indices is used.)

The constant β is determined by introducing the incremental octahedral shear strains and stresses from Eq. 11 into Eq. 6b which provides:

$$\beta = \frac{\gamma_o}{2G_T \tau_o} \cdot \frac{S_{k\ell} \cdot dS_{k\ell}}{e_{ij} \cdot dS_{ij}} \tag{12}$$

With the constant β thus defined and after introducing Eq. 6a for the hydrostatic strains into Eq. 10a, the incremental strain-stress relation expressed in the coordinate directions takes the form:

$$d\varepsilon_{ij} = \beta \cdot dS_{ij} + \delta_{ij}\, \frac{1}{3K_T}\, d\sigma_o + \delta_{ij}\, \frac{1}{H_T}\, d\tau_o \tag{13}$$

After replacing the octahedral stresses with coordinate stresses:

$$d\sigma_o = \frac{1}{3}\, \delta_{k\ell}\, d\sigma_{k\ell}$$

$$dS_{ij} = (\delta_{ik} \cdot \delta_{j\ell} - \frac{1}{3}\, \delta_{ij} \cdot \delta_{k\ell}) \cdot d\sigma_{k\ell}$$

$$d\tau_o = \frac{1}{3\tau_o} \cdot S_{rs} \cdot (\delta_{rk} \cdot \delta_{s\ell} - \frac{1}{3}\, \delta_{rs} \cdot \delta_{k\ell}) \cdot d\sigma_{k\ell}$$

and by use of the fact that $S_{rr} = 0$, the final incremental flexibility relation is:

$$d\varepsilon_{ij} = [\frac{1}{3} \cdot \frac{1}{3K_T}\, \delta_{ij} \cdot \delta_{k\ell} + \beta \cdot (\delta_{ik} \cdot \delta_{j\ell} - \frac{1}{3}\, \delta_{ij}\, \delta_{k\ell})$$

$$+ \frac{1}{H_T 3\tau_o}\, \delta_{ij}\, S_{k\ell}] \cdot d\sigma_{k\ell} \tag{14}$$

or in matrix notation:

$$\{d\varepsilon\} = [D^{sc}] \cdot \{d\sigma\} \tag{15}$$

$[D^{sc}]$ is the incremental flexibility matrix of solid concrete.

The incremental flexibility matrix is singular if the initial values of the octahedral shear stress $\tau_o = 0$. This occurs during the first load step or for pure hydrostatic pressure. It is appropriate in both cases to replace the incremental flexibility by the secant relation, neglecting coupling effects.

$$d\varepsilon_{ij} = [\frac{1}{3}\frac{1}{3K_T}\delta_{ij} \cdot \delta_{k\ell} + \frac{1}{2G_T} \cdot (\delta_{ik} \cdot \delta_{j\ell} - \frac{1}{3}\delta_{ij} \cdot \delta_{k\ell})] \cdot d\sigma_{k\ell} \tag{16}$$

3.2 Failure Criterion for Concrete

The four parameter failure criterion, proposed by Hsieh/Ting/Chen [8], is used to define the ultimate state of stresses. This criterion is applicable for plain concrete with isotropic behaviour subjected to multiaxial short-time loading. It covers the total area of stress states from uniaxial tension beyond biaxial stresses to the general triaxial stress state.

Written in terms of octahedral stresses, the failure function reads

$$A \cdot \frac{3}{2}\left(\frac{\tau_o}{f'_c}\right)^2 + B\sqrt{\frac{3}{2}}\frac{\tau_o}{f'_c} + C\frac{\sigma_{11}}{f'_c} + D \cdot 3\frac{\sigma_o}{f'_c} - 1 = 0 \tag{17}$$

with the parameters A, B, C and D as established by Hsieh [8].

As proposed by Hsieh/Ting/Chen [8] the modes of failure are categorized into three types, namely cracking, crushing, and a mixture of cracking and crushing. The pure cracking zone is defined by the maximum tensile stress condition, $\sigma_{11} > 0$. For this mode it is assumed that the element keeps its stiffness tangential to the crack plane. The pure crushing mode occurs if all three principal strain components will be compressive strains, $\varepsilon_{11} < 0$. In this case the residual stiffness and residual strength of the crushed concrete is neglected. The post failure behaviour in the mixed mode is regarded as linear interpolation of the perfectly deformable behaviour in the purely crushed mode and the anisotropic behaviour in the purely cracked mode.

The failure envelope is open in the direction of the negative hydrostatic axis. For numerical reasons (namely to limit the stress envelope for the calculation of σ_{ou} and τ_{ou}, see below), the maximum hydrostatic compression stress, related to the uniaxial compression strength f_c', is limited by a factor c_f. (In the numerical examples, described in section 5, the factor $c_f = 100$ is used.)

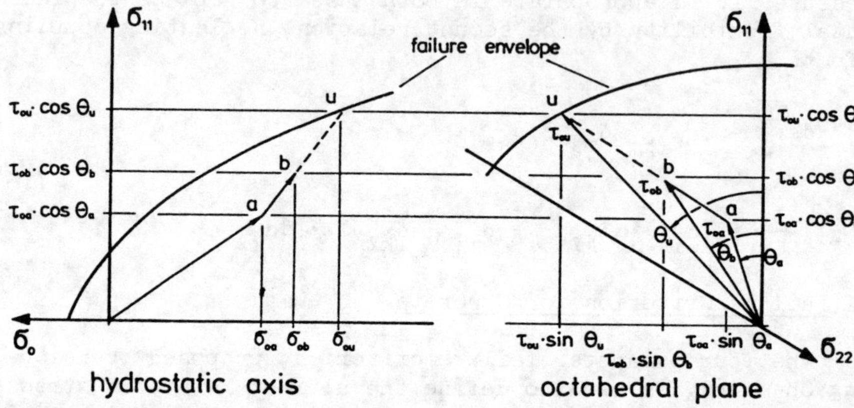

Fig. 3: Limit Stresses σ_{ou} and τ_{ou}

The ultimate hydrostatic stress σ_{ou} and deviatoric stress τ_{ou}, required for calculation of the tangent values of the material characteristics, Eqs. 7, are defined as that point on the failure surface, where the linear extension of the stress increment vector penetrates the stress envelope. Defining point "a" as the initial stress state in the octahedral stress space, point "b" as the new stress state (after adding the stress increments to the stresses at point a), and point "u" as the penetration point of the linear extension of the increment vector from "a" through "b" to the failure envelope (see Fig. 3), the following relations hold:

$$\frac{\Delta\sigma_{au}}{\Delta\sigma_{ab}} = \frac{\sigma_{ou} - \sigma_{oa}}{\sigma_{ob} - \sigma_{oa}} \tag{18a}$$

$$= \frac{\tau_{ou} \cdot \cos\Theta_u - \tau_{oa} \cdot \cos\Theta_a}{\tau_{ob} \cdot \cos\Theta_b - \tau_{oa} \cdot \cos\Theta_a} \tag{18b}$$

$$= \frac{\tau_{ou} \cdot \sin\Theta_u - \tau_{oa} \cdot \sin\Theta_a}{\tau_{ob} \cdot \sin\Theta_b - \tau_{oa} \cdot \sin\Theta_a} \tag{18c}$$

$\Delta\sigma_{au}$ and $\Delta\sigma_{ab}$ are the length of the vector increments from point a to u and a to b respectively.

Introducing the relation $p = \Delta\sigma_{au}/\Delta\sigma_{ab}$ and the abbreviations

$$X = \tau_{oa} \cdot \sin\Theta_a + (\tau_{ob} \cdot \sin\Theta_b - \tau_{oa} \cdot \sin\Theta_a) \cdot p \qquad (19a)$$

$$Y = \tau_{oa} \cdot \cos\Theta_a + (\tau_{ob} \cdot \cos\Theta_b - \tau_{oa} \cdot \cos\Theta_a) \cdot p \qquad (19b)$$

the unknown stresses σ_{ou} and τ_{ou} and the angle Θ_u can be calculated:

$$\sigma_{ou} = \sigma_{oa} + (\sigma_{ob} - \sigma_{oa}) \cdot p \qquad (20a)$$

$$\tau_{ou} = \sqrt{X^2 + Y^2} \qquad (20b)$$

$$\tan\Theta_u = \frac{X}{Y} \qquad (20c)$$

Insertion of Eqs. 20 into the failure condition, Eq. 17, yields an equation of fourth order in the unknown p that can be solved only iteratively. During iteration the relation p is used as parameter. Convergence is assumed to be sufficient, if the difference of the length of stress vector $\Delta\sigma_{au}$ is less than $f'_c/100$. Thus the convergence criteria in terms of the parameter p reads:

$$|p^{i+1} - p^i| < \frac{f'_c}{100 \cdot \Delta\sigma_{ab}} \qquad (21)$$

The failure envelope is defined only within the limits of Θ between 0° and 60°. Beyond these limits the failure envelope is the symmetric extension of Eq. 17 with respect to the axis $\Theta = 0°$ (σ_{11} - axis) or $\Theta = 60°$ ($-\sigma_{22}$ - axis) within the deviatoric plane. This must be carefully observed during calculation of the proper value of Θ_u.

3.3 Rough Crack Model

The nonlinear behaviour of concrete structures is mainly influenced by the brittle cracking in tension which causes an abrupt loss of strength and stiffness. It is essential to find an appropriate model for the description of cracking and post cracking behaviour of concrete to get realistic results in any analysis of concrete structures.

Although cracking is a discontinuous attribute of concrete, the smeared crack approach is used here. This approach allows an equivalent continuum treatment with localized material anisotropy and thus fits well into the approximate model of the finite element method without the laborious changing of structure topology.

A new crack starts to open if the failure criterion is violated in the pure cracking or mixed mode. It is assumed that the crack is planar and that the plane of the crack is normal to the direction of the main tensile strain ε_{11}. At first crack opening the stress normal to the crack is set to zero. The crack opens in normal direction while the initial slip (relative tangential displacement of the crack surfaces) is zero. The increase of further crack opening may be accompanied by shear slip of the crack surfaces. Due to aggregate interlock cracks sustain substantial shear forces even at open cracks. In addition a compressive normal stress is induced on the crack surface if during increasing shear slip the opening of the crack is restrained by reinforcement across the crack or by boundary constraints of the surrounding concrete.

The rough crack model of Bazant and Gambarova [9], [10] is used to calculate the stress transfer accross the crack surfaces. This model is based on the assumption that crack properties may be considered as a material property similar to stress-strain relations for solid concrete and may be generally written in the incremental form of Eq. 22. Eq. 22 is an extension of the planar formulation of Bazant and Gambarova [9], [10] to a three-dimensional state with independent and similar behaviour in the two directions s and t tangential to the crack surface.

$$\begin{Bmatrix} d\sigma_{nn} \\ d\sigma_{ns} \\ d\sigma_{nt} \end{Bmatrix} = \begin{bmatrix} C_{nn} & C_{ns} & C_{nt} \\ C_{sn} & C_{ss} & C_{st} \\ C_{tn} & C_{ts} & C_{tt} \end{bmatrix} \begin{Bmatrix} d\delta_{n} \\ d\delta_{s} \\ d\delta_{t} \end{Bmatrix} \tag{22}$$

δ_n, δ_s and δ_t represent the relative normal and tangential displacements of the crack surfaces. Within the smeared crack approach the relative displacements are considered to be continuously distributed. The average strains due to the smeared cracks then are

$$\varepsilon_{nn}^{cr} = \delta_n/s \tag{23a}$$

$$\varepsilon_{ns}^{cr} = \delta_s/s \tag{23b}$$

$$\varepsilon_{nt}^{cr} = \delta_t/s \tag{23c}$$

with s as mean crack spacing. The crack spacing depends on diameter and arrangement of the reinforcement, any bond slip of the reinforcement, composition of the concrete, stress distribution, gradient and much more. The determination of the crack spacing is beyond the scope of this work. It is introduced as constant parameter.

To calculate the stiffness coefficients C_{nn}, C_{ns}, C_{nt}, ..., a path-independent total stress total displacement formulation is used as derived by Bazant and Gambarova [9], [10].

$$\sigma_{nn} = f_n(\delta_n, \delta_s, \delta_t), \ \sigma_{ns} = f_s(\delta_n, \delta_s), \ \sigma_{nt} = f_t(\delta_n, \delta_t) \quad (24)$$

Differentiation of the functions (24) together with the definitions (23) for the strains due to cracking yields the components of the incremental crack stiffness matrix $[C^{cr}]$.

The average over all strains of the reinforced concrete are the sum of the strains of the solid concrete between the cracks $\{\varepsilon^{sc}\}$ and the strains due to the cracks $\{\varepsilon^{cr}\}$ so that:

$$\{d\varepsilon\} = \{d\varepsilon^{sc}\} + \{d\varepsilon^{cr}\} \quad (25)$$

The strains of the solid concrete are related to the stresses by the incremental flexibility matrix as derived in sub-section 3.1

$$\{d\varepsilon^{sc}\} = [D^{sc}]\ \{d\sigma\} \quad (26)$$

It remains to invert the crack stiffness matrix $[C^{cr}]$ and to transform the resulting matrix from the (n, s, t) coordinate system to the global (x, y, z)-system for deriving the incremental relation between the strains due to cracking and the stresses

$$\{d\varepsilon^{cr}\} = [D^{cr}] \cdot \{d\sigma\} \quad (27)$$

with $[D^{cr}] = [R]^T \cdot [C^{cr}]^{-1} \cdot [R]$

[R] is the matrix to transform the stresses, as well as the corresponding strains, from the (x, y, z)-system to the (n, s, t)-system with n being normal to the crack surface.

Substituting Eqs. 26 and 27 into Eq. 25 and by regard of the fact, that the stresses in the solid concrete between the cracks are equal to the stresses on the crack surface, we obtain

$$\{d\varepsilon\} = \Big[[D^{sc}] + [D^{cr}]\Big] \cdot \{d\sigma\} \quad (28)$$

The sum $[D] = [D^{sc}] + [D^{cr}]$ is the final incremental flexibility matrix of the cracked concrete element.

The direction of the main tensile stresses and strains in the solid concrete between the cracks may change with further increases in the load because of stress transfer across the cracks. As a result a second crack system at an arbitrary angle to the first crack system is possible. To prevent numerical inaccuracy, however, the development of a new crack system is allowed only if the deviation of the vector normal to the new crack from the normal vector to the existing crack is larger than 20°.

The derivation of the stiffness relation Eq. 28 for concrete with one crack is easily extented to concrete with two crack systems of arbitrary directions. The incremental stiffness matrix $[C^{cr}]$ must then be written for each of the crack systems. After applying transformations $[R_1]$ and $[R_2]$, where subscripts 1 and 2 refer to crack system 1 and 2, respectively, and adding the deformations, the combined flexibility matrix due to the existence of both crack systems is

$$[D^{cr}] = [R_1]^T \cdot [C_1^{cr}]^{-1} \cdot [R_1] + [R_2]^T \cdot [C_2^{cr}]^{-1} \cdot [R_2] \quad (29)$$

Within the described approach the development of a third crack system is possible. At the present state of the program, however, the stiffness and strength of a concrete element with three cracks are neglected.

The total stress total displacement relations (24), as proposed by Bazant and Gambarova [9], [10], are singular, if $\delta_n = 0$. This singularity is avoided by the use of a special procedure at first crack opening (see section 4). If for an existing crack δ_n is negative or very small again (e. g. if $\delta_n < 0.001$ mm) the crack is defined as closed and the concrete element is treated under subsequent load as an uncracked element, that is the incremental flexibility matrix $[D^{cr}]$ is zero. On the other hand no contact at the crack surface is possible, that is $\sigma_{nn} = \sigma_{ns} = \sigma_{nt} = 0$, if at an open crack ($\delta_n > 0.001$ mm) the relative tangential displacements are both zero ($\delta_s = \delta_t = 0$). In this case the incremental crack stiffness matrix $[C^{cr}]$ is neglected.

4. SOLUTION ALGORITHM

An incremental iterative technique is adopted to solve the nonlinear equations. The total load expected is divided into a number of increments. Changes of nodal displacements within a load step are computed by an iterative process, based on the Newton-Raphson method with constant stiffness and initial stress. Because the labourious, time intensive computation of the stiffness matrix for the element used, the stiffness matrix of the system is calculated and triangulized only once with initial material characteristics and kept for all subsequent load steps. During the iteration procedure, the nodal displacements are corrigated by applying additional corrective nodal loads, equivalent to initial stress increments that are required for coincidence of the elastic solution with the true stress increments corresponding to the computed strain increments. As a convergence criterion for the iteration process

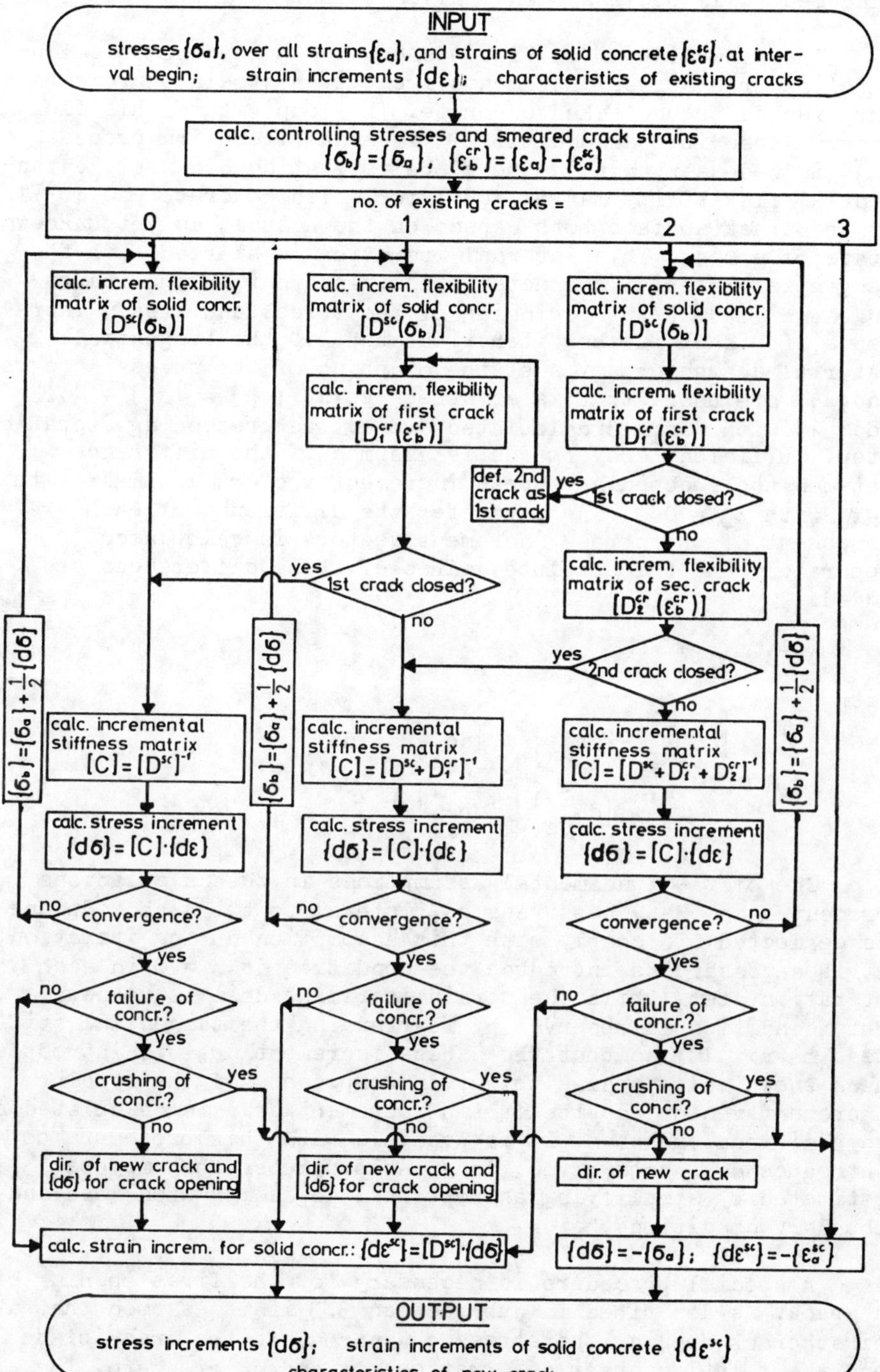

Fig. 4: Calculation of True Stress Increments

the maximum deformation criterion is used. Convergence is evident if the relative magnitude of the additional maximum displacement will be less than a given value.

The flow diagram in Fig. 4 shows the procedure to calculate the true stress increments from given strain increments for cracked or uncracked concrete. The loop within this flow diagram describes an additional, internal iteration process that is necessary within the global iteration because the tangential flexibility matrix $[D^{sc}]$ for solid concrete and $[D^{cr}]$ of the crack surface both depend on the actual, as yet unknown state of stress. This internal iteration is started with the tangential material characteristics, inherent to the values $\{\sigma_a\}$ and $\{\varepsilon_a\}$ at the beginning of the stress and strain increments. For each further iteration step "i" the tangential material characteristics at the midpoint of the stress interval $\{d\sigma_{i-1}\}$ are used, that is at stress point $\{\sigma_b\} = \{\sigma_a\} + 1/2 \{d\sigma_{i-1}\}$ with $\{d\sigma_{i-1}\}$ calculated during the preceeding iteration step. Sufficient convergence is assumed if the difference between two succeeding stress increment vectors is smaller than $\lambda_s f'_c$ with $\lambda_s = 0.01$. As parameter the difference of each component of the stress increment vectors is calculated separately and the absolute values of these differences are added.

$$\Delta\sigma^i < \lambda_s f'_c \tag{30}$$

with

$$\Delta\sigma^i = |d\sigma_{xx}^{i-1} - d\sigma_{xx}^i| + |d\sigma_{yy}^{i-1} - d\sigma_{yy}^i| + + \dots + |d\sigma_{yz}^{i-1} - d\sigma_{yz}^i|$$

One of the fundamental assumptions in the calculations presented is that the stress σ_{zz} lateral to the slab midplane is neglected. To comply with this assumption during iteration, it is necessary to introduce the condition $d\sigma_{zz} = 0$ in each iteration step, while the strain increment $d\varepsilon_{zz}$ is unknown. This condition is observed by rearranging the tangential stiffness matrix [C] so that the stress increment $d\sigma_{zz}$ is introduced as a known value beside the given components of the strain increment vector and the strain increment $d\varepsilon_{zz}$ is calculated as an additional unknown. It is possible with this procedure to introduce $d\sigma_{zz}$ as a given, constant parameter that may be estimated by simplifying assumptions, depending upon load and boundary conditions.

A special procedure is necessary for the first opening of a crack. As described in sub-section 3.3 it is assumed that at first crack opening the stress σ_{nn} normal to the crack plane and the relative shear deformations δ_s and δ_t are zero. The final stresses in the concrete after initial crack opening hence are:

$$\{\sigma_b\} = \{\sigma_a\} + \{d\sigma\} = \{\sigma_a\} + [D^{sc}]^{-1}\ \{d\bar{\varepsilon}\} \qquad (31)$$

$$\text{with } \sigma_{bnn} = 0 = \sigma_{ann} + C_{11}^{sc} d\bar{\varepsilon}_{nn} + C_{12}^{sc} d\bar{\varepsilon}_{ss} + \ldots + C_{16}^{sc} d\bar{\gamma}_{st} \qquad (32)$$

C_{11}^{sc}, C_{12}^{sc}, ... are the components of the tangential stiffness matrix $[D^{sc}]^{-1}$ of solid concrete. Because of the assumption of isotropic behaviour for solid concrete, the matrix $[D^{sc}]^{-1}$ is invariant during stress transformation. Therefore a transformation of this matrix into the (n, s, t)-system of the crack direction is not necessary.

As consequence of condition $\delta_s = \delta_t = 0$, the components of the incremental strain vector $\{d\bar{\varepsilon}\}$ of the solid concrete after cracking, excluding $d\bar{\varepsilon}_{nn}$, are equal to the components of $\{d\varepsilon^{sc}\}$ before cracking:

$$\begin{aligned} d\bar{\varepsilon}_{ss} &= d\varepsilon_{ss}^{sc},\ d\bar{\varepsilon}_{tt} = d\varepsilon_{tt}^{sc},\ d\bar{\gamma}_{ns} = d\gamma_{ns}^{sc}, \\ d\bar{\gamma}_{nt} &= d\gamma_{nt}^{sc},\ d\bar{\gamma}_{st} = d\gamma_{st}^{sc} \end{aligned} \qquad (33)$$

With this condition and Eq. 32 we can write the incremental strain vector in matrix notation

$$\{d\bar{\varepsilon}\} = \frac{1}{C_{11}^{sc}}\ [A] \cdot \{d\varepsilon^{sc}\} - \frac{1}{C_{11}^{sc}}\ \{\sigma_a^o\} \qquad (34)$$

with

$$[A] = \begin{bmatrix} 0 & -C_{12}^{sc} & -C_{13}^{sc} & \ldots & -C_{16}^{sc} \\ 0 & 1 & 0 & \ldots & 0 \\ 0 & 0 & 1 & \ldots & 0 \\ . & \ldots & \ldots & \ldots & \ldots \\ 0 & 0 & 0 & \ldots & 1 \end{bmatrix}$$

and $\{\sigma_a^o\}^T = \{\sigma_{ann}, 0, 0, 0, 0, 0\}$

After transformation from the (n, s, t)-system into the global (x, y, z)-system the incremental strains of the solid concrete between the new cracks are

$$\{d\bar{\varepsilon}\} = \frac{1}{C_{11}^{sc}} \cdot [B] \cdot \{d\varepsilon^{sc}\} - \frac{1}{C_{11}^{sc}}\ \{\bar{\sigma}_a\} \qquad (35)$$

with $[B] = [R]^T \cdot [A] \cdot [R]$ and $\{\bar{\sigma}_a\} = [R]^T \cdot \{\sigma_a^o\}$

The desired stress increments are calculated after multiplying $\{d\bar{\varepsilon}\}$ with the tangential stiffness matrix $[D^{sc}]^{-1}$ of solid concrete.

$$\{d\sigma\} = \frac{1}{C_{11}^{sc}} \cdot [D^{sc}]^{-1} \cdot [B] \cdot \{d\varepsilon^{sc}\} - \frac{1}{C_{11}^{sc}} \cdot [D^{sc}]^{-1} \cdot \{\bar{\sigma}_a\} \tag{36}$$

In Eq. 36 $\{d\varepsilon^{sc}\}$ is identical with the over all strain increment vector $\{d\varepsilon\}$ if the first crack opens. For the second crack it is necessary to multiply

$$\{d\varepsilon^{sc}\} = [D^{sc}] \cdot [C] \cdot \{d\varepsilon\} \tag{37}$$

to find the strains in the solid concrete between the cracks with [C] as incremental stiffness matrix of concrete with one crack.

For three cracks or pure crushing it is assumed that the concrete loses its strength and stiffness. In this case, the stress increments are simply the negative values of the starting stresses

$$\{d\sigma\} = - \{\sigma_a\} \tag{38}$$

As proposed in sub-section 3.2, the stress increments for the mixed failure mode are calculated as linear interpolation between the pure cracking mode and the pure crushing mode.

For simplicity the derivation of Hsieh/Ting/Chen [8] is used. They employ Hooke's law of elasticity to the dual failure criterion $\sigma_{11} > 0$ and $\varepsilon_{11} < 0$ to identify the mixed failure mode as:

$$1 \leqq \chi \leqq \frac{1 + \nu}{(1 - 2\nu)} \tag{39}$$

with the crusing coefficient

$$\chi = - \frac{\sigma_o}{(\sigma_{11} - \sigma_o)} \tag{40}$$

With $\{d\sigma_1\}$ as strain increments for pure cracking and $\{d\sigma_2\} = - \{\sigma_a\}$ for pure crushing, the stress increments for the mixed mode then are:

$$\{d\sigma\} = \{d\sigma_1\} + (\{d\sigma_2\} - \{d\sigma_1\})\,(\chi - 1)\frac{1 - 2\nu}{3\nu} \tag{41}$$

5. NUMERICAL EXAMPLES

5.1 System and Element Idealisation

To examine the program system, the tests S1 and S2 from a series carried out by H. Sunidja [11] are analytically verified. The objective of the tests was to study the strength and behaviour of unbonded prestressed concrete plate-edge column connections representative of those used in prestressed flat plate buildings at static vertical loadings.

The test specimens consist of square prestressed concrete slabs, clamped in square columns which are located adjacent to and centered along one edge of the slabs. The columns have hinged ends above and below the slabs. The system and dimensions of the slabs are shown in Fig. 5.

The slabs are prestressed in both directions to represent a two-way flat slab with concentration of the tendons in a narrow strip over the columns. The arrangement of the tendons is shown in Fig. 6. The tendons consist of 7-wire stress-relieved strands 9.52 mm (3/8 in) in diameter. In addition, No. 3 deformed bars (71 mm cross-sectional area) are used as bonded reinforcement of the slab.

The live load of the slabs is simulated by four concentrated loads placed at different distances from the column face for slab S1 and slab S2 (see Fig. 5). The load is increased incrementally and after each load step the load is held constant and data on deflection, prestressing forces, strains and cracks are recorded. During the first test, slab S1, the load was released at special load steps and reloaded back to the previous level again. It was found, however, that the influence of these load cycles on the behaviour of the slab was neglectable. For all other tests the load was simply increased monotonically.

The idealization of the slab into elements and layers is shown in Fig. 7. The slab is assumed as clamped at the adjacent column. The displacements and rotations of the nodal points surrounding the column are set to zero. This assumption is justified because the heavily reinforced column is very stiff compared to the slab and because additional measurements at the column during the tests showed no significant deformations.

From the bonded reinforcement only the top reinforcement within the column strips is considered. The distribution of the reinforcement as smeared reinforcement layers is approximately taken into account by use of different thicknesses for the layers.

The pretension is idealized as a horizontal load at the slab edges. The eccentric anchorage of the tendons is approxi-

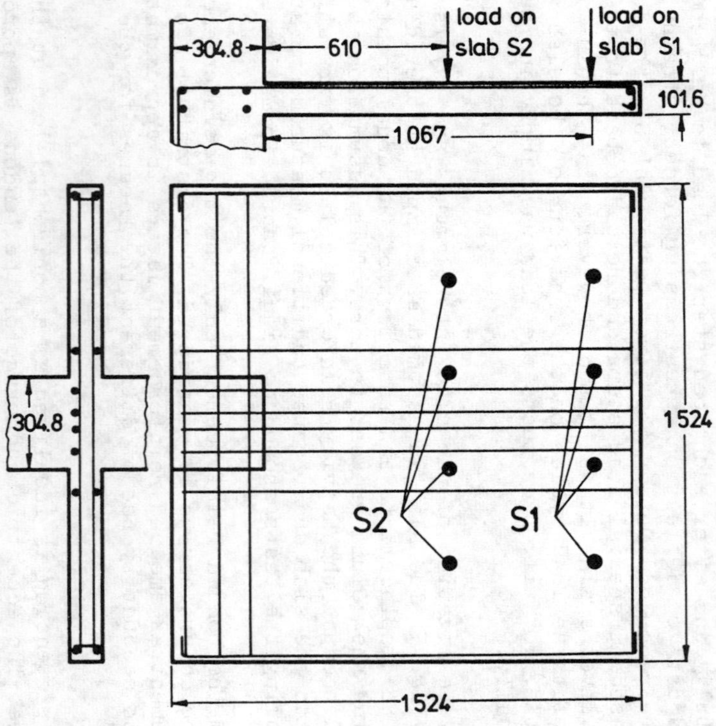

Fig. 5: Test Specimens with Bonded Reinforcement

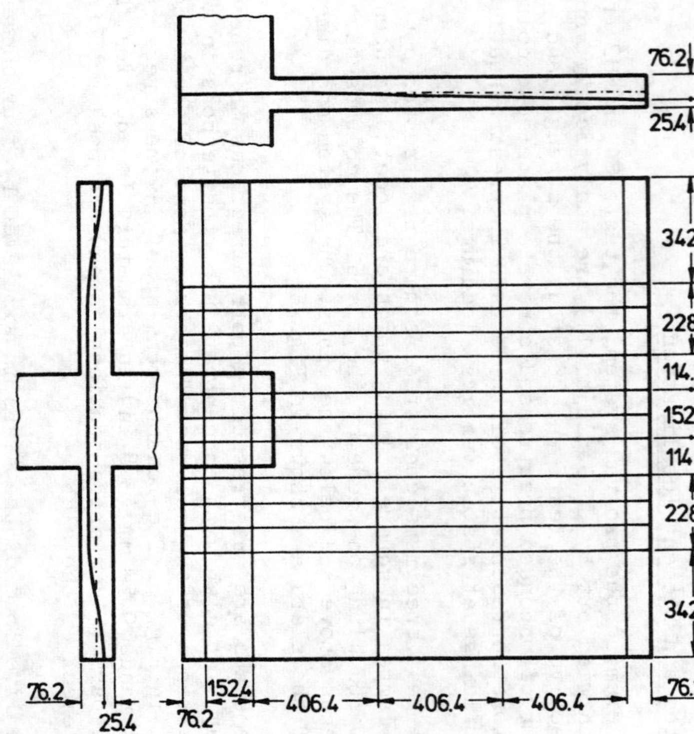

Fig. 6: Unbonded Tendons

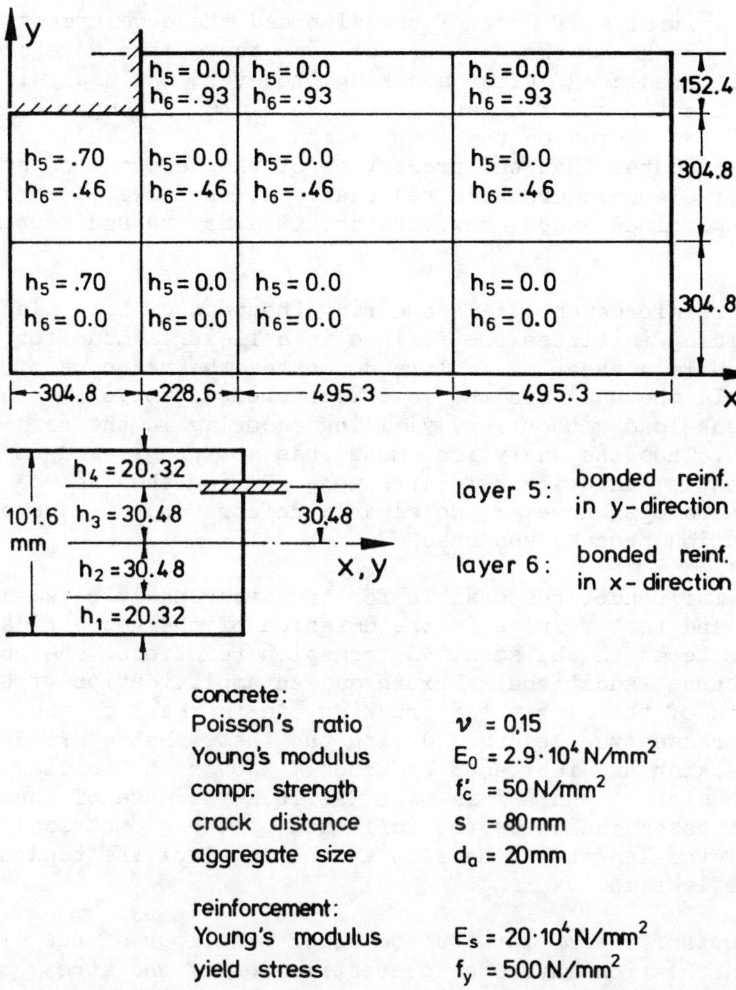

Fig. 7: Element Idealization and Material Properties

mately considered by the distribution of the loads as stresses over the two mid layers. To account for the diversion of the pretension forces because the curvature of the tendons additional lateral forces are applied to the slabs. In the calculation the pretension and the self-weight load of the slabs are applied as initial loads and kept constant during the stepwise increase of the lateral, concentrated loads.

5.2 Comparison of Analysis with Test Results

In Fig. 8 the moment-deflection relation of the analysis is compared with the test results. The moment is obtained by multiplying the total applied load by the distance from the load to the face of the column.

The initial stiffness of the slab and the development of the first cracks at the top surface near the column edge are quite well predicted. After cracking, however, the analysis results in a too stiff behaviour of the slabs. A comparison of the concrete strains on the compression side of the slab, Fig. 9, indicates that the prediction of the deformation behaviour of plain concrete is reasonable at the lower load steps. At the upper load steps, however, the strains are underestimated.

The reinforcement yielded during the test of both slabs. That yielding initiates the failure in a flexural mode for slab S1 while a shearing failure dominates the response of slab S2. In the analysis the load is increased above the actual peak load without any yielding occuring in the reinforcement. Thus the analytical model has not progressed to the development of an evident failure mode. During the last calculated load steps, however, no clearly defined convergence of the iteration process was reached.

One influence, responsible for the discrepancy between analysis and test results is the omission of the geometrically nonlinear terms in the strain-deformation relations. The pretension causes additional flexure and an amplification of the deflection of the slab after cracking. An increase of the tendon stresses was observed during the tests that started after cracking and increased by about 7 % over the initial stress in slab S1 and 4 % in slab S2. This increase of the tendon stresses indicates the influence of the geometrical change of the length of the slab at the level of the tendons during deflection.

A further reason for the too stiff behaviour of the slab during calculation may be an overestimation of the stress transfer across the cracks. The stiffness of the cracks depends on the crack width and thus is largely influenced by the assumed distance between the cracks in the smeared crack approach. In Fig. 10 and 11 the calculated crack patterns at the slab top surfaces are compared with the test results. At lower load steps, namely before yielding of reinforcement, parallel cracks with an approximately constant spacing developed so that the smeared crack approach seems applicable. With increasing load, however, the crack opening is concentrated into only a few cracks, consequently with larger relative displacements and less transfer of stress across those cracks than assumed in the calculations.

To demonstrate the calculated amount of stress transfer across a crack, the distribution of the normal stresses and the lateral shear stresses in a cross-section of slab S2 near the column edge are plotted in Fig. 12 for different load steps and

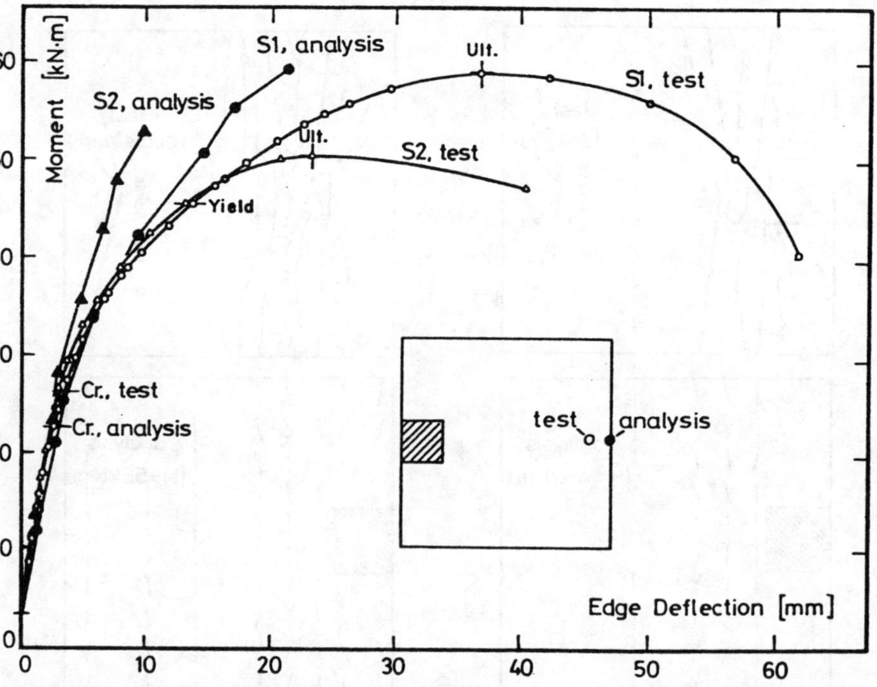

Fig. 8: Moment-Deflection-Relation

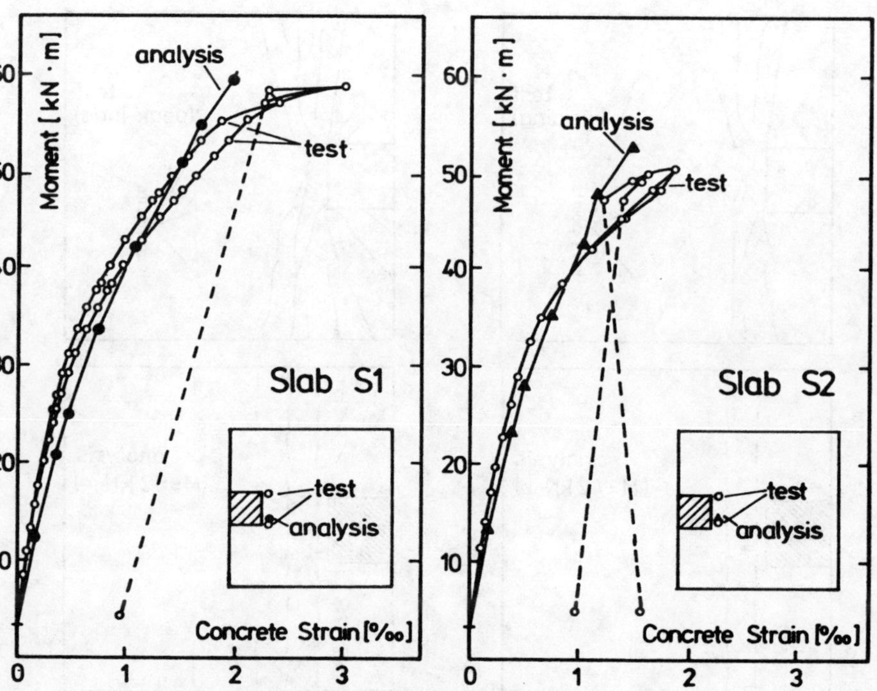

Fig. 9: Concrete Strains at Bottom Surface

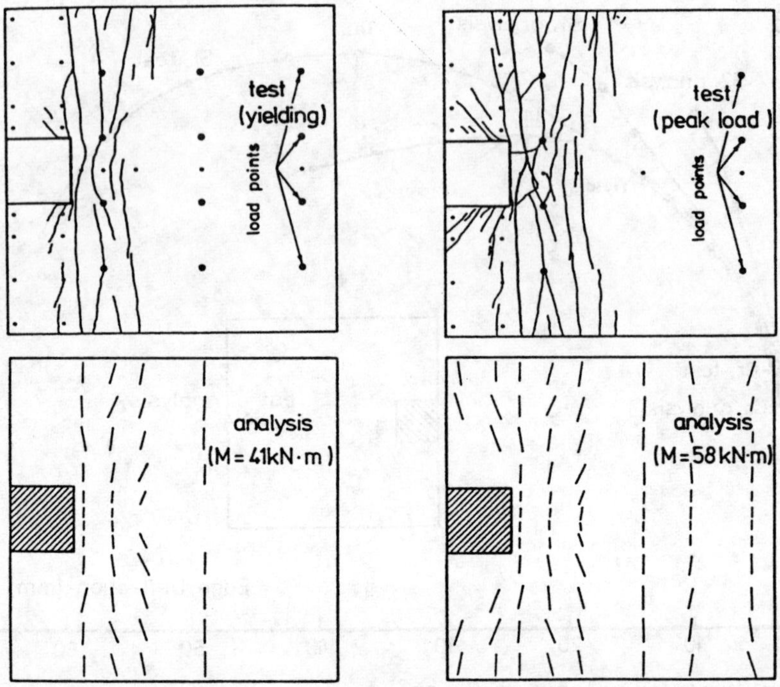

Fig. 10: Crack Patterns on Top Surface of Slab S1

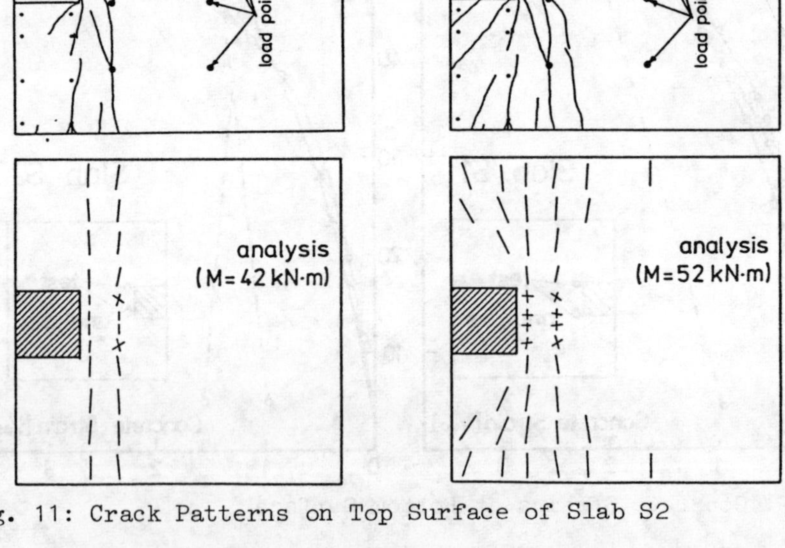

Fig. 11: Crack Patterns on Top Surface of Slab S2

compared with the formation of the crack through the slab. It is evident that the proposed model enables the calculation of nonlinear stress distribution across the slab thickness and the development of inclined cracks.

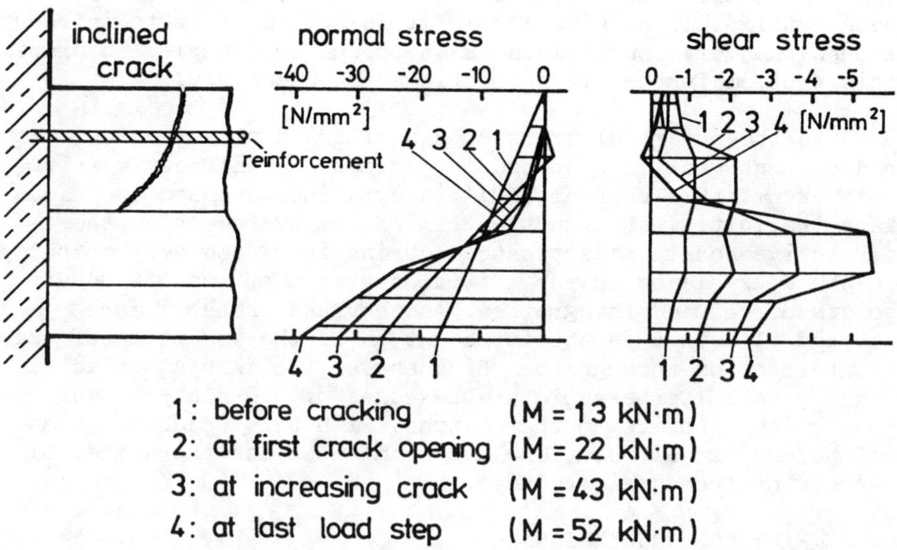

Fig. 12: Crack and Stress Distribution Over Slab Thickness Near Column Edge

5.3 Some Remarks to Numerical Application of the Used Element

The Ahmad thick shell element with full integration (3*3*2 Gauss points for a quadratic element) has proven to be very successful in "thick" shell situations. It was soon discovered, however, that this element is much too stiff when the thickness is reduced. The stiffening effect is produced by parasitic shear stresses present in the lower order elements. This shear "locking" could be reduced with reduction or selective reduction of the numerical integration (see for example Zienkiewicz et al. [12]). A negative feature of the reduced integration techniques, however, is rank defiency as Hughes et al. [13] indicate. This translates into the possibility of zero energy modes which in turn causes singularity of the stiffness matrix unless adequate constraints exist for the boundary nodes.

Another problem arises with the application of reduced integration for the layered element model used. As Zienkiewicz [14] points out, the stiffness matrix must be singular if the number of unknowns exceeds the number of independent relations supplied at all integration points. With each layer two additional unknowns, the rotations α and β, are added to the element at each nodal point and correspondingly two additional independent relations should be added at each integration point. The additional conditions are the relation of the displacements u and v of the layer to the deformation of the reference layer, Eqs. 1a and 1b. For reduced integration, the number of unknowns may exceed the number of independent relations and the stiffness matrix will be singular if it is not possible to introduce enough boundary constraints in the peculiar example. To clarify this problem further investigations are necessary.

In the numerical example, described in this section, reduced integration resulted in a singular stiffness matrix. To overcome this problem, full integration was used to calculate the initial stiffness matrix of the system while the corrective nodal loads produced during iteration were computed on the equivalence to the initial stresses at the 2*2 Gauss points of reduced integration. Another and probably more successful method is the introduction of the condition of constant rotation through the thickness of the slab at selected nodal points (for example at the edges of the slab-column connection opposite to the column). With this condition some of the additional rotations are eliminated and the number of degrees of freedom is reduced.

6. SUMMARY AND CONCLUSION

A finite element model for calculation of reinforced concrete slabs including lateral shear forces is introduced. To give a realistic description of the behaviour of reinforced concrete slabs, nonlinear deformation and failure behaviour of concrete under three dimensional stresses, inclined, progressive cracks of concrete through the slab thickness, and stress transfer across the crack surfaces because aggregate interlock are considered. The method is verified by comparison of results with those from two tests of slab-column connections, representative for the edge fields of prestressed, flat plates. The analysis results in a too stiff behaviour of the slabs. It is assumed that the main reasons for deviation of the analysis from the test results stems from lack of consideration of the geometric nonlinearities and an overestimation of the stress transfer across the cracks with the smeared crack concept. To justify these conjectures, calculation of more examples with variation of the different parameters is necessary.

7. ACKNOWLEDGEMENTS

This investigation was carried out, while the author was a Visiting Research Scholar at the University of Illinois at Urbana-Champaign, with support of the Deutsche Forschungsgemeinschaft, West Germany. The author greatly appreciates the help of Professor W. C. Schnobrich and acknowledges the facilities provided by the Nathan M. Newmark Civil Engineering Laboratory.

8. REFERENCES

[1] Ahmad, S., Irons, B.M. and Zienkiewicz, O.C. "Analysis of Thick and Thin Shell Structures by Curved Finite Elements", Int. J. for Num. Methods in Eng., Vol. 2, 419-451, 1970.

[2] Klein, D., Kristjansson, R., Link, J., Mehlhorn, G. and Schäfer, H. Berechnung von dünnen Stahlbetonplatten bei Berücksichtigung eines wirklichkeitsnahen Werkstoffverhaltens, Forschungsberichte aus dem Institut für Massivbau der TH Darmstadt, Nr. 15/1974, Nr. 25/1975, Nr. 32/1976.

[3] Kristjansson, R. Physikalisch und geometrisch nichtlineare Berechnung von Stahlbetonplatten mit Hilfe Finiter Elemente, Ph.D.Thesis TH Darmstadt, 1977.

[4] Mawenya, A.S. and Davis, J.D. "Finite Element Bending Analysis of Multilayer Plates", Int. J. for Num. Methods in Eng, Vol. 8, 215-225, 1974.

[5] Gerstle, K.H. et al. "Behaviour of Concrete under Multiaxial Stress States", J. of the Eng. Mech. Div., ASCE, Vol. 106, 1383-1403, 12/1980.

[6] Gerstle, K.H. "Simple Formulation of Triaxial Concrete Behaviour", ACI-Journal, No. 5, Proc. V.78, 382-387, 9-10/1981.

[7] Cornelius, V. Report on Simple Theory, Preliminary Report, Univ. of Colorado in Boulder, Aug. 1981.

[8] Hsieh, S.S., Ting, E.C. and Chen, W.F. "A Plastic Fracture Model for Concrete"in Fracture in Concrete, W.F. Chen and E.C. Ting, Editors, Proc. of a Session at the ASCE National Convention in Hollywood, Florida, Oct. 1980.

[9] Bažant, Z.P., Gambarova, P. "Rough Cracks in Reinforced Concrete", J. of the Struct. Div., ASCE, Vol. 106, 819-842, April 1980.

[10] Gambarova, P. "On Aggregate Interlock Mechanism in Reinforced Concrete Plates with Extensive Cracking", IABSE-Colloquium on Advanced Mechanics of Reinforced Concrete in Delft, Netherlands, 1981.

[11] Sunidja, H., Foutch, D.A. and Gamble, W.L. Response of Prestressed Concrete Plate-Edge Column Connections, Civil Eng. Studies, Structural Research Series No. 498, Univ. of Illinois in Urbana-Champaign, March 1982.

[12] Zienkiewicz, O.C., Taylor, R.L. and Too, J.M. "Reduced Integration Technique in General Analysis of Plates and Shells", Int. J. of Num. Methods in Eng., Vol. 3, 275-290, 1971.

[13] Hughes, J.R., Cohen, M. and Haroun, M. "Reduced and Selective Integration Techniques in the Finite Element Analysis of Plates", Nuclear Eng. and Design, 46, 203-222, 1978.

[14] Zienkiewicz, O.C. The Finite Element Method in Engineering Science, McGraw-Hill, London, 3. Edition.

CHAPTER 8

PRACTICAL APPROACH FOR MODELLING THE NONLINEAR RESPONSE OF RC SHELLS

J. A. FIGUEIRAS
Prof. Auxiliar, Department of Civil Engineering,
University of Porto, Portugal.

SUMMARY

This paper describes the application of finite element techniques to the nonlinear analysis of reinforced concrete shell problems under short-term loading. Material and geometric nonlinearities are considered in the numerical model which is concerned with the deformational characteristics and ultimate load prediction. Degenerate quadratic thick shell elements employing a layered discretization through the thickness are adopted. Some aspects of the nonlinear solution procedures employed in the model are discussed. Numerical examples are presented and comparison is made with experimental results whenever possible. A practical problem of a liquid storage tank is analysed using the present model.

1. INTRODUCTION

The development of adequate computational models to analyse the complex structural behaviour of reinforced concrete shell problems has been increasingly demanded in order to assess current design procedures. The primary objective of the finite element models developed for global analysis and design purposes is the accurate prediction of the overall deformational characteristics of structures and their limit loads. The modelling of the complex behaviour of reinforced concrete is frequently simplified in the local context (crack discontinuity, bond slip) with a twofold objective:

(i) to avoid the use of material parameters whose knowledge is insufficient, and

(ii) to make the numerical process more efficient.

Despite this, the fundamental nonlinear behaviour up to collapse must be captured in the analysis of structures for which the model is employed.

The present article is concerned with the deformational characteristics and ultimate load analysis of reinforced concrete shell structures under quasi-static short-term loading. To model the compressive behaviour of the concrete, both a perfect and a strain hardening plasticity approach are employed. In bending problems involving normal or under-reinforced structures little difference can be observed between both approaches, but the strain hardening plasticity model must be employed in problems wherein the compressive concrete can play a major role in the collapse. The yield criterion is defined in function of the first two stress invariants taking into account the five nonzero stress components. For the crushing condition a similar function but in terms of the strain components is considered. Although, the transverse shear type of failure cannot be accurately predict by the present model, the inclusion of the transverse shear stress components in the yield and crushing conditions make possible that compression-shear failures (namely, in the vicinity of concentrated loads) may be detected. These conditions for compression are complemented with a tension cut-off representation. A smeared or distributed model for cracked concrete is assumed and an average shear modulus, varying with tensile strain, is used in cracked zones. Full bond is assumed at the reinforcement/concrete interface, but a tension stiffening diagram is adopted to simulate gradual bond deterioration with progressive cracking. The behaviour of cracked reinforced concrete zones can have a significant influence in the deformational response of shell problems, and the material parameters defining this behaviour should be correctly assessed. Results of the application of the present model to a simple panel under uniaxial tension will be given to illustrate the definition of the material parameters used to simulate the tensile behaviour.

For structural discretization degenerate quadratic thick shell elements employing a layered representation through the thickness are used. Both the 8-node Serendipity and heterosis elements are used with reduced and selective integration rules. The concrete characteristics are specified for each layer through the thickness and reinforcing steel is represented as a smeared layer of equivalent thickness with uniaxial strength and rigidity properties. Nonlinear geometric effects which are of major importance in the deformational response and ultimate load analysis of reinforced concrete shell problems are accounted for. An incremental and iterative process is adopted in the nonlinear finite element analysis in order to trace out the response for increasing loads and to guarantee convergence towards the correct solution path. The tangential stiffness matrix is recalculated in the second iteration of each load increment and immediately after a local unloading has been detected in any sampling point. This algorithm has shown good convergence characteristics for the examples studied.

The finite element computational model adopted is briefly described and results of simple numerical applications are presented to illustrate the influence of some parameters and different formulations on the structural response. Three numerical examples are analysed and the results discussed to illustrate the capabilities of the present code. In the first two problems the numerical solution is compared with available experimental results. The final application refers to a practical axisymmetric shell used as a wine storage tank.

2. MATERIAL MODELLING

Description of the highly complex behaviour of concrete is a difficult task and to date generally accepted constitutive equations do not exist. A variety of models have been proposed to characterize the stress-strain relations and failure behaviour of concrete material under multi-axial stress states [1,2]. All these models have certain inherent advantages or disadvantages which depend to a large degree on the particular application considered. The use of a simplified formulation (appropriate to specific purposes) is computationally more efficient and results in a more comprehensive analysis. The formulation adopted here [3,4] is directed at the ultimate load analysis of thick and thin plate and shell problems. Full details of the approach adopted are given in ref. [4] and only a brief summary is provided below.

2.1 Compressive behaviour of concrete

Based on the flow theory of plasticity, the nonlinear compressive behaviour of concrete is characterized by the yield condition, the flow and hardening rules and the crushing condition.

a) The yield condition

In the present analysis of thick plates and shells the following yield criterion in terms of the first two stress invariants is employed.

$$f(I_1,J_2) = \left[\beta(3J_2) + \alpha I_1\right]^{\frac{1}{2}} = \sigma_o \tag{1}$$

where α and β are material parameters and σ_o is the equivalent effective stress taken as the compressive stress from a uniaxial test. The uniaxial compression test and the biaxial test, under equal compression stress ($\sigma_1=\sigma_2$) are used to determine α and β. For practical purposes a relation can be assumed between the equal biaxial yield stress (or strength), f_{cb}, and the uniaxial yield stress (strength), f_c,

$$f_{cb} = 1.16(\text{to } 1.20) f_c \tag{2}$$

If Kupfer's results [5] are employed ($f_{cb}=1.16f_c$), the material constants are:

$$\alpha = 0.355\,\sigma_o \;;\; \beta = 1.355 \tag{3}$$

and the yield condition can be written in terms of the stress components as:

$$f(\sigma) = \left[1.355\left[(\sigma_x^2+\sigma_y^2-\sigma_x\sigma_y) + 3(\zeta_{xy}^2+\zeta_{xz}^2+\zeta_{yz}^2)\right] + 0.355\sigma_o(\sigma_x+\sigma_y)\right]^{\frac{1}{2}} = \sigma_o \tag{4}$$

being a function of the concrete cylinder strength, f_c, only. This expression is compared in Figure 1 with the experimental results of Kupfer et al. [5] in biaxial stress space.

Both a perfect plastic and a strain-hardening plasticity approach are employed, which are illustrated, for one-dimension in Figure 2. The results performed by the two different approaches in the analysis of the single span beam, specified as B-3, tested by Bresler and Scordelis [6] are compared in Figure 3. The beam failed by crushing of the concrete compression zone near midspan where a concentrated load was applied. In Figure 3, the experimental and the two numerical load-deflection curves and the corresponding collapse loads are illustrated (full details can be found in ref. [4]). A significant improvement in the results is observed using the hardening plasticity model, not only in the deformational behaviour but also for the limit load, in which the crushing of the concrete plays a primary role.

b) The flow and hardening rules

The associated flow rule is considered for concrete predominantly for practical reasons, since there is very little supportive experimental evidence. Using the concept of normality, the plastic strain increment is defined [7] as,

$$d\varepsilon_{ij}^p = d\lambda \frac{\partial f(\sigma)}{\partial \sigma_{ij}} \tag{5}$$

where $d\lambda$ is a proportional constant which determines the magnitude of the plastic strain increment, and $\partial f(\sigma)/\partial\sigma_{ij}$ defines its direction to be perpendicular to the yield surface. The current stress function $f(\sigma)$ is the yield condition, or the subsequent loading functions in the strain hardening model.

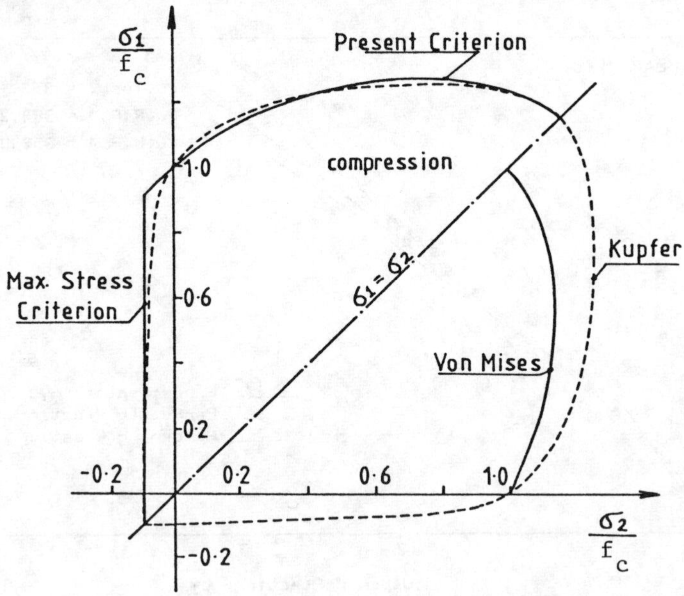

Fig. 1 Compressive yield and tensile cracking criterion for concrete (biaxial stress space).

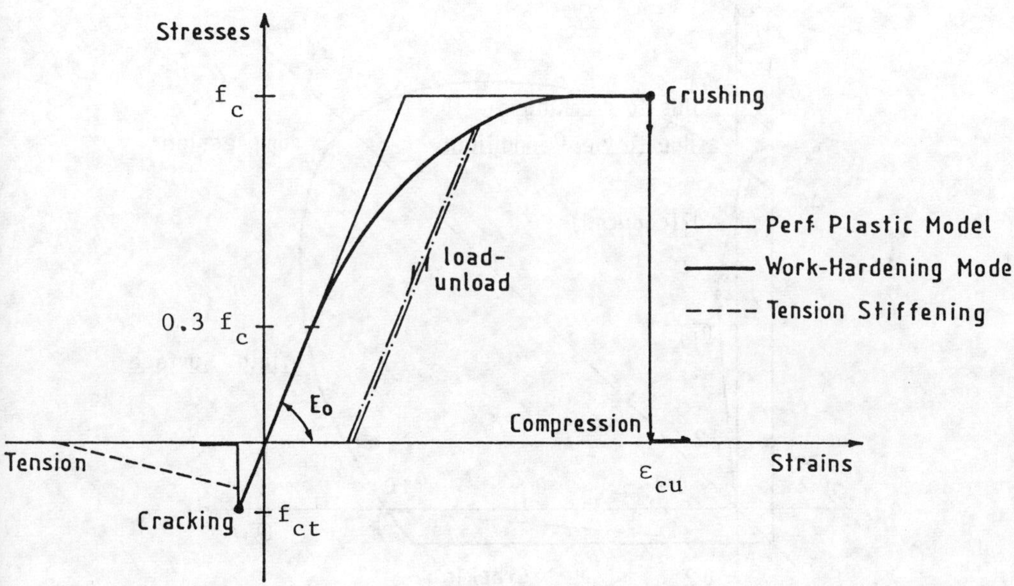

Fig. 2 One-dimensional representation of the concrete constitutive model.

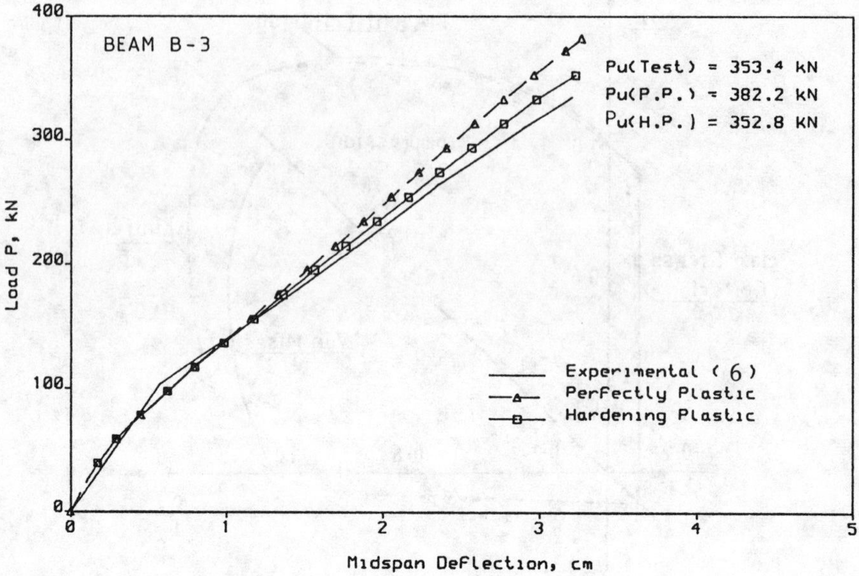

Fig. 3 Numerical and experimental load-deflection curves for beam B-3.

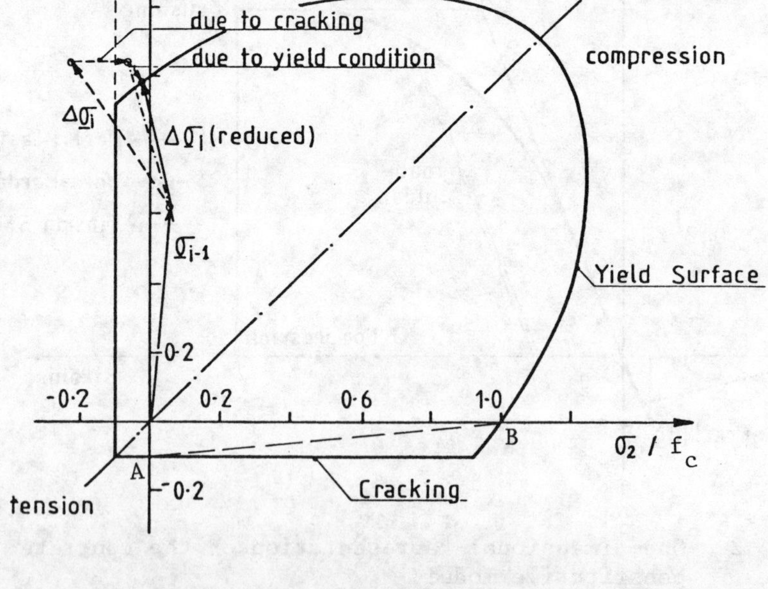

Fig. 4 Reduction of incremental stresses to the yield surface for cracked concrete in one direction.

The hardening rule defines the relation between the loading surfaces $f(\sigma)$ (or the effective stress) and the accumulated plastic strain (or the effective plastic strain). In the present work the relation between effective stress and effective plastic strain is extrapolated from the uniaxial stress-strain relationship given by the CEB parabola [8].

c) <u>The crushing condition</u>

Lack of available experimental data on concrete ultimate-deformation capacity under multiaxial stress states has resulted in the appropriate limiting strain criterion being developed by simply converting the yield criterion described in terms of stresses directly into strains. Thus,

$$\beta(3J_2') + \alpha I_1' = \varepsilon_{cu}^2 \qquad (6)$$

where I_1' and J_2' are strain invariants [2], and ε_{cu} is an ultimate total strain extrapolated from uniaxial test results, taken in practical examples equal to 3.5‰. The previous values of material parameters α and β obtained from Kupfer's results can be adopted.

d) <u>Compressive behaviour of cracked concrete</u>

Cracking of concrete is a partial and directional fracture, since the stiffness and strength characteristics in directions parallel to the cracking plane are not (directly) affected. In such directions, plastic yielding as well as the crushing type of fracture can occur. The elasto-plastic analysis of a point where the concrete is cracked in one direction is performed as previously specified for uncracked points. Firstly, for each iteration of the nonlinear procedure, the stress level is checked against possible cracking and subsequently the yield condition is verified. If cracking has occurred the total stress components are accordingly modified, that is, the stress level is reduced to the domain of the yield condition. The modified stress components are then employed to calculate the effective stress and reduced to the yield surface [7], if it is reached. Figure 4 illustrates the behaviour described in the biaxial principal stress space. The maximum tensile stress criterion (tension cut-off) is a simple and effective condition to be coupled with the yield and crushing criterion in this type of problems. Note that, in the cracking condition represented by the dashed line AB (Fig. 4) any numerical perturbation may cause 'spurious' paralled cracking when concrete is about to yield in one direction. If concrete is cracked in one direction, the fictitious concrete strains used by the smeared crack model must be taken equal to zero when the crushing condition is applied.

2.2 Tensile idealization of concrete

The deformational characteristics of plates and shells are significantly influenced by the behaviour of cracked concrete zones. Thus, the tensile modelling of reinforced concrete plays an important role in the analysis, specially when the serviceability limit states are to be verified. However, the variability of characteristics presented by tension concrete in shell structures, due to in-situ factors, reinforcing and loading conditions, makes the definition a rigorous model for cracked concrete impracticable. In the present approach, the influence of the cracked zones on the structural response is described by a simple smeared crack model with material parameters obtained from experimental test results.

The response of concrete under tensile stresses is assumed to be linear elastic until the fracture surface is reached. This tensile type of fracture or cracking is governed by a maximum tensile stress criterion (tension cut-off). Cracks are assumed to form in planes perpendicular to the direction of maximum principal tensile stress as soon as this stress reaches the specified concrete strength, f_{ct}. In order to avoid further complexities, cracks are assumed to form only in planes perpendicular to the shell mid-surface. A sudden and total release of the normal stress in the cracked direction, or its gradual relaxation according to a tension stiffening diagram (see next section) is adopted after cracking has occurred. Before cracking, concrete is assumed to be an isotropic material and the stress-strain relations are expressed in terms of five components of stresses ($\sigma_z=\sigma_3=$ $=0$) and strain [9]. If a cracked plane has formed, the elasticity modulus and the Poisson's ratio are reduced to zero in the direction perpendicular to the cracked plane, and a reduced shear modulus is employed to simulate the aggregate interlock and dowel shear effects. The stress-strain relationships for cracked concrete are expressed in terms of total stresses and strains, and the cracked shear moduli are assumed to be a linear function of the current tensile strain [10].

Taking 1 and 2 as the two principal stress directions in the plane of the structure, the cracked shear moduli, for concrete cracked in the 1-direction (crack plane normal to the 1-direction), are:

$$G^c_{12} = 0.25G(1-\varepsilon_1/0.0045), \quad G^c_{12} = 0 \quad \text{if} \quad \varepsilon_1 > 0.0045$$
$$G^c_{13} = G^c_{12} \; ; \quad G_{23} = 5/6\ G \qquad (7)$$

where G is the uncracked shear modulus and ε_1 is the tensile strain in the 1-direction. When the tensile stress in

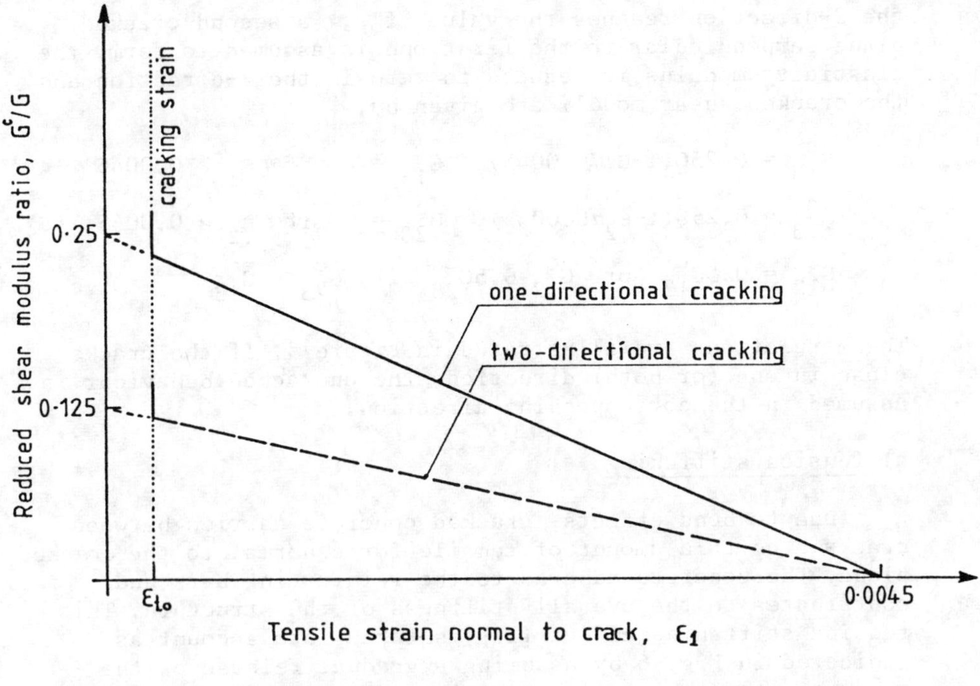

Fig. 5 Reduced shear modulus for cracked concrete.

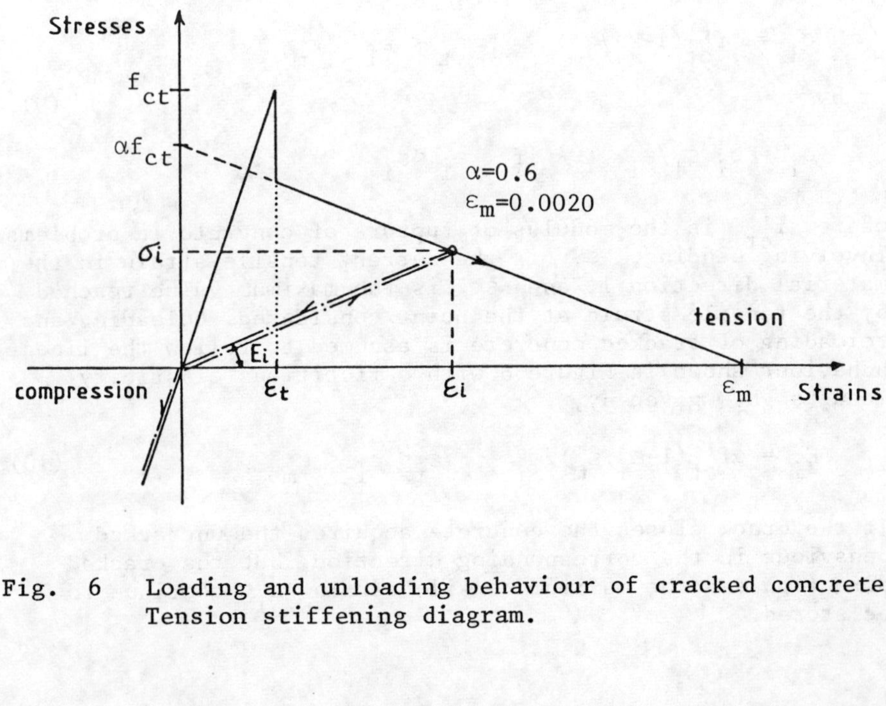

Fig. 6 Loading and unloading behaviour of cracked concrete. Tension stiffening diagram.

the 2-direction reaches the value f_{ct}, a second cracked plane perpendicular to the first one is assumed to form, the elasticity modulus is reduced to zero in the 2-direction and the cracked shear moduli are given by,

$$
\begin{aligned}
G^c_{13} &= 0.25G(1-\varepsilon_1/0.0045), \quad G^c_{13} = 0 \quad \text{if} \quad \varepsilon_1 > 0.0045 \\
G^c_{23} &= 0.25G(1-\varepsilon_2/0.0045), \quad G^c_{23} = 0 \quad \text{if} \quad \varepsilon_2 > 0.0045 \qquad (8) \\
G^c_{12} &= 0.5G^c_{13} \quad \text{or} \quad G^c_{12}=0.5G^c_{23} \quad \text{if} \quad G^c_{23} < G^c_{13}
\end{aligned}
$$

These relations are illustrated in Figure 5. If the cracks close in one (or both) directions the uncracked behaviour is assumed in the corresponding direction.

a) Tension stiffening

Due to bond effects, cracked concrete carries between cracks a certain amount of tensile force normal to the cracked plane. The concrete adheres to the reinforcing bars and contributes to the overall stiffness of the structure. This tension stiffening effect [11] is taken into account as indicated in Fig. 6 by assuming a gradual release of the concrete stress component normal to the cracked plane for the concrete layers in which reinforcement can supply tensile stresses. The previous stress-strain relations for cracked concrete are then modified according to the tension stiffening diagram. The normal stress σ_1 (and, or σ_2) is obtained as (Fig. 6),

$$\sigma_1 = f'_{ct}(1-\varepsilon_1/\varepsilon_m) \qquad \varepsilon_t < \varepsilon_1 < \varepsilon_m$$

or by (9)

$$\sigma_1 = \sigma_i\ \varepsilon_1/\varepsilon_i \qquad \text{if} \quad \varepsilon_1 < \varepsilon_i$$

where f'_{ct} is the modulus of rupture of concrete in problems involving bending, ε_1 is the current tensile strain in the material direction 1, and ε_i is the maximum value reached by the tensile strain at the point considered. Unloading and reloading of cracked concrete is assumed to follow the linear behaviour shown in Figure 6 with a fictitious elasticity modulus E_i given by,

$$E_i = \alpha f'_{ct}(1-\varepsilon_i/\varepsilon_m)/\varepsilon_i \qquad \varepsilon_t < \varepsilon_i < \varepsilon_m \qquad (10)$$

If the crack closes the concrete acquires the uncracked behaviour in the corresponding direction, but the cracked direction and the maximum tensile strain ε_i continue to be stored.

b) Analysis of a simple panel under uniaxial tension

The coefficients to be used in the definition of the tension stiffening diagram and the cracked shear modulus have been assessed by using experimental results. A rectangular panel with variable orientation of reinforcement and subjected to a uniform uniaxial tension, tested by Peter [12], can be used as a first assessment for the numerical models simulating the cracked concrete behaviour. Results of these tests reported by Cervenka and Gerstle [13] were compared in ref. [4] with the numerical results given by the present model, and some of these comparisons are now illustrated. The test specimens are square panels 160×160 cm^2 with 8 cm thickness, and orthogonal reinforcement uniformly distributed and inclined of an angle θ with respect to the axis of applied tension. A square central area, 100×100 cm^2, of each panel was utilized to mesure the displacement d in the direction of the applied force and the transverse displacement d_h. The direct tensile strength, f_{ct} = 0.15 kN/cm^2, was used in the analysis instead of the modulus of rupture, f'_{ct} = 0.22 kN/cm^2, therefore a value of $\alpha = 0.8$, $(0.8f_{ct} \simeq 0.6\ f'_{ct})$, has been used in the tension stiffening diagram. To take into account the effect of the variable angle between the reinforcement and the principal stress direction, β, which is approximately paralell to the direction of the applied load, a correction is performed on the tension stiffening coefficient, ε_m. The parameter ε_m takes a value of 0.0025 for $\beta < 15^o$, $\varepsilon_m = 2\times0.0025$ for $\beta = 45^o$ and for intermediate values of $\beta(15-45)$ a linear interpolation is used to obtain the current value of ε_m [14].

Figures 7 and 8 compare the experimental and numerical load-longitudinal displacement curves for different reinforcing angles, θ. A fairly good correlation of results can be observed in this example, wherein the response is essencially controlled by the behaviour of the cracked concrete. The dashed lines illustrate the behaviour when the action of the cracked concrete is neglected. Figure 9 shows the results of three different tests performed with the $\theta = 0^o$ - panel varying the tension stiffening coefficients. The transverse displacement, d_h, representing the shear deformation, is plotted in Figure 10 for different reinforcing angles, θ, and for a total applied load of N = 343 kN. The analysis performed with the present model presents good correlation with the experimental results. The displacements obtained with different values of G^c for $\theta = 20^o$ are also represented in Fig. 10. The full line designated by 'theory' corresponds to neglecting the shear stiffness of cracked concrete.

2.3 Behaviour of steel in tension and compression

The reinforcing bars are considered as steel layers of equivalent thickness in the present model. Each steel layer

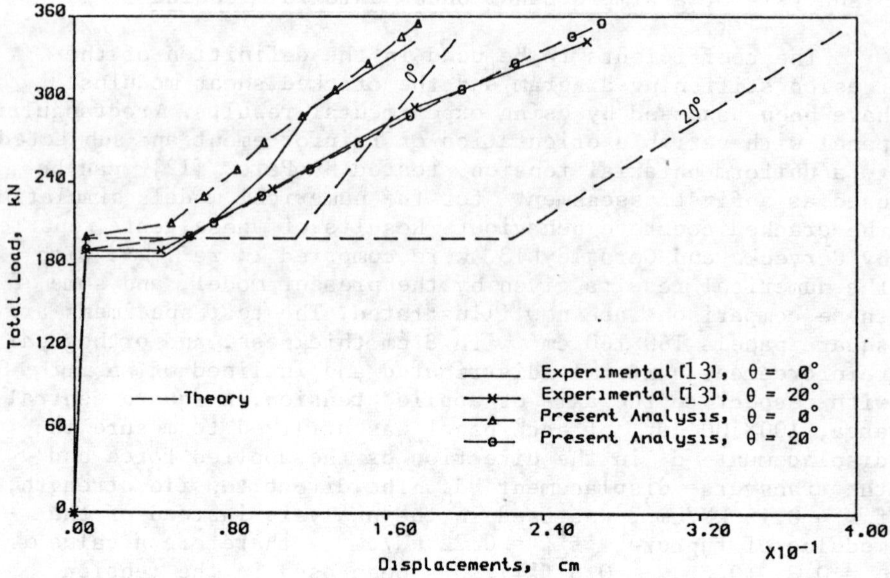

Fig. 7 Comparison of load-longitudinal displacement curves for Peter's panels.

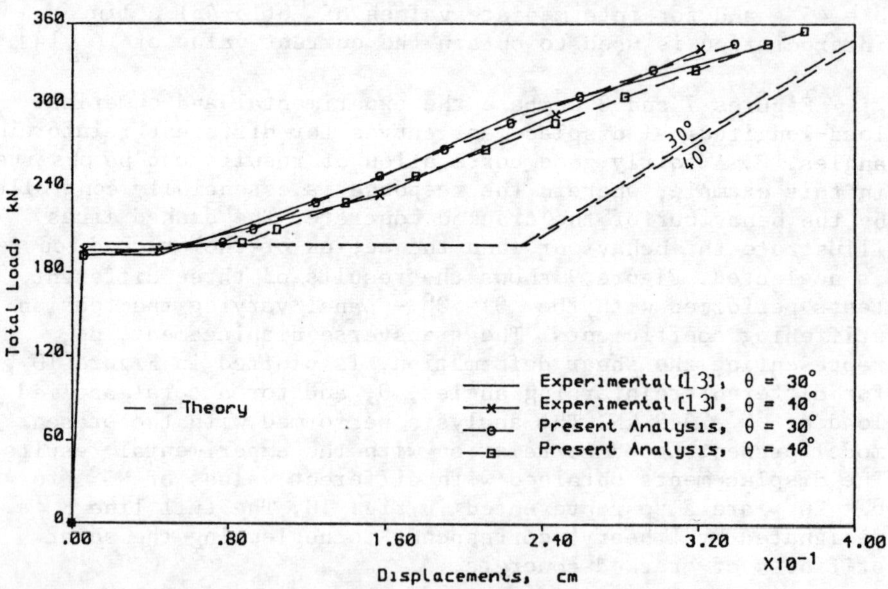

Fig. 8 Comparison of load-longitudinal displacement curves for Peter's panels.

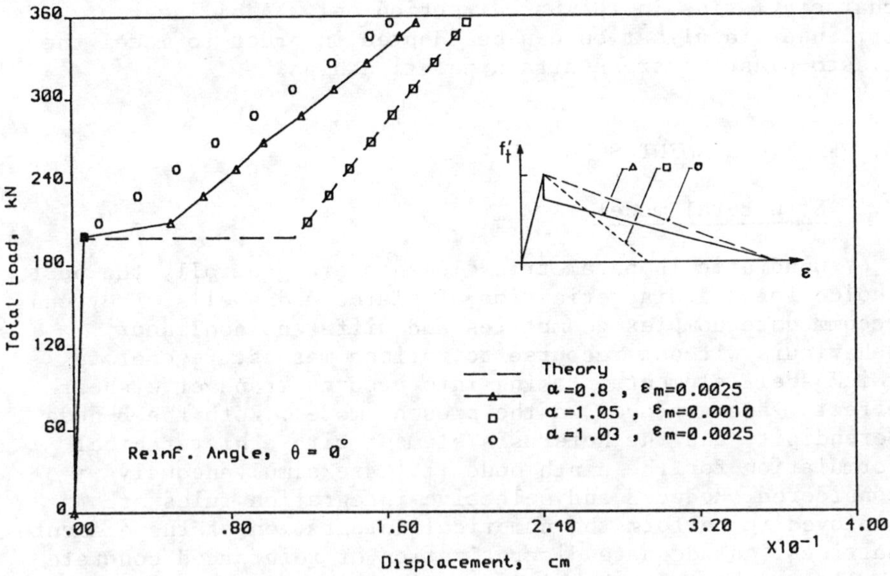

Fig. 9 Load-displacement diagrams for different values of the tension stiffening parameters.

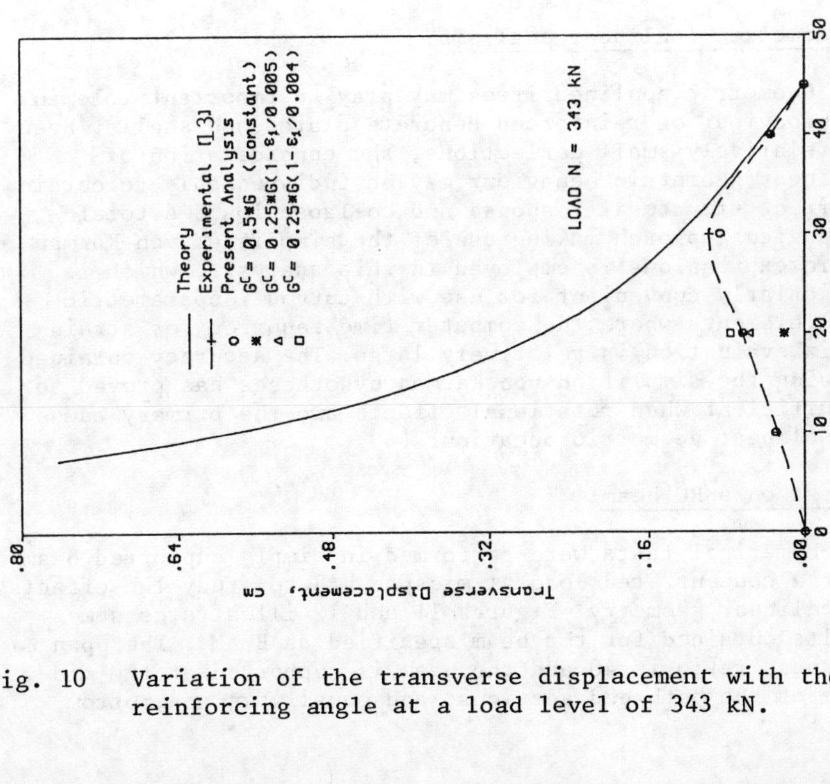

Fig. 10 Variation of the transverse displacement with the reinforcing angle at a load level of 343 kN.

exhibits a uniaxial response, having strength and stiffness characteristics in the bar direction only. A bilinear or a trilinear idealization can be adopted in order to model the elasto-plastic stress-strain relationships.

3. FINITE ELEMENT SOLUTION

3.1 Structural model

Quadratic isoparametric elements are generally the best choice for the discretization of plates and shells. They can accommodate complex geometries and different nonlinear behaviours without recourse to refined meshes. Degenerate thick shell elements, taking into account transverse shear effects, are employed in the present model. Both the 8-node Serendipity and the heterosis element with a hierarchical formulation for the ninth node [15] are simultaneously considered. Reduced and selective integration rules are employed to perform the numerical integration of the element matrices. An adequate discretization of reinforced concrete shells in the through-thickness direction has proved to be very important when flexural forces are presented. A layered approach is employed to both represent the steel reinforcement and to discretize the concrete behaviour through the thickness. The progress of cracked zones, as well as the compressive behaviour of the concrete is analysed and monitored for each layer and for each element Gauss point.

a) Geometric nonlinear analysis

Geometric nonlinearities may play an important role in the behaviour of reinforced concrete plates and shells. Even for relatively small deflections, the consideration of nonlinear geometric behaviour may be indispensable to obtain a correct structural response and collapse load. A total Lagrangian approach making use of the simplified von Karman strain expressions is employed in this analysis, which is particularly convenient for use with curved isoparametric shell elements where the computer time required for strain matrix evaluation is relatively large. The accuracy obtained by using the simplified von Karman hypotheses has proved to be sufficient when rotational effects are the primary cause of nonlinear geometric behaviour [4].

b) Test on a RC beam

Numerical tests were performed in simply supported beams under a concentrated load at midspan [4] to study the effects of nonlinear geometry. Figures 11 and 12 illustrates some results obtained for the beam specified as BEAM2. The span to thickness ratio is 40 and the beam is supported at the mid--line of the both end sections, and equally top and bottom

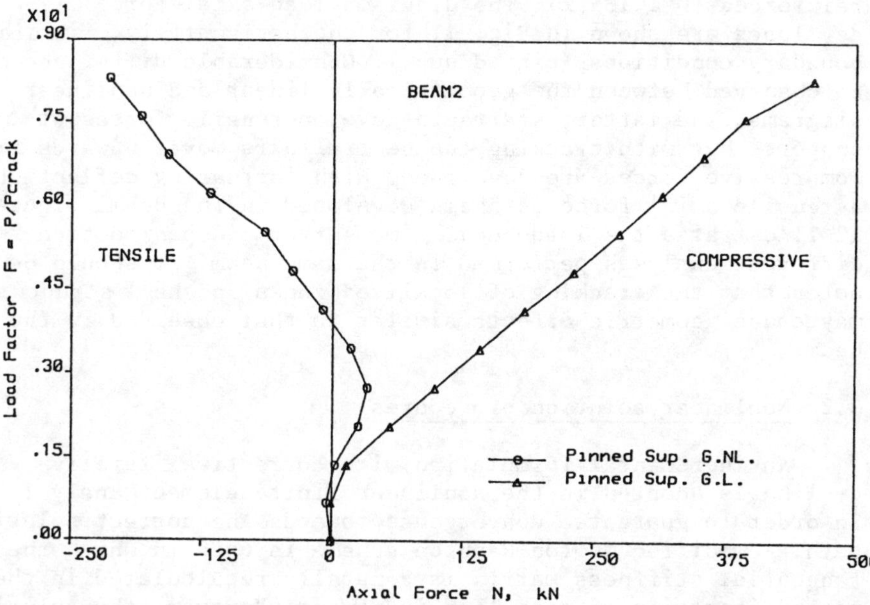

Fig. 11 Load-axial force curves for BEAM2.

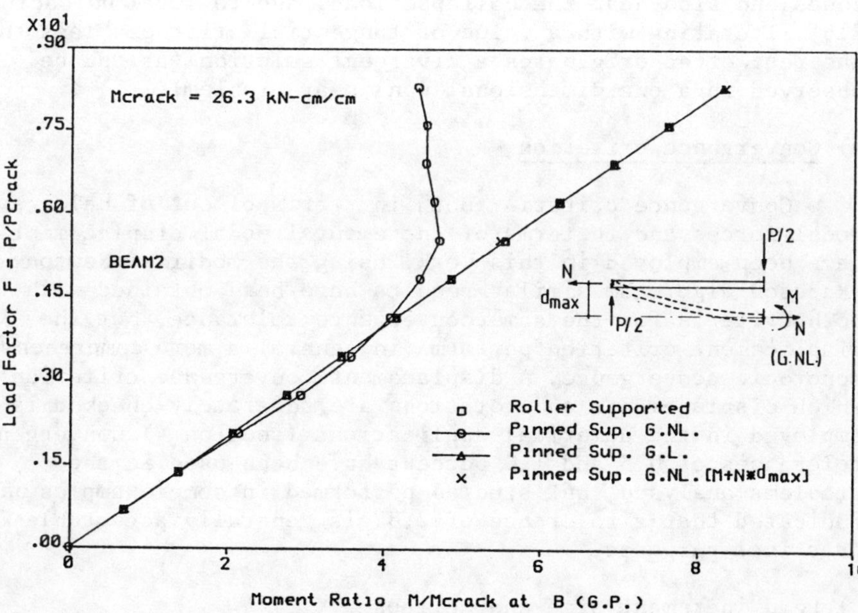

Fig. 12 Load-bending moment curves for BEAM2 at midspan.

reinforced (ρ=1.135 %). The diagrams load-axial force developed are shown in Fig. 11 for the horizontally restrained boundary conditions (pinned sup.). Considerable difference can be observed between the geometrically linear and nonlinear diagrams. The latter, starts to develop tensile forces at supports but with cracking the neutral axis moves upwards and compressive forces are developed. With increasing deflections a tensile axial force is again developed in the beam. Figure 12 illustrates the load-bending moment at midspan section for different analyses performed in the same beam. It should be noted that the cracking of localized zones in shell structures may cause geometric effects similar to that observed in the beam.

3.2 Nonlinear solution procedures

An incremental formulation with corrective iterative cycling is adopted in the nonlinear finite element analysis in order to guarantee convergence towards the correct solution path. A modified Newton-Raphson scheme is used in which the tangential stiffness matrix is generally recalculated in the second iteration of each load increment. However, the solution stability has been greatly improved by evaluating the stiffness matrix at stages within a load increment when the change of material characteristics implies a local increase of stiffness. For increasing monotonic loading, this recalculation has to be performed more frequently during the first spread of cracked zones and also near the collapse load, due to local unloadings [16]. Iterating with a value of tangential stiffness less than the real, often originates a divergent solution (as can be observed in a one-dimensional nonlinear problem).

a) Convergence criterion

Convergence criteria both in terms of out of balance nodal forces and in terms of incremental nodal displacements have been employed in this work. Using the modified Newton--Raphson algorithm similar results have been obtained with both criteria for the same convergence tolerance, but the displacement criterion performs in general a more comprehensive monotonic convergence. A displacement convergence criterion in which displacements and rotations are separately checked is employed in the numerical applications (section 4). Convergence tolerances of 0.5 and 1.0 percent have been used in the problems analysed, but studies performed in some examples have indicated that a tolerance of 2.5% is generally acceptable for practical purposes.

b) Load increment size and collapse

To establish the entire deformational response of structures up to collapse load an incremental procedure is usually employed. In the particular case of reinforced concrete structures a reasonable number of load increments must be used, since these structures are generally path dependent. The size of the load increments employed in the present work has ranged from 0.1 to 0.2 of the structure cracking load. However, values of increment size approximately 0.5 times the cracking load have produced reasonable results for the cases analysed. To obtain the collapse load within narrow limits the size of the load increments must be refined when the structure approaches collapse. This is performed by using the restart facilities included in the present code.

When convergence is not reached for a specified number of iterations, usually fifteen, the state of the material is then investigated to check if a real collapse load has been achieved. In general, the crushing of concrete layers and a subsequent divergent solution announces the limit load. The formation of a total or a parcial collapse mechanism can in general be observed using the plastic strains in steel at the Gauss points. The system failure is attained when the stiffness and (or) strength of some influent Gauss integration points are lost. The stiffness at nodal points can be positive but the equilibrium between applied loads and equivalent nodal forces may not be achieved.

c) <u>Nonlinear solution test</u>

Results of the application of this model to a simply supported two span beam, experimentally tested by Duddeck et al. and reported by Mueller [17], are shown. Full details of the corresponding numerical analysis can be found in ref. [4]. The results of four different nonlinear algorithms are compared in Fig. 13 where the curves load factor versus point load deflection are illustrated, and the meaning of the abbreviations is:

D.C.C. - displacement convergence criterion
F.C.C. - force convergence criterion
K.I. - initial stiffness matrix
K.T. - tangential stiffness matrix calculating by using the modified N-R method described.
Ref.Analysis - refined analysis obtained with D.C.C.--K.T. and a tolerance of 0.02%.

Both the D.C.C. and the F.C.C. with K.T. show good results for the specified tolerance of 1.0%. Figure 13 presents as well the number of iterations per load increment for the D.C.C.-K.T. algorithm. Figure 14 illustrates the response and ultimate load obtained with the D.C.C.-K.T. algorithm considering different tolerances. The stresses in the concrete bottom layer and in the steel top layer, which suffer a major influence from the nonlinear solution procedure, are shown in Fig. 15. These stresses, which refer to the mid-supported section, were obtained with the K.T. algorithm. Good agreement

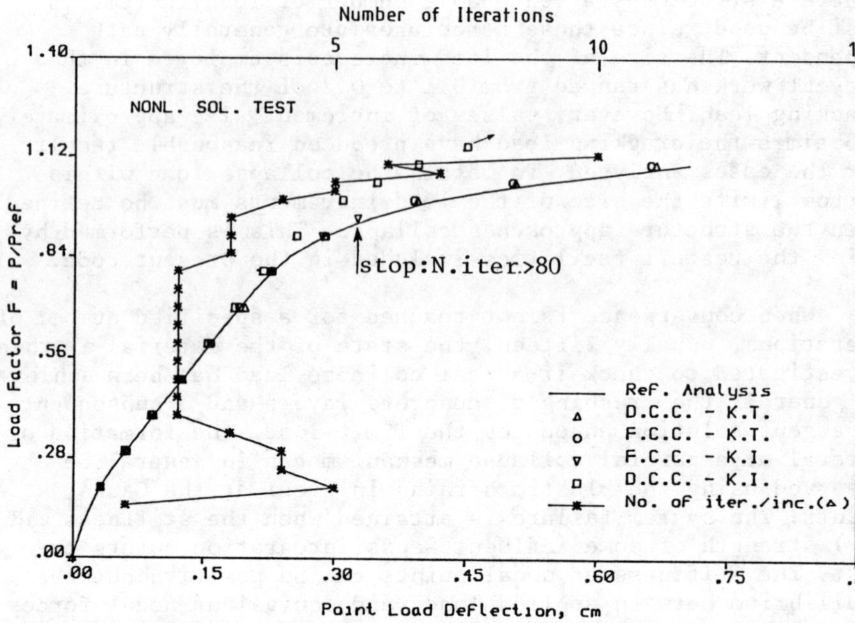

Fig. 13 Comparison of the solution algorithm performance for the two-span-beam.

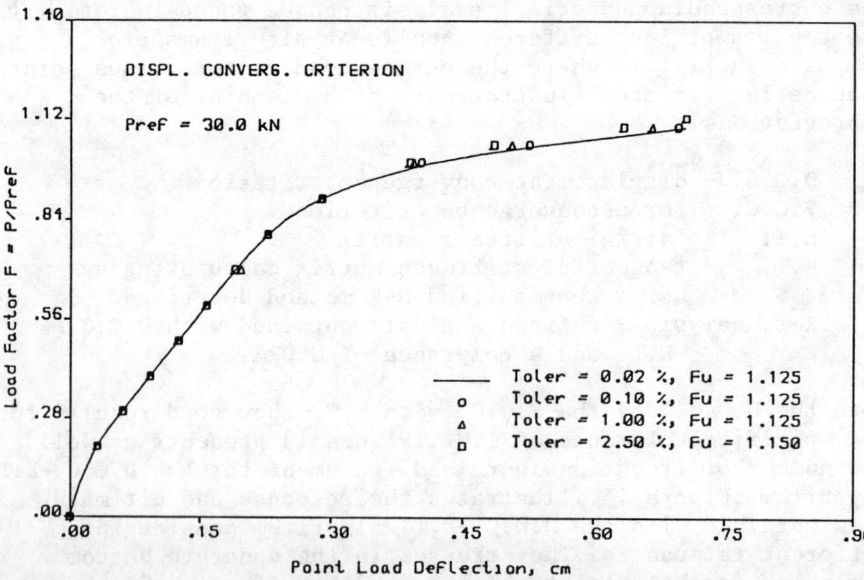

Fig. 14 Load-dflection curves using variable convergence tolerances for the two-span-beam.

is exhibited even for the loosest tolerance, 2.5%.

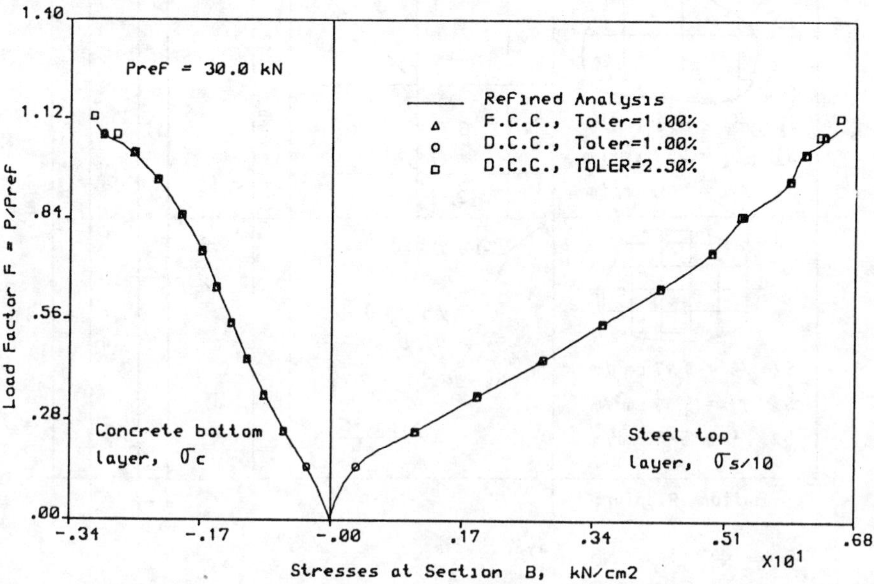

Fig. 15 Concrete and steel stress diagrams using variable convergence tolerances for the two-span-beam.

4. NUMERICAL APPLICATIONS

4.1 Corner supported square plate

A corner supported square plate loaded by a concentrated force at the centre, tested by Duddeck et al. [18] and also reported in ref. [17] is considered. This model, constructed from expanded slate lightweight concrete, has the advantage of well defined boundary conditions (horizontally free). Three plates were tested (S1, S2, S3) with different amounts of reinforcement in each direction, but with the total amount of reinforcement in the plates kept constant. The plate geometry, amount of reinforcement and finite element discretization are shown in Fig. 16 and the material properties are specified in Table 1. Numerical results for the three plates can be found in ref. [3] and here results are presented for the plate S3 only.

Figure 17 compares the numerical and experimental load--deflection curves and collapse loads for plate S3 (with anisotropic reinforcement). An excellent agreement can be observed between both analyses. The verification of the material constitutive relations and the local variations of

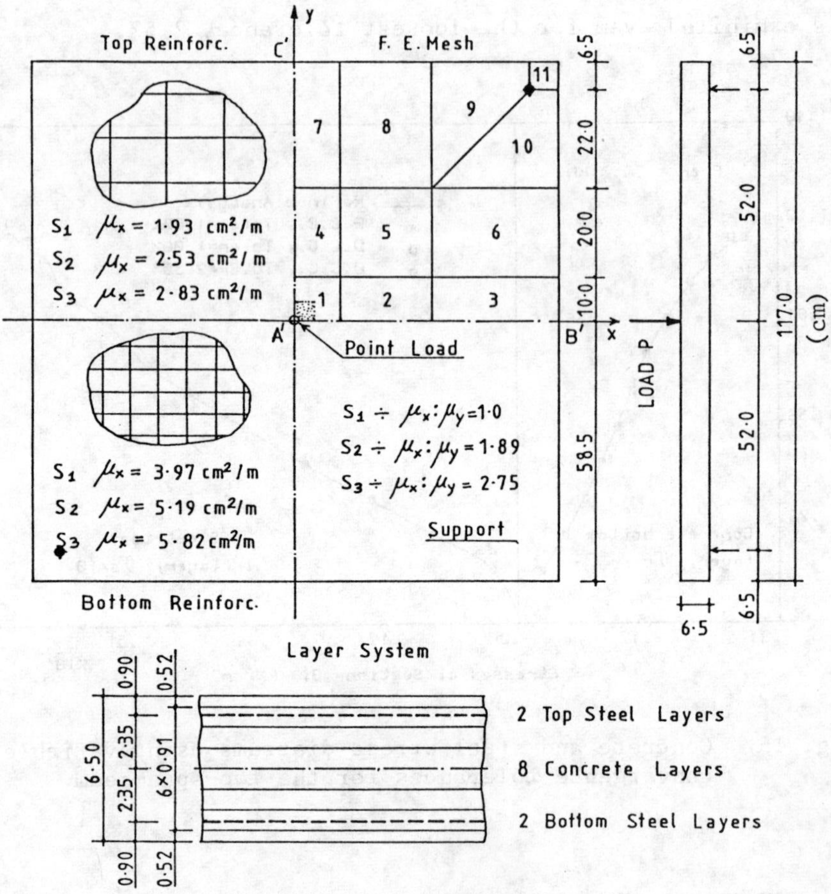

Fig. 16 Corner supported centrally loaded plates: geometry, reinforcement and finite element idealization.

MATERIAL PROPERTIES [cm; KN]	
CONCRETE	STEEL
Youngs Modulus, E_c = 1640.0	Youngs Modulus, E_s = 20100.0
Poisson's Ratio, ν = 0.0	Yield stress, f_y = 60.0
Ult. Comp. st., f'_c = 4.30	Youngs Modulus, E'_s = 700.0
Ult. Tens. st., f_t = 0.30	Ult. stress, f'_y = 70.0
Ult. Comp. stn., ε_u = 0.0035	Youngs Modulus, E''_s = 0.0
Tension Stiff.Coeff. α = 0.7	Angle x-axis, θ_x = 0.0
Tension Stiff.Coeff. ε_m = 0.0020	Angle x-axis, θ_y = 1.57

Table 1 Problem parameters for corner supported square plate (Fig. 16).

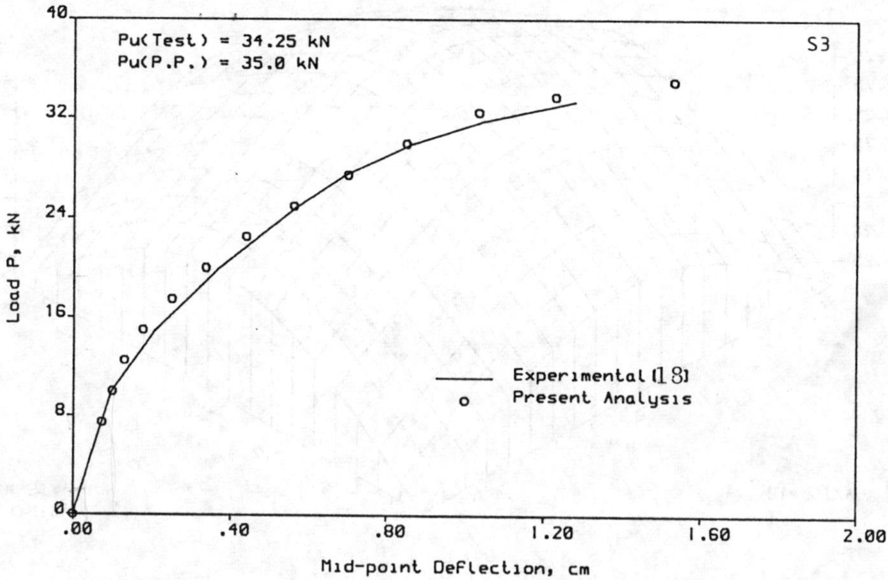

Fig. 17 Comparison of the numerical and experimental load--central deflection curves for plate S3.

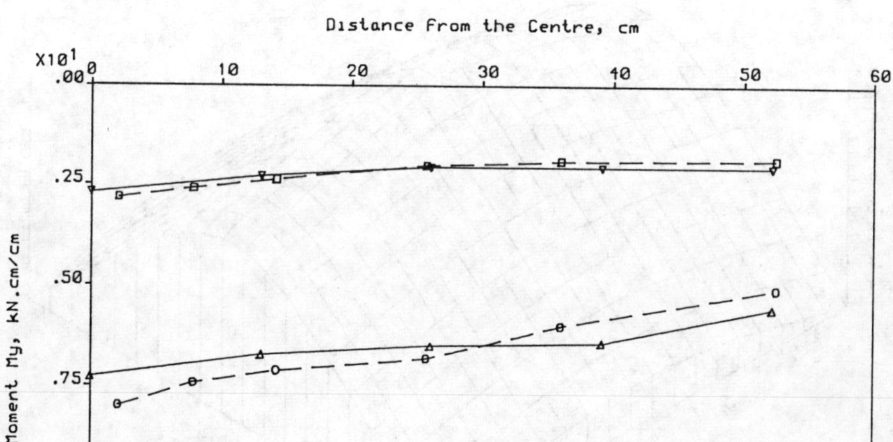

Fig. 18 Comparison of numerical and experimental moments, My along the x-axis for plate S3.

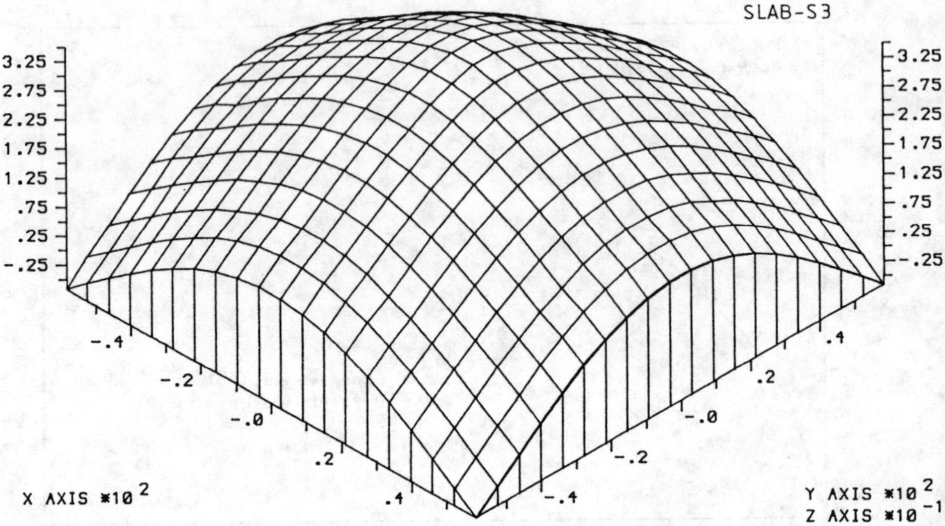

Fig. 19 Deflectionfield for plate S3 assuming a linear elastic behaviour (P=35 kN).

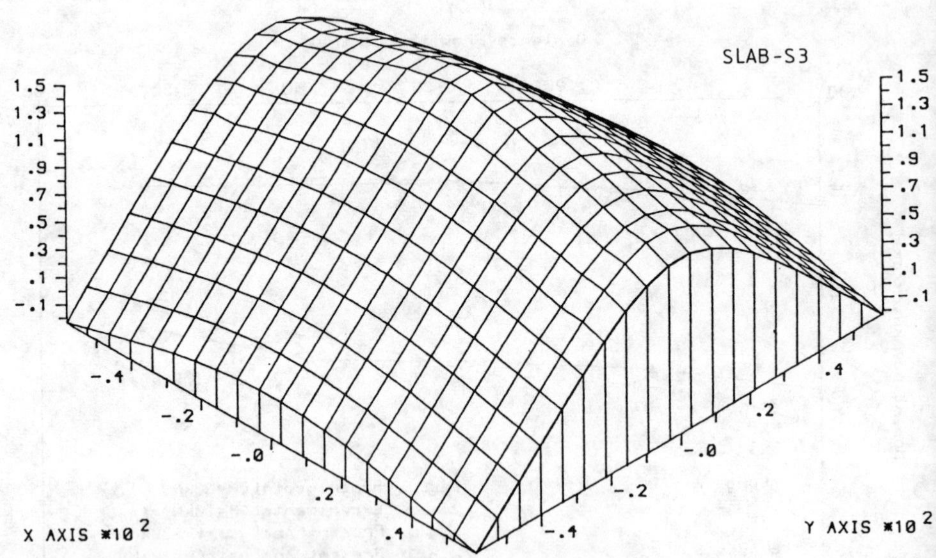

Fig. 20 Deflection field for plate S3 corresponding to the actual nonlinear response (P=35 kN).

stiffness demands the comparison of internal moments in addition to the displacement check. The experimental and numerical bending moments along the symmetry axis A'-B' (Fig. 16) are compared in Fig. 18, exhibiting a good correlation. Figures 19 and 20 illustrate the deflection field of plate S3, assuming either an uncracked elastic behaviour (Fig. 19) or the real nonlinear response at collapse load (Fig. 20). The deflections, corresponding to P = 35 kN, are represented in the z-direction (vertical) and all dimensions are in centimeters. Comparing the linear and the final nonlinear deformational configurations, the collapse mechanism that is being developed can be easily detected.

4.2 Parabolic cylindrical shell

A parabolic cylindrical shell with variable thickness, subjected to uniformly distributed pressure, which was tested by Hedgren[19] is analysed with the present finite element code. The shell was tested with end support diaphragms and free longitudinal edges. The shell geometry, finite element mesh, and zones with different layer patterns are shown in Figure 21. Thirty-six finite elements are used to model one quarter of the shell, each of which is divided into 8 equal concrete layers. The material properties are specified in Table 2. The shell is considered simply supported on both curved ends, assuming free longitudinal displacements (x-direction) and free rotation on a plane normal to the shell surface.

The design load of the shell was 0.358 N/cm^2 (0.52 lb/in^2) [19] and in the following presentation of results the applied load is expressed in terms of a load factor F, which is defined as the ratio of the applied load to the design load. In Figure 22 the experimental load-vertical deflection curve at midspan of the free edge is compared with the present results using a linear and a nonlinear geometric model. The experimental curve is given only up to F = 2.8 and an experimental extrapolated curve presented in ref. [20] is also included in Figure 22. The collapse load found for the non-linear geometric analysis, F = 4.35, agrees quite well with the reported experimental ultimate load, Fu = 4.40, and both were reached for identical collapse mechanisms. The geometrically linear analysis exhibits a quite different response with softer behaviour and lower load-carrying capacity (Fu=3.20). The importance of nonlinear geometric effects on the behaviour of this shell was pointed out by Hedgren[19]. An increase of the shell stiffness and its load-carrying capacity, due to the vertical uplift of the shell crown and the downward displacement of the free edges, is observed. An increase of the lever arm of the internal forces is consequently obtained, when the deformed geometry is taken into consideration. Figure 23 compares the linear and nonlinear geometric results (using a lower value for E'_s = 2000 kN/cm^2) with the results of ref.[20]

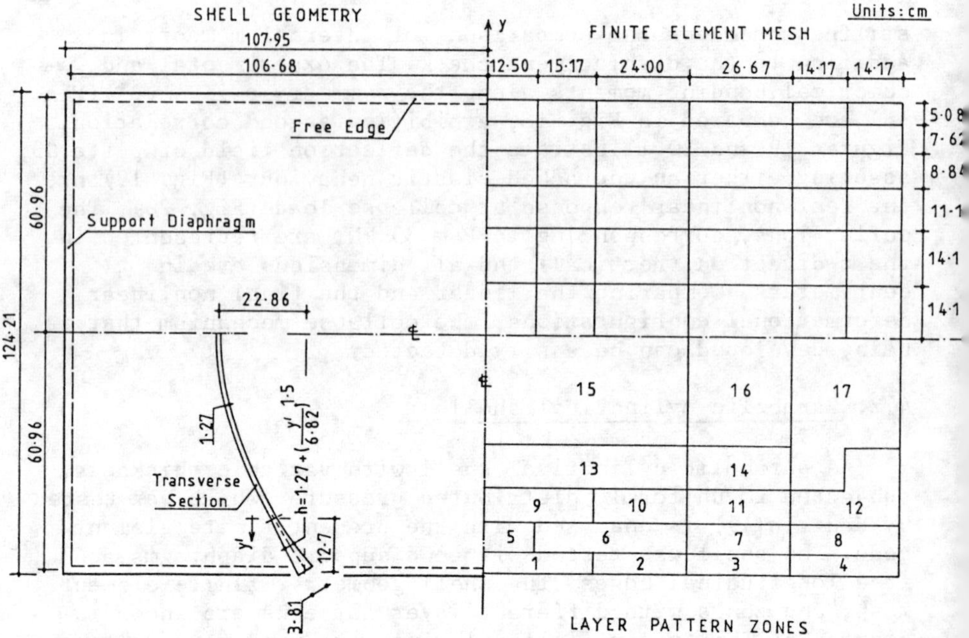

Fig. 21 Geometry and finite element idealization for parabolic cylindrical shell.

MATERIAL PROPERTIES (cm; KN)		
CONCRETE	STEEL	
Youngs Modulus, E_c = 2069.0 Poisson's Ratio, ν = 0.145 Ult. comp. st., f'_c = 3.02	Youngs Modulus, E_s = 20000.0 Yield stress, f_s = 30.66 Youngs Modulus, E'_s = 4000.0	#6
Ult. tens. st., f'_t = 0.48 Ult. comp.stn., ε_u = 0.0035 Tension Stiff.Coeff., α = 0.6	Youngs Modulus, E_s = 20000.0 Yield stress, f_s = 25.29 Youngs Modulus, E'_s = 4000.0	#3
Tension Stiff.Coeff., ε_m = 0.002	Youngs Modulus, E_s = 20000.0 Yield stress, f_s = 21.91 Youngs Modulus, E'_s = 4000.0	#4

Table 2 Problem parameters for the parabolic cylindrical shell of Fig. 21.

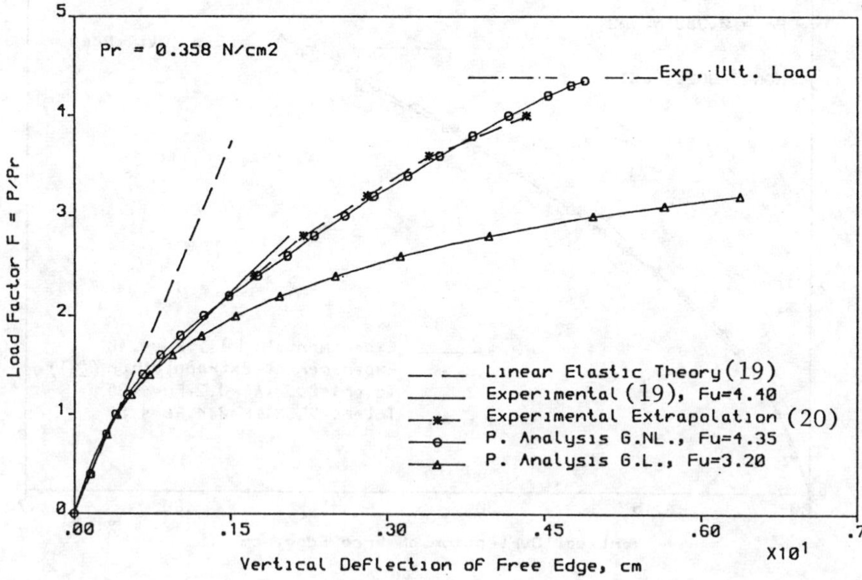

Fig. 22 Comparison of the experimental and numerical load--vertical deflection response for cylindrical shell.

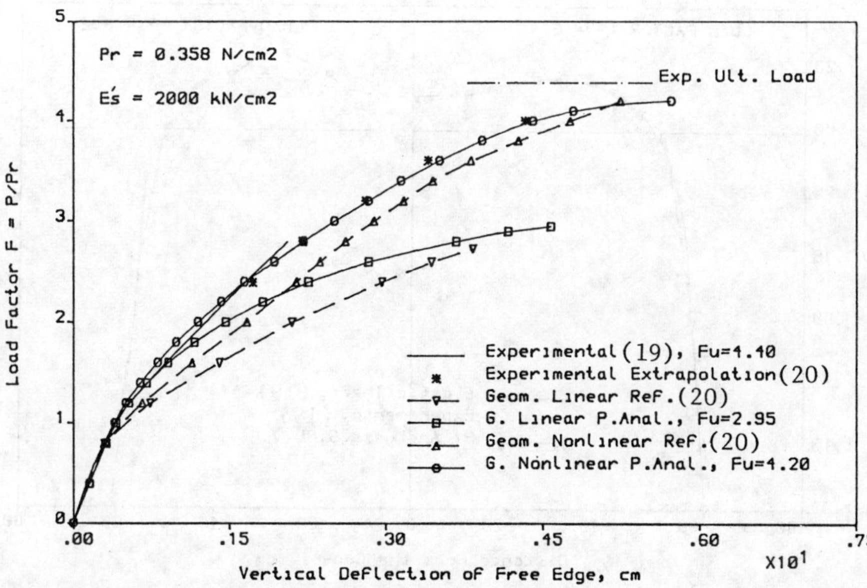

Fig. 23 Comparison of numerical load-deflection responses for cylindrical shell.

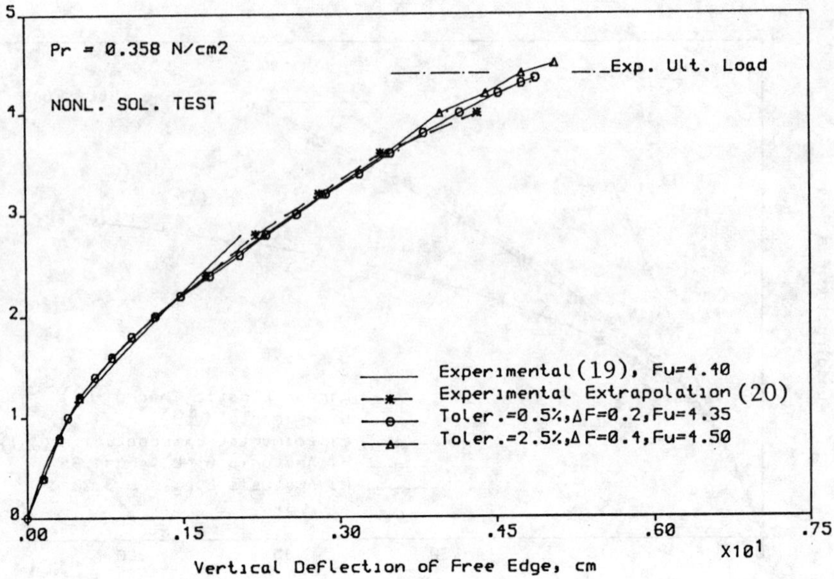

Fig. 24 Effects of the increment size and convergence tolerance on the nonlinear response.

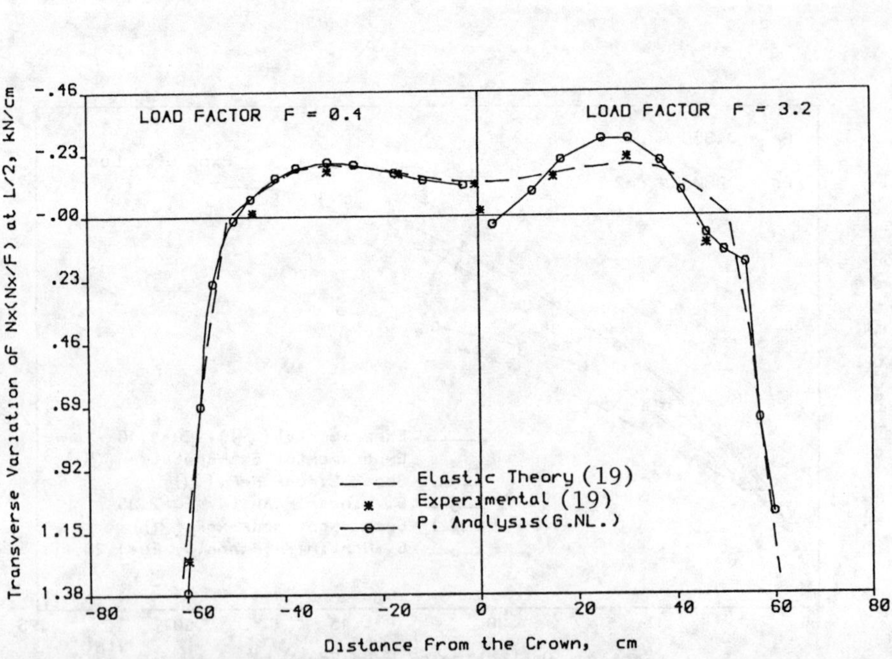

Fig. 25 Tranverse variation of the stress resultant Nx at midspan. Comparison with experimental results and elastic theory.

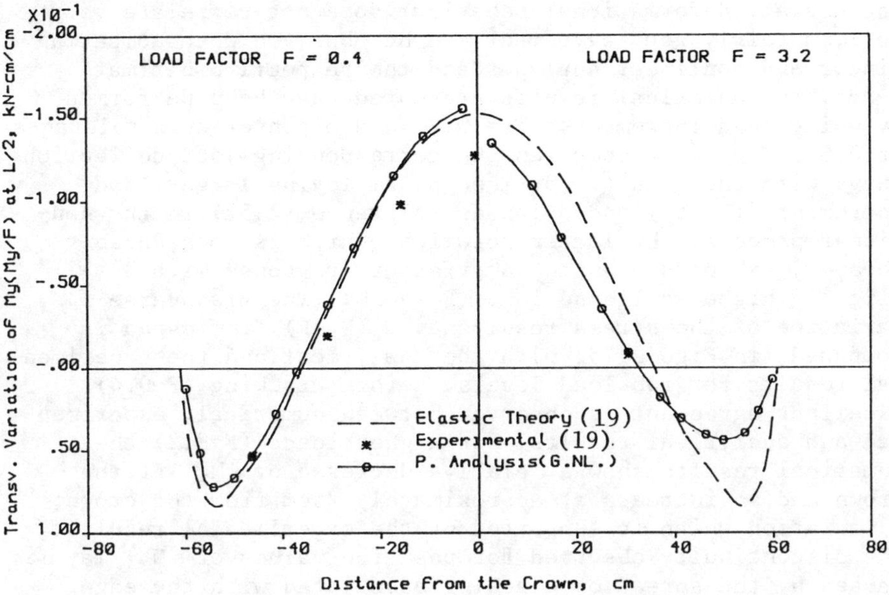

Fig. 26 Transverse variation of the moment My at midspan. Comparison with experimental results and elastic theory.

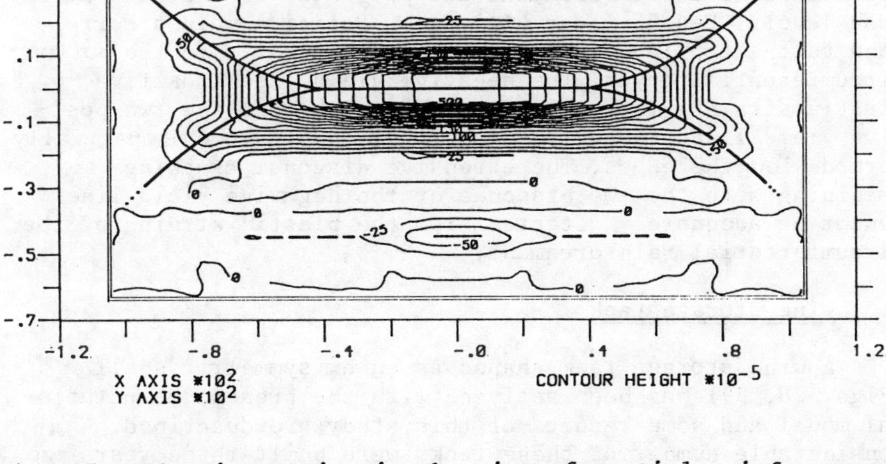

Fig. 27 Plastic strains in the circumferential reinforcement at failure.

which were obtained with a different numerical model. Although the overall deformational behaviour does not correlate very well, a fairly good agreement can be observed when comparing linear and nonlinear analyses and the respective ultimate loads. The numerical results presented have been performed by using load increments F = 0.2 and a convergence tolerance of 0.5%. Figure 24 compares the corresponding load-deflection curve with the results obtained by employing larger load increments (F=0.4) and a looser tolerance (2.5%) in the non-linear process. The latter solution, which is considerably cheaper, exhibits a quite satisfactory response with a slightly higher collapse load (Fu=4.50). The transverse variation of the stress resultant $N_x(N_x/F)$ at midspan is compared, in Figure 25, with the analytical and the experimental results for two load levels. Before cracking (F=0.4) excellent agreement is observed between numerical, experimental and analytical results. For higher loads (F=3.2) the numerical results show a relative decrease of N_x at the crown and an increase at approximately 25cm from the crown; a variation which is supported by the experimental results. The discontinuity observed for positive values of N_x may be caused by the spread of cracking associated with the edge reinforcement and the variable thickness in the vicinity of the free edge. Another stress resultant having a major influence on the load-carrying capacity of this shell is the moment M_y, whose variation is illustrated in Figure 26 for two different load levels. In the elastic domain (F=0.4) the numerical and the analytical results show good agreement, whereas the experimental values are smaller in the vicinity of the crown; probably due to the diaphragm interaction. For high loads (F=3.2) an extensive longitudinal cracking is developed at the crown top surface, which originates a redistribution of the M_y-moment, decreasing the maximum relative values. The numerical failure mechanism can be clarified by reference to Figure 27, where the plastic strains in the circumferential reinforcement are presented as contours at a load level, F=4.35 (near ultimate). Positive values correspond to the top layer and negative values refer to the bottom circumferential layer. The negative yield line (positive plastic strains) is completely formed, whereas the two positive yield lines (negative plastic strains) are only partially formed along the edges. The extensive diagonal cracking combining with the two branches of the negative yield line cannot be adequately detected from the plastic strains of the circumferential reinforcement.

4.3 Wine storage tank

A wine storage tank shaped as an axisymmetric shell (Figs. 28, 29) has been analysed with the present computational model and some results of this study are described. A considerable number of these tanks were built three years ago and several are presently out of service due to extensive

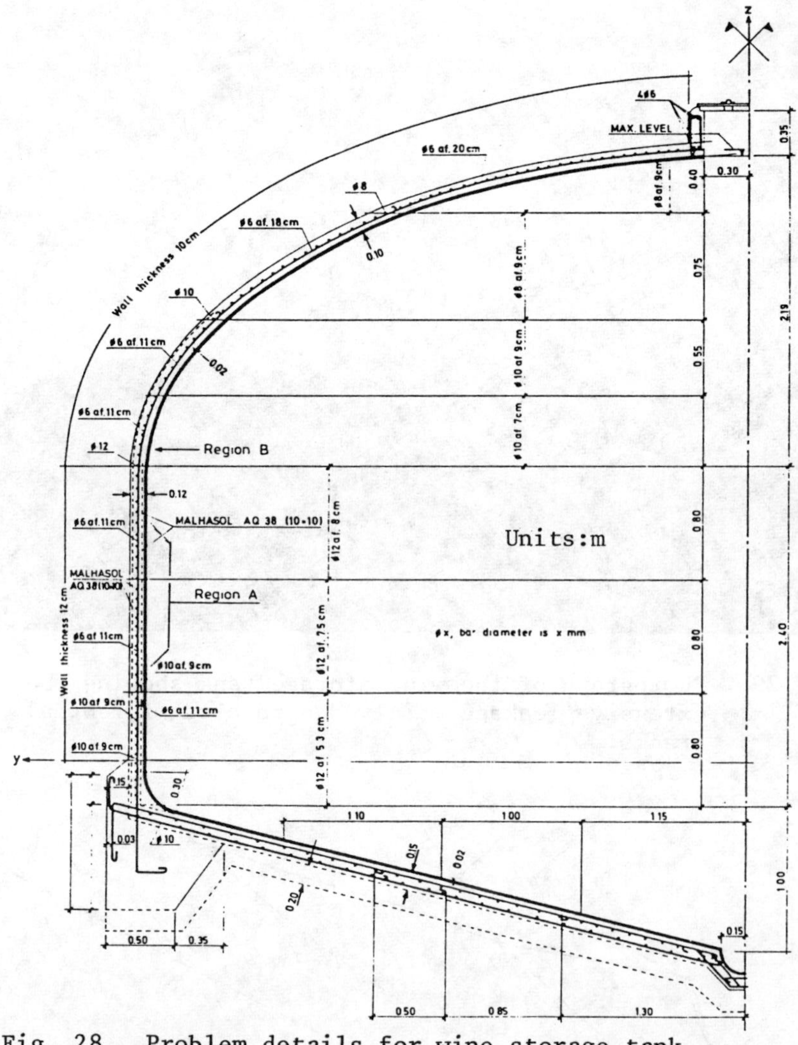

Fig. 28 Problem details for wine storage tank.

MATERIAL PROPERTIES [m, MN]		
CONCRETE	STEEL	
Young's Modulus, E_c = 20000.0 Poisson's Ratio, ν = 0.20 Ult. Comp. st. f_c = 16.0 Ult. Tens. st. f_{ct} = 0.8 Ult. Comp. stn. ε_u = 0.0035 Tens. stiff. Coef., α = 0.6 (*) Tens. Stiff. Coef., ε_m = 0.0015	Young's Modulus, E_s = 200000.0 Yield stress, f_y = 230.0 Young's Modulus, E'_s = 2000.0	Reinforcing bars
	Young's Modulus, E_s = 200000.0 Yield stress, f_y = 460.0 Young's Modulus, E'_s = 5000.0 (*)α = 0.0 for the two inner and outer layers without wire mesh	Wire mesh (Malhasol AQ38)

Table 3 Problem parameters for wine storage tank (Fig. 28).

Fig. 29 Photograph of the wine storage tank showing the extensive leakage mainly due to circumferential cracks.

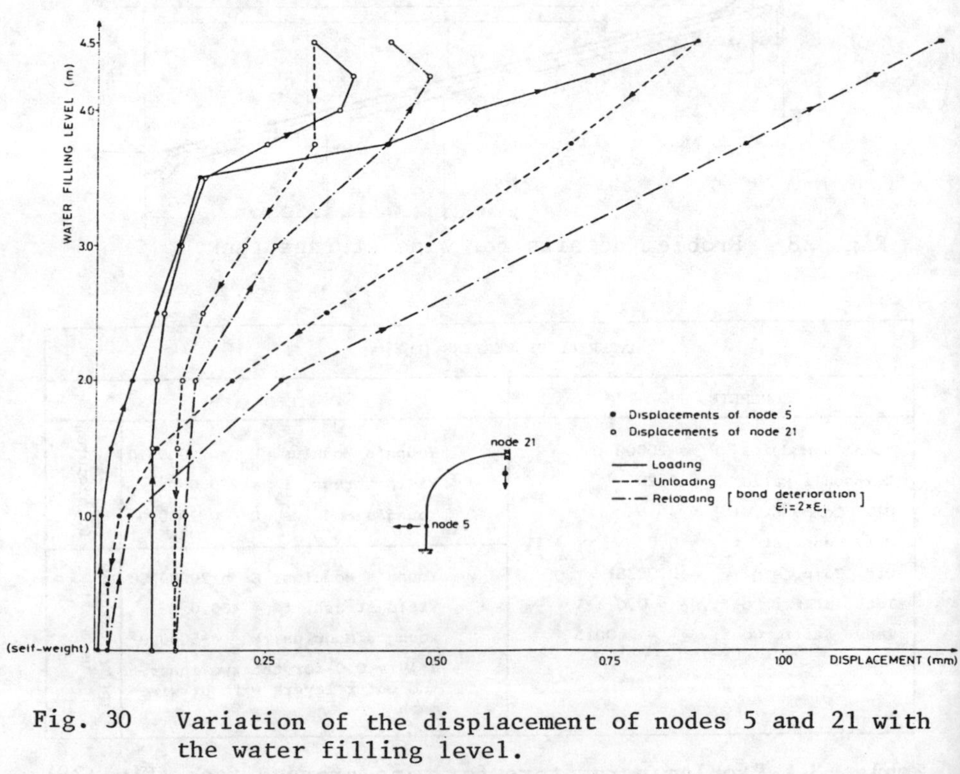

Fig. 30 Variation of the displacement of nodes 5 and 21 with the water filling level.

cracking resulting in leakage of wine. The tanks were constructed using a pneumatic membrane to model the internal surface shape, and then with the reinforcement in the correct position, concrete was mechanically sprayed onto the membrane until the desired thickness was achieved. A waterproof sealer was finally applied to the inner surface. With the first loading, leakage was soon observed in a few tanks, but several became unserviceable when reloaded. All the storage tanks show extensive vertical and horizontal (circumferential) cracking on the external surface of region A (see Fig. 28) and a circumferential crack on the internal surface at region B. Leakage of the contained liquid has been detected mainly at region A due to circumferential cracks (see Fig. 29) which are relatively wide. The numerical analysis is undertaken to investigate the possible structural causes of the damage.

The geometry and the reinforcement of the shell are illustrated in Fig. 28. Taking advantage of the axial symmetry of the shell, a sector of 15 degrees is considered for analysis which is divided into ten finite elements in the z-direction. The corresponding mesh is implicity defined by the nodal point positions shown in Figure 31. The wall thickness is discretized by six equal concrete layers and the necessary number of steel layers to represent the reinforcing bars in their correct location. The boundary conditions assumed in the analysis correspond to a built-in edge at the base and a vertically-free only condition at the top. A hydrostatic pressure load corresponding to the tank filled with wine up to the maximum level (Fig. 28) and the tank self-weight ($=0.0245$ MN/m^3) is the service loading considered. The material properties assumed in the analysis are specified in Table 3. The reinforcing bars placed at the mid-surface of the shell are of mild steel with a yield stress, $f_y=235$ MPa, whereas the 10×10 cm^2 of 3.8 mm diameter wire mesh (Malhasol AQ38 - - 1.13 cm^2/m) with a 2 cm concrete cover is of cold-worked steel for which $f_y=490$ MPa. Tension stiffening effect is considered for the two inner and the two outer concrete layers only in the region where Malhasol wire mesh exists (see Fig.28).

In the analysis, the self-weight loading was first considered and then the hydrostatic pressure was gradually applied up to the maximum filling level, of 4.5 meters. The gradual unloading and reloading of the water pressure was subsequently considered; to simulate the bond deterioration of cracked concrete with repeated loading a value of $\varepsilon_i=2\times\varepsilon_i$ was taken in the tension stiffening diagram (see Fig. 6). Figure 30 shows the variation of the displacement of two characteristic points with the water filling level in this process. The displacement profiles for a filling level of 3.0 m are illustrated in Fig. 31 and in Fig. 32 are shown the in-plane force, N_x, and moment, M_y, diagrams corresponding to the first loading process for which the concrete is uncracked. Concrete cracking, which is the only factor of nonlinearity,

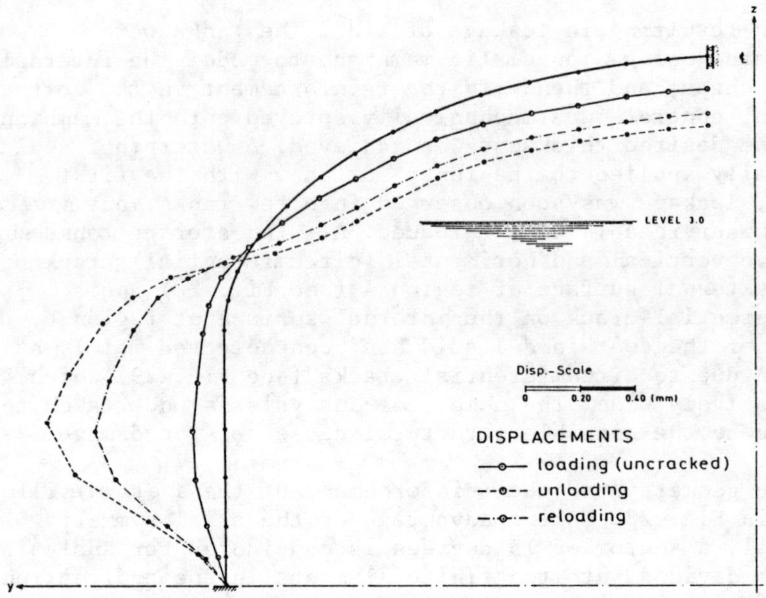

Fig. 31 Displacement profiles for the wine storage tank at a filling level of 3.0 m.

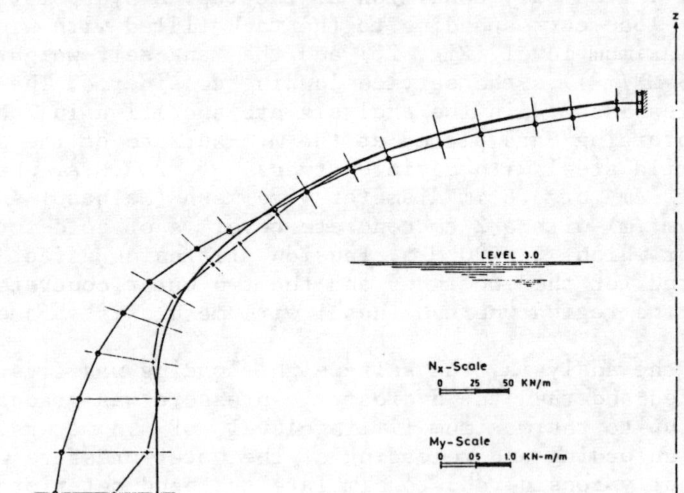

Fig. 32 Variation of the in-plane circumferential force Nx and the bending moment My at a filling level of 3.0m.

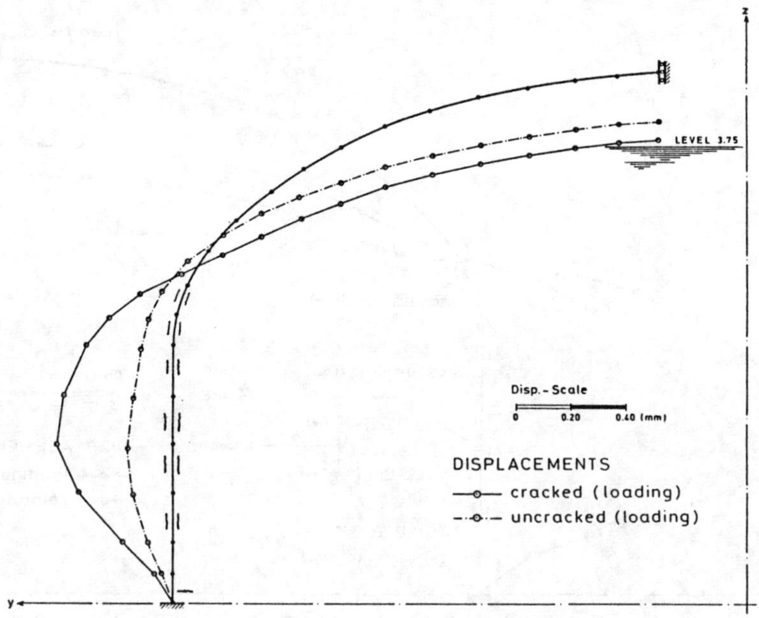

Fig. 33 Displacement profiles and cracking for the wine storage tank at a filling level of 3.75m.

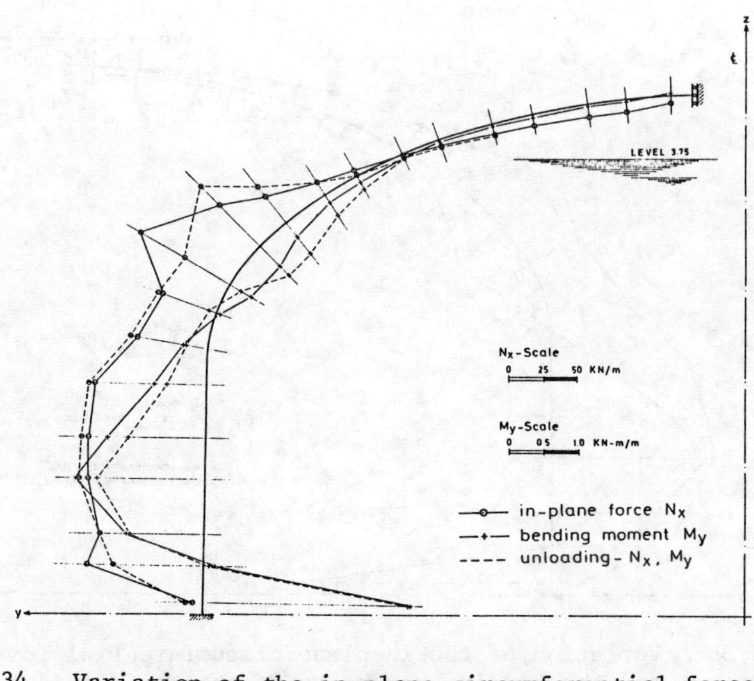

Fig. 34 Variation of the in-plane circumferential force Nx and the bending moment My at a filling level Of 3.75m.

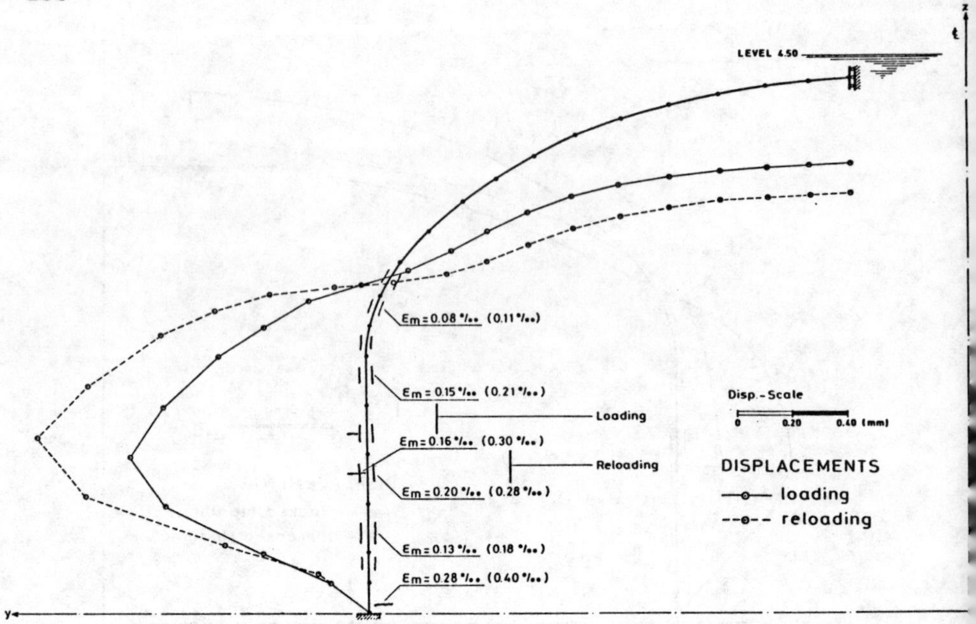

Fig. 35 Displacement profiles and cracking for the wine storage tank at the maximum filling level of 4.50m.

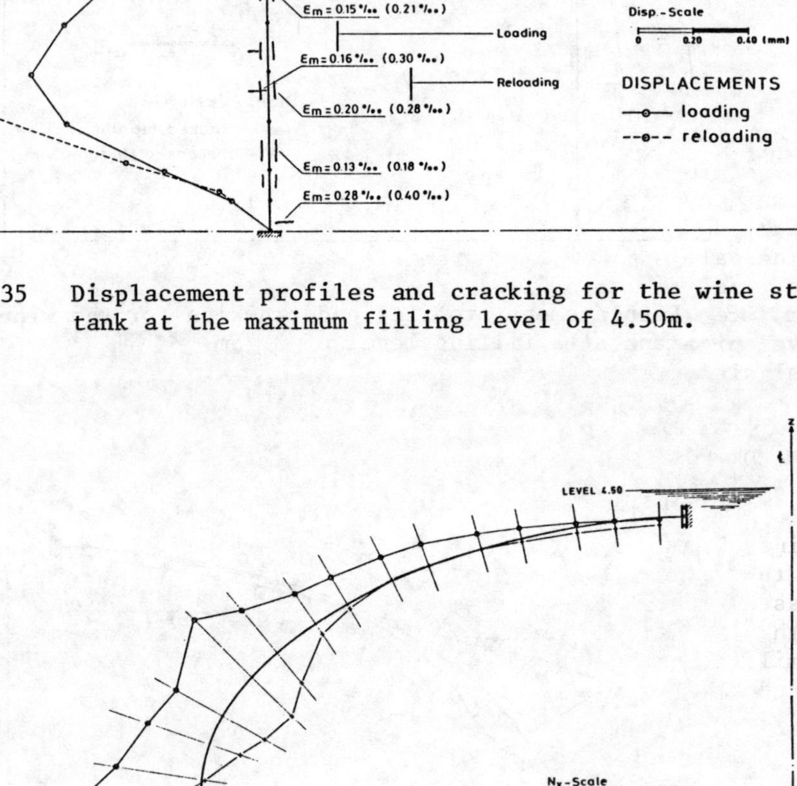

Fig. 36 Variation of the in-plane circumferential force Nx and the bending moment My at the maximum level of 4.50m.

starts at a filling level of 3.75 m. The displacement profiles, before and after cracking has occurred, are illustrated in Figure 33. Cracking starts in the inner concrete layer at the bottom (horizontal crack) and vertical cracking through the complete thickness commences immediatley in region A (about node 5) spreading upwards and downwards with progressive loading. The decrease of circumferential stiffness near node 5 due to vertical cracking leads to a substantial and sudden increase of the displacements in region A, which is responsible for a sudden increase of the bending moment, M_y, in this region causing circumferential (horizontal) cracks for higher loading in the outer concrete layers. The variation of the two more significant stress resultants, the in-plane force N_x and the bending moment M_y, at a filling level of 3.75m are shown in Figure 34. Figure 35 illustrates the displacement profiles and the values obtained for the mean strains in cracked concrete for the maximum water pressure, while, in Figure 36 are presented the corresponding in-plane force N_x and bending moment M_y. Taking into account the small amount of vertical reinforcement (Fig. 28), the circunferential cracks due to M_y are few and consequently wide, being the principal responsible for the leakage. The recuperation of the wine storage tanks has been carried out by applying to the external surface, near node 5, a pre-stressed ring developing a circunferential force of 135 kN, being the analysis performed by the present code.

5. CONCLUSIONS

A finite element computational model for the analysis of reinforced concrete shells has been described and the results of three numerical applications have been presented and discussed. The 8-node Serendipity and the heterosis element used with reduced and selective integration produced excellent and similar results for the cases analysed. Good correlation was found between numerical and experimental results throughout the entire structural response which demonstrates the effectiveness of the present code. Geometric nonlinear effects need to be taken into consideration in order to obtain realistic numerical solutions. The present numerical model and the resulting computational code has been applied to the structural analysis of several shell problems being an important and effective tool for research and design of reinforced concrete structures.

6. REFERENCES

1. ASCE Committee on Concrete and Masonry Structures, Task Committee on Finite Element Analysis of Reinforced Concrete Structures: "A state-of-the-art report on finite element analysis of reinforced concrete structures",ASCE Special Publication, 1981.

2. Chen, W. F. Plasticity in reinforced concrete, McGraw-Hill Book Company, New York, 1982.

3. Owen, D. R. J., Figueiras, J. A. and Damjanic, F. "Finite element analysis of reinforced and prestressed concrete structures including termal loading", in Comp. Meth. in Appl. Mech. and Eng., vol. 41, pp.323-366, 1983.

4. Figueiras, J. A. "Ultimate load analysis of anisotropic and reinforced concrete plates and shells, Ph. D. Thesis, C/Ph/72/83, Dept. of Civil Eng., University College of Swansea, U.K., 1983.

5. Kupfer, H., Hilsdorf, K. H. and Rush, H. "Behaviour of concrete under biaxial stresses", Proceedings, ACI, vol. 66, Nº 8, pp. 656-666, 1969.

6. Bresler, B. and Scordelis, A. C. "Shear strength of reinforced concrete beams", ACI Journal, Proceedings, vol. 60 Nº 1, pp. 51-73, 1963.

7. Owen, D. R. J. and Hinton, E.,Finite elements in plasticity: theory and practice, Pineridge Press Ltd., Swansea, U.K., 1980.

8. Comité Euro-International du Béton, CEB-FIP model code for concrete structures, CEB-FIP International Recommendations, 3rd. Ed., 1978.

9. Owen, D. R. J. and Figueiras J. A., "Ultimate load analysis of reinforced concrete plates and shells including geometric nonlinear effects", in Finite element software for plates and shells, (eds. E. Hinton and D. R. J. Owen), Pineridge Press, Swansea, U.K., 327-388, 1984.

10. Kolmar, W. and Mehlhorn, G. "Comparison of shear stiffness formulations for cracked reinforced concrete elements", in Proceeding of the International Conference on Computer aided analysis and design of concrete structures, Part I, (eds. F. Damjanic et al.), Pineridge Press, Swansea, U.K., 133-147, 1984.

11. Cope, R. J. "Material modelling of real, reinforced concrete slabs" in Proceeding of the International Conference on Computer aided analysis and design of concrete structures, Part I, (eds. F. Damjanic et al.), Pineridge Press, Swansea, U.K. 85-117, 1984.

12. Peter, J., "Zur bewehrung von scheiben und schalen fur hauptspannungen schiefwinklig zur bewehrungsrichtung", Die Bautechnik, 43, 5, 149-154, 1966.

13. Cervenka, V. and Gerstle, K. H., "Inelastic analysis of reinforced concrete panels: theory", Pubs. IABSE, 31-II, 1971.

14. Leonhardt, F., Construções de concreto, vol. 4, Editora Interciência, Rio de Janeiro, Brasil, 1979.

15. Figueiras, J. A. and Owen, D. R. J., "Analysis of elasto--plastic and geometrically nonlinear anisotropic plates and shells", in Finite element software for plates and shells, (eds. E. Hinton and D. R. J. Owen), Pineridge Press, Swansea, U.K., 235-326, 1984.

16. Crisfield, M. A. "Difficulties with current numerical models for reinforced concrete and some tentative solutions" in Proceedings of the International Conference on Computer aided analysis and design of concrete structures, Part I (eds. F. Damjanic et al.), Pineridge Press, Swansea, U.K., 331-357, 1984.

17. Mueller, G. "Numerical problems in nonlinear analysis of reinforced concrete "Report Nº UC SESM 77-5, Dept. of Civil Eng., Univ. of California, Berkeley, Sept. 1977.

18. Duddeck, H., Griebenow, G. and Shaper, G., "Material and time dependent nonlinear behaviour of cracked reinforced concrete slabs", in Nonlinear behaviour of reinforced concrete spatial structures, vol. 1, Preliminary Report IASS Symposium held in Darmstadt (eds. G. Mehlhorn et al.), 101-113, 1978.

19. Hedgren, A. W. and Billington, D. P., "Mortar model test on a cylindrical shell of varying curvature and thickness", ACI Journal, 73-83, 1967.

20. Floegl, H. and Mang, H., "On tension stiffening in cracked reinforced concrete slabs and shells considering geometric and physical nonlinearity", Inginieur-Archiv, vol. 51, 215-242, 1981.

CHAPTER 9

FINITE ELEMENT TECHNIQUES FOR THE ANALYSIS OF HYPERBOLIC COOLING TOWER SHELLS

S. S. J. Moy
Department of Civil Engineering
University of Southampton
Southampton, England.

SUMMARY

Finite element methods for the analysis of cooling tower shells are presented. It is shown that the semi-analytical method using axisymmetric elements is more economic than using general shell elements for linear analysis.

A method for obtaining directly the stresses in the ring beam is presented in detail together with typical results. Two methods are given for analysing geometrically imperfect shells and their relative merits compared.

Other aspects of cooling tower behaviour, cracking, buckling and non-linear analysis, which have been studied using finite elements are also outlined.

NOTATION

A	Cross sectional area of a column
c	Number of column pairs
E,G	Young's and shear moduli for a column
$\tilde{F}_\lambda$	Vector of reactions in the λth column pair
H (superscript)	Number of harmonics in Fourier series
j (superscript)	Indicates Fourier series 'constants' for the jth harmonic. Unbarred constants, e.g. $\overline{P}(Z)^j$, refer to cosine terms, barred constants, e.g. $\overline{P(Z)}^j$, refer to sine terms.
I	Second moment of area of a column
J	Polar moment of area of a column
m_ϕ, m_θ, $m_{\phi\theta}$	Changes in bending stress resultants
M_ϕ, M_θ, $M_{\phi\theta}$	Bending stress resultants
n_ϕ, n_θ, $n_{\phi\theta}$	Changes in membrane stress resultants
N_ϕ, N_θ, $N_{\phi\theta}$	Membrane stress resultants

$\underset{\sim}{N}_b$	Vector of stress resultants at the base of the shell from continuous flexible boundary analysis.
$\underset{\sim}{N}_c$	Vector of actual stress resultants at base of shell
$\underset{\sim}{N}_s$	Vector of self equilibrating base loads
$\underset{\sim}{N}_\lambda$	Vector of actual stress resultants across the λth column pair
p (superscript)	Refers to perfect shell
$p(Z,\theta)$	General function of Z and θ
$\underset{\sim}{P}^j$	Vector of nodal forces for the jth harmonic
q_n	Total normal pressure
$q_{n\phi}$	Normal pressure due to N_ϕ
R_b	Horizontal radius at the base of the shell
$[S_b]$	Stiffness matrix equivalent to the continuous flexible boundary
$[S_{cp}]$	Stiffness matrix for a column pair
$[S^j]$	Shell stiffness matrix for the jth harmonic
T (superscript)	Transpose matrix
Z	Vertical elevation above the base of the shell
$\underset{\sim}{\delta}^j$	Vector of nodal displacements for the jth harmonic
$\underset{\sim}{\delta}_\lambda$	Vector of nodal displacements at the centreline of the λth column pair
$\Delta F_{n\phi}$	Normal force due to N_ϕ
θ	Horizontal angle around the shell circumference
θ,φ (subscripts)	Indicate circumferential and meridional variables respectively e.g. N_θ, N_ϕ
Ω	Circumferential extent of half column head

Terms not shown here are defined in the text or in the figures.

1. INTRODUCTION

Reinforced concrete cooling towers are particularly interesting structures for the analyst. They are very large structures. In the U.K., typically, they are 114m high with a base diameter of about 79m and they weigh about 9000T. At the same time they are very slender structures whose wall height to thickness ratio is about the same as a sheet of thin card. Figure 1 shows the geometry of a typical tower.

The top of the shell is stiffened by a thickened cornice and the base by a ring beam. The ring beam is required because the base of the shell is supported on discrete columns which provide a gap for the entry of cooling air to the tower.

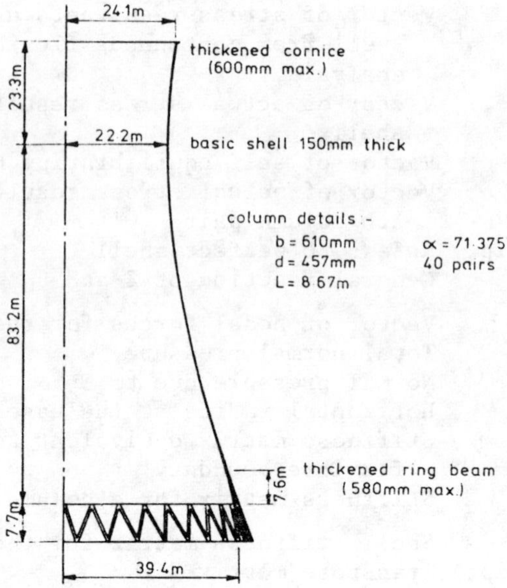

Figure 1: Typical Cooling Tower Geometry

Such structures are difficult to construct accurately. The problems of setting out a large diameter circular shutter and pouring concrete tens of metres above ground level are obvious. Consequently it is not uncommon for the shell to have significant imperfections which will affect the distribution of stress in the shell. Cooling towers are also prone to cracking, particularly vertical cracks, as would be expected in such large areas of concrete. Again these cracks will affect the behaviour of the tower.

The need for a deeper understanding of cooling tower shell behaviour has been vividly brought home on several occasions. In the U.K. there were no serious structural problems with concrete cooling towers until 1965. On November 7th of that year three almost completed towers (only the internals were incomplete) at Ferrybridge power station in Yorkshire collapsed during a South Westerly gale. In 1974 a tower of similar size collapsed at the I.C.I. Nylon works at Ardeer, Scotland. That tower was known to have imperfections in its shell. The U.K. has not been alone in having such problems!

The geometry of the shell and its possible problems have generated considerable interest amongst researchers. Much of the analytical work done has made use of the finite element method. The purpose of this article is to present a summary of the application of finite element techniques to the analysis of cooling tower shells, showing what has been achieved and also what is being attempted.

2. LOADING ON COOLING TOWERS

There are two main loads on cooling tower shells. These are the weight of the reinforced concrete (dead load) and wind load. Dead load is simply a vertical force applied to the shell.

Wind load is considerably more complex. It is usual to represent the wind load on the tower as a pseudo-static load which allows for the dynamic effect of the wind. (This may not be satisfactory on towers larger than those in use today, but that is another problem). This wind loading has been derived from very extensive wind tunnel testing. This testing has shown that the wind pressure varies around the circumference of the shell. Figure 2a shows the so-called "Most Adverse" wind pressure distribution which is used in

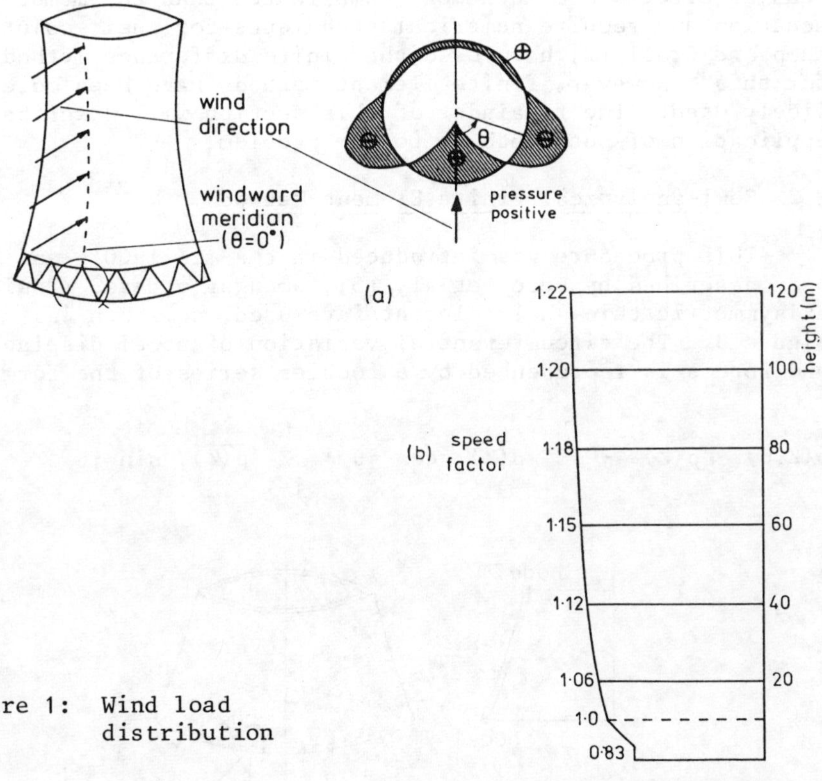

Figure 1: Wind load distribution

the U.K. In addition the wind pressure varies with height above the ground, Figure 2b shows the distribution used in the U.K. Wind blowing across the top of the tower creates an internal suction which adds to the wind pressure on the outside of the shell due to the hot water (which is being cooled) inside the shell. The reactions at the top of the columns are concentrated forces applied to the base of the shell. These forces are important in the design of the ring beam. It is possible to visualise other loadings such as concentrated forces at prestressing anchors fixed to the shell. (This is not fanciful, a scheme to prestress a tower has been considered).

3. ANALYSIS TECHNIQUES

3.1 Background

The earliest analyses of cooling towers treated the shell as a pure membrane. Martin and Scriven [1] have described an effective procedure for this type of analysis. Unfortunately, the towers do not choose to behave as pure membranes, and significant bending moments can occur, particularly around the ring beam.

The governing equations of shell behaviour which include bending effects are much more complicated than the membrane equation and require numerical techniques for their solution. Kemp and Croll [2] have used the finite difference method for this. However, finite element methods have been more widely used. The remainder of this section will describe the application of such methods to the problem.

3.2 Semi-analytical Finite Element Method

This procedure was introduced in the mid 1960's and has been described by Percy et al, [3], amongst others. Basically axisymmetric thin shell elements are used, as shown in Figure 3. The circumferential variation of nodal displacements and forces is represented by a fourier series of the form

$$p(Z,\theta) = p(Z)_o + \sum_{j=1}^{H} p(Z)_j \cos j\theta + \sum_{j=1}^{H} \overline{p(Z)}_j \sin j\theta \qquad (1)$$

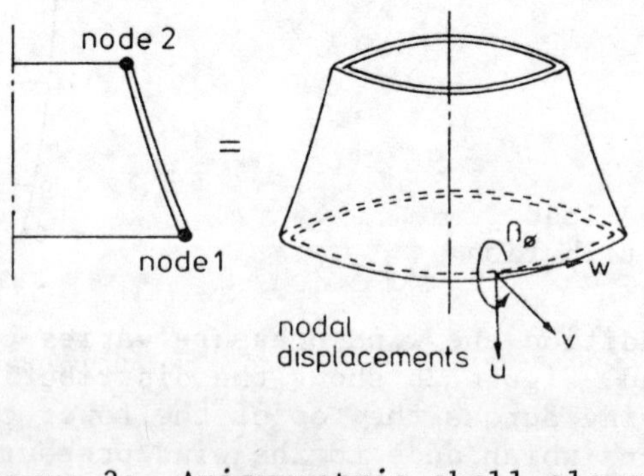

Figure 3: Axisymmetric shell element

Dead load requires only the constant term of the series and the "Most Adverse" wind loading can be represented by an eight term series. Because the wind loading is symmetric about the $\theta = 0°$ axis the series contains only cosine terms. Due to the symmetry of the loading, the individual series for each nodal variable contain only either the cosine or the sine terms. From these series and the assumed geometry and shape functions of the element the displacement approach gives the element stiffness matrix. In linear analysis the harmonics (terms) of the Fourier series decouple to give

$$[\theta^j]\, \underset{\sim}{P}^j = [\theta^j][S^j]\, \underset{\sim}{\delta}^j \tag{2}$$

where $\underset{\sim}{P}^j$ and $\underset{\sim}{\delta}^j$ are vectors of the Fourier series constants of the jth harmonic for the nodal forces and displacements respectively, $[S^j]$ is the element stiffness matrix for the jth harmonic and $[\theta^j]$ is a diagonal matrix containing the terms $\cos j\theta$ or $\sin j\theta$. $\underset{\sim}{P}^j$, $\underset{\sim}{\delta}^j$ and $[S^j]$ are totally independent of θ (hence decoupling). From equation (2)

$$\underset{\sim}{P}^j = [S^j]\, \underset{\sim}{\delta}^j \tag{3}$$

Solution of equation (3) for each harmonic in turn gives the Fourier series constants for the nodal displacements. From these and the strain-displacement and stress-strain relationships the Fourier series constants for the stress resultants within each element are calculated. Figure 4 shows the stress resultants which are obtained. By summing the Fourier series using the appropriate angle the stress resultants and displacements can be found at any location on the circumference.

The complete shell is represented by the two noded axisymmetric shell elements and effectively the problem is reduced to one dimension. Relatively few elements are required and thus only a small number of simultaneous equations have to be solved, although there is a set of equations for each harmonic. The approach is extremely economic in terms of both computer time and storage.

The earliest elements were straight sided cone-toroids as shown in Figure 5a. The nodal displacements u and v were represented by linear shape functions and w by a cubic function. The nodal rotation β was simply dw/ds. The element was easy to use and automatically fulfilled rigid body motion requirements. It had the disadvantage that when representing a doubly curved shell (such as a cooling tower) there had to be a slope discontinuity at each node (see Figure 5b). This introduced additional bending stresses into the shell, which would not occur in reality. Modern elements have curved meridians, as in Figure 6, and their geometry is defined by

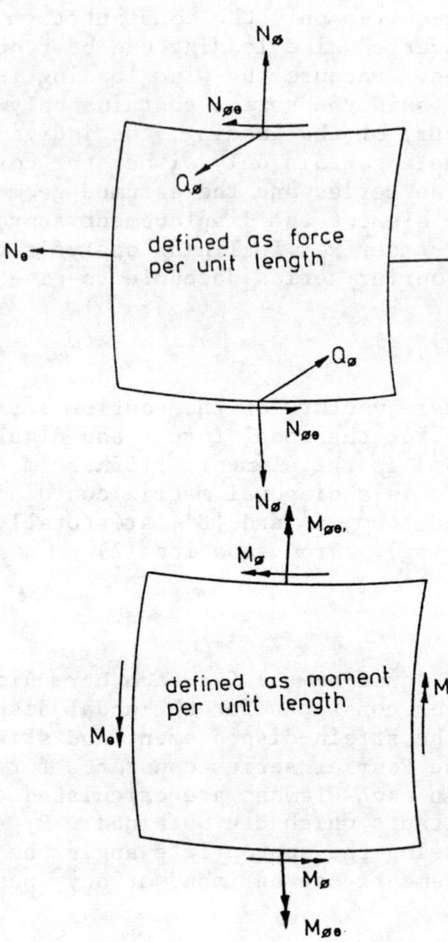

Figure 4: Stress resultants

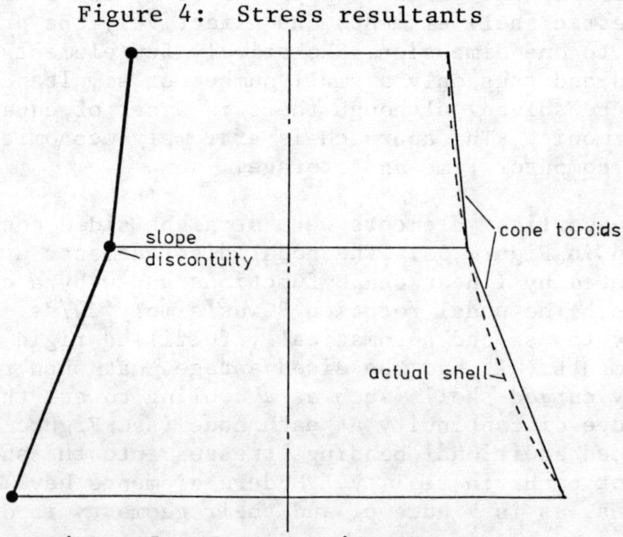

Figure 5: Cone toroid elements

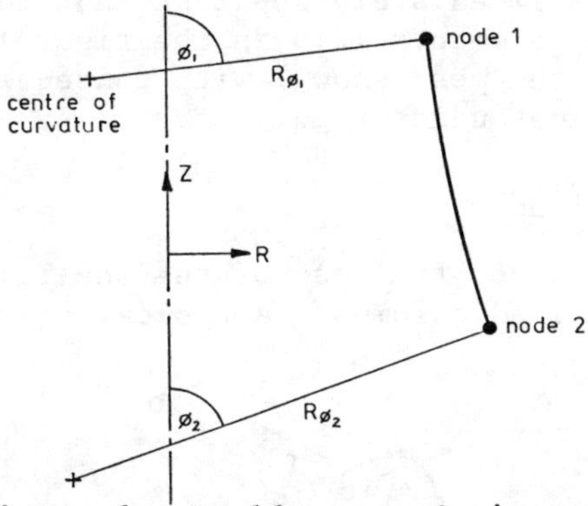

Figure 6: Doubly curved ring element

interpolation between nodal values of radius, slope and curvature [4]. There are problems in satisfying rigid body motion requirements with these elements. One means of overcoming these problems is to introduce internal degress of freedom into the element [4], and then to use static condensation to express the element stiffness matrix in terms of nodal variables only. Attempts are still being made to produce curved elements which explicitly satisfy all requirements without recourse to internal degrees of freedom [5].

The technique which has been described in this section is similar in many respects to the finite strip method.

3.3 General Shell Finite Elements

An alternative approach is to use general shell elements to represent the complete shell. Elements used have ranged from flat triangles and quadrilaterals [6] to sophisticated 9-node degenerated Lagrangian elements [7] which can allow material non-linearities. A typical mesh of triangular elements is shown in Figure 7. Only half the shell need be represented because of the symmetry of the loading about the

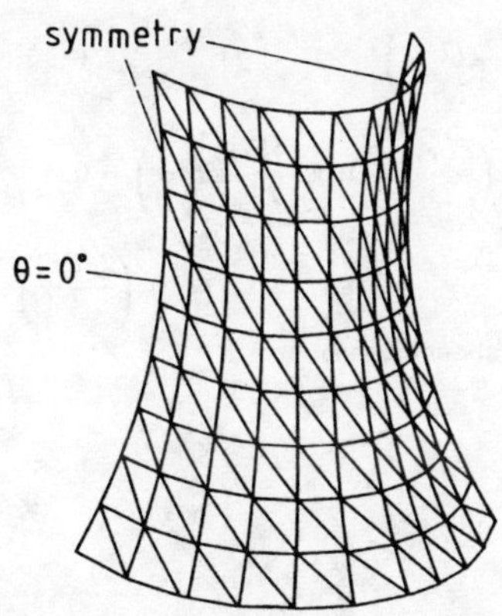

Figure 7: Typical mesh of flat triangular elements

$\theta = 0°$ axis. It is immediately apparent that the size of the tower requires many elements (216 in the figure) and degrees of freedom (780 in the mesh shown) with consequent penalties on both computer time and storage.

4. BOUNDARY CONDITIONS

As explained above, the base of the shell is supported on equi-spaced inclined columns. A typical pair of columns is shown in Figure 8.

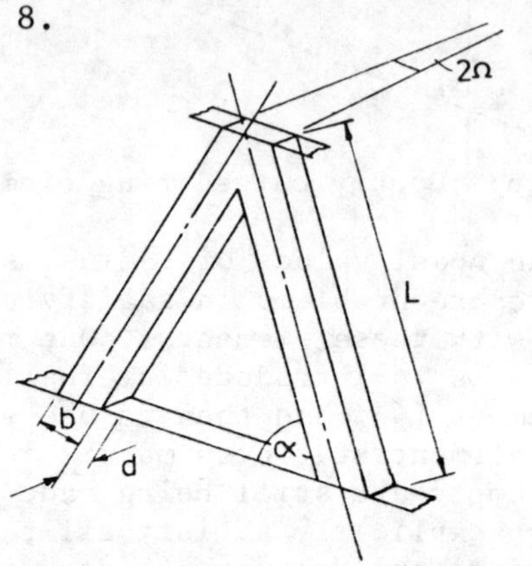

Figure 8: Typical Column Pair

In many analyses the column flexibility has been ignored. Extreme boundary conditions of complete base fixity or hinged support were chosen. The columns produce a flexible boundary. A stiffness matrix for a column pair can be determined by considering each column in the pair as a prismatic member in bending and applying appropriate transformations of axes to bring the column stiffness matrix into the local shell co-ordinate system. When the matrices for the individual columns are added they give the stiffness matrix for the column pair.

$$[S_{cp}] = \begin{bmatrix} \left(\frac{2EA}{L}\sin^2\alpha + \frac{24EI_y}{L^3}\cos^2\alpha\right) & 0 & 0 & 0 \\ & \left(\frac{2EA}{L}\cos^2\alpha + \frac{24EI_y}{L^3}\sin^2\alpha\right) & 0 & 0 \\ & & \left(\frac{24EI_z}{L^3}\right) & \left(-\frac{12EI_z}{L^2}\sin\alpha\right) \\ \text{Matrix symmetric about diagonal} & & & \left(\frac{2GJ}{L}\cos^2\alpha + \frac{8EI_z}{L}\sin^2\alpha\right) \end{bmatrix} \qquad (4)$$

Referring to Figure 8, $A = bd$, $Iy = (db^3)/12$, $Iz = (bd^3)/12$ and $J \cong (bd^3)/5$.

Since the shell is supported on a large number of column pairs it can be assumed that the stiffness of the column pairs is uniformly distributed around the base of the shell. The average stiffness per unit length is

$$[S_b] = \frac{c}{2\pi R_b} [S_{cp}] \tag{5}$$

Where c is the number of column pairs and R_b is the horizontal radius at the base of the shell. With appropriate transformations of axes, $[S_b]$ can be added into the overall stiffness matrix to represent the flexible boundary. In the semi-analytical method further minor modifications are required for each harmonic considered [8].

5. THE APPLICATION OF FINITE ELEMENT METHODS

5.1 The Perfect Shell

The shell in Figure 1 has been analysed using both finite element techniques. Figures 9(a), (b), (c) and (d) show plots of the membrane stress resultants N_ϕ (meridional) and N_θ (circumferential) and the bending stress resultants M_ϕ and M_θ at the 67.5° meridian, which is close to where wind suction is greatest. The stress resultants are due to wind load only, with a free stream wind pressure of 1 kN/m². (For simplicity the variation of wind pressure with height was ignored). The agreement between axisymmetric, flat triangle and flat quadrilateral elements is excellent for N_ϕ, N_θ and M_θ but is not good for M_ϕ, particularly near the base of the tower. However, M_ϕ has a small magnitude compared to the other stress resultants.

Element type	*Number of elements*	*Total degrees of freedom*	*Computer time (ICL 2970 units)*
Axisymmetric	34	140	220
Flat triangle	1156	3780	1600
Flat quadrilateral	578	3780	3200

Table 1: Comparison of analyses and perfect shell

It is worth comparing the computational effort to obtain these resultants. Table 1 shows number of elements, degrees of freedom and computer time for the three types of element. (Computer time is machine dependent and is only included for comparison). The economy of the axisymmetric elements is

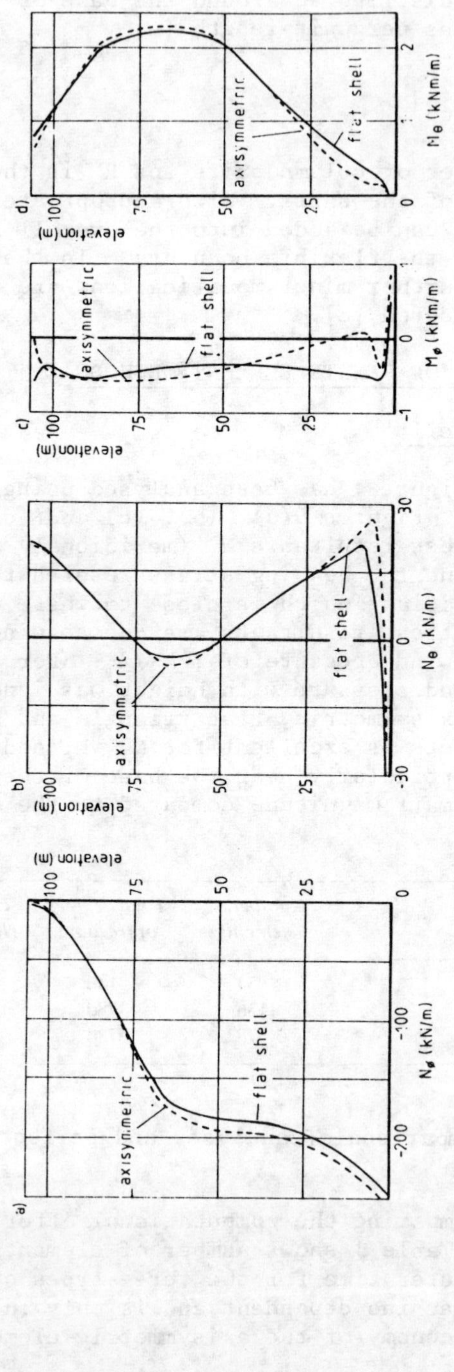

Figure 9: Stress resultants in the perfect shell

obvious. Due to this economy the semi-analytical technique has been widely adopted for the analysis and design of perfect cooling tower shells and there are various programs available for this purpose. Reference [9] is typical.

5.2 Analysis of the Ring Beam

The thickened ring beam at the base of the shell is required to span between the pairs of columns. The perfect shell analysis with the fixed, hinged or flexible base gives the stress resultants in the bulk of the shell but not in the ring beam because of the approximation in the boundary conditions at the base of the shell. Until recently a separate analysis of the ring beam has been necessary, with the attendant data preparation and a considerable degree of approximation. However, it has been possible to develop the semi-analytical method [8] to give stress resultants in the ring beam.

The continuous flexible base described above is only an approximation to the actual base support. The analysis using the continuous flexible base gives Fourier series for a system of stress resultants $\underset{\sim}{N}_b$ at the base of the shell. Thus

$$\underset{\sim}{N}_b = \sum_{j=0}^{H} [\theta^j]\, \underset{\sim}{N}_b^j \qquad (6)$$

where $\underset{\sim}{N}_b = [N_\phi, N_{\phi\theta}, Q_\phi, M_\phi]^T$

$$[\theta^j] = \begin{bmatrix} \cos j\theta & & & 0 \\ & \sin j\theta & & \\ & & \cos j\theta & \\ 0 & & & \cos j\theta \end{bmatrix} \qquad (7)$$

and $\underset{\sim}{N}_b^j$ is the vector of Fourier constants for the jth harmonic for each of the stress resultants in $\underset{\sim}{N}_b$. Figure 10(a) shows approximately the membrane stress resultant N_ϕ at the base. In reality, N_ϕ should be zero at the base of the shell, except where the shell rests on a column pair, as shown in Figure 10(b). This is also the case for the other stress resultants in equation (6). To produce the actual base conditions these stress resultants must be made to have the required distribution. (The other resultants N_θ and M_θ are not forced to be zero at the base and are not treated in the same manner).

The actual stress resultants are the reactions at the top of each column pair and can be calculated from the flexible boundary analysis. In any column pair the concentrated reactions (at the centreline of the column pair) are given by

$$\underset{\sim}{F}_\lambda = [S_{cp}]\, \underset{\sim}{\delta}_\lambda \qquad (8)$$

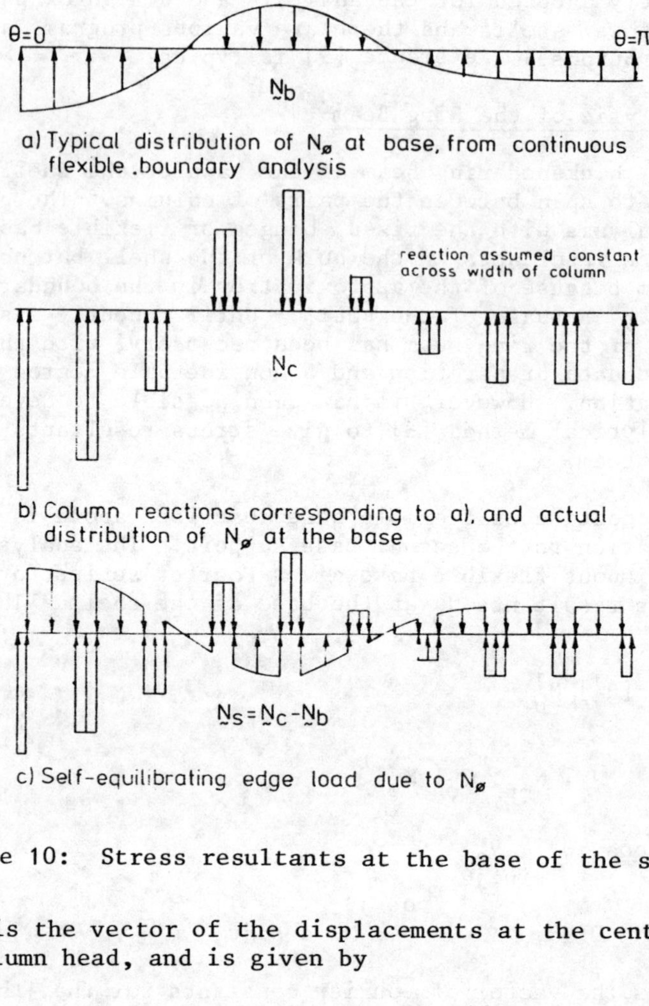

Figure 10: Stress resultants at the base of the shell

where $\underset{\sim}{\delta}_\lambda$ is the vector of the displacements at the centreline of the column head, and is given by

$$\underset{\sim}{\delta}_\lambda = \sum_{j=0}^{H} [\theta_\lambda^j] \, \underset{\sim}{\delta}_b^j \tag{9}$$

where $\underset{\sim}{\delta}_b^j$ is the vector of constants for the displacements at the base of the shell for the jth harmonic and $[\theta_\lambda^j]$ is the matrix in equation (7) with θ replaced by θ_λ, the angular position of the centre of the column head.

Since the width of the column head is small compared to the circumference, it can be assumed that the stress resultants are uniformly distributed across the column head so that the actual stress resultants $\underset{\sim}{N}_\lambda$ are given by

$$\underset{\sim}{N}_\lambda = \frac{1}{2\Omega R_b} [S_{cp}] \sum_{j=0}^{H} [\theta_\lambda^j] \underset{\sim}{\delta}_b^j \tag{10}$$

where 2Ω is the total angular extent of the column head. The stress resultants at all the column pairs can be written

$$\underset{\sim}{N}_c = \sum_{\lambda=1}^{C} \underset{\sim}{N}_\lambda = \frac{1}{2\Omega R_b} [S_{cp}] \sum_{\lambda=1}^{C} \sum_{j=0}^{H} [\theta_\lambda^j] \, \underset{\sim}{\delta}_b^j \qquad (11)$$

$\underset{\sim}{N}_c$ is effectively a very discontinuous function of θ. These stress resultants $\underset{\sim}{N}_c$ must be in overall equilibrium with the resultants $\underset{\sim}{N}_b$.

If the self-equilibrating load $\underset{\sim}{N}_s$, given by

$$\underset{\sim}{N}_s = \underset{\sim}{N}_c - \underset{\sim}{N}_b \qquad (12)$$

and illustrated in Figure 10(c) is applied at the base of the shell, the stress resultants caused by this load are the changes which correct the continuous flexible boundary solution to give the actual discrete column base conditions.

It is necessary to express $\underset{\sim}{N}_s$ by a Fourier series. Since $\underset{\sim}{N}_s$ is such a discontinuous function of θ a large number of harmonics are necessary to express $\underset{\sim}{N}_s$ accurately. However, a careful study of the series [8] has shown that many of the terms are zero, the only non-zero terms being

$$\left.\begin{array}{l} 0 \leqslant j \leqslant H \\ nc - H \leqslant j \leqslant nc + H \end{array}\right\} \qquad (13)$$

where n takes the values 1, 2 etc. This has given a considerable saving in computation because the non-zero harmonics (typically about 35% of all harmonics for wind load and only 2.5% for dead load) can be identified directly and the corresponding series constants calculated.

The truncation of the series has also been analysed. Table 2 shows the loading around one column pair for various numbers of terms in the series. Twenty column pairs, each with an angular extent of 0.02 rad and carrying a load of 1 MN were used. This was equivalent to an N_ϕ of 50 MN/rad over the column heads. It can be seen that after $n = 25$ the approximation is sufficiently accurate. It was also confirmed, see Figure 11, for column reactions corresponding to the 'most adverse' wind loading, that $n = 25$ was a suitable truncation of the series.

The tower in Figure 1 has been reanalysed to determine the effect of the discrete columns. Figures 12 and 13 show stress resultants from dead load only. Since dead load is constant around the circumference it is sufficient to present results at the centre of one column pair and midway between pairs. Figure 12(a) shows that the meridional stress

Length of series 'n'	θ (degrees)								% of reaction in width of column
	0	0·286	0·573	1·8	3·6	5·4	7·2	9·0	
10	52·50	45·21	27·35	−0·09	1·04	1·23	1·29	1·31	86·8
25	48·73	52·83	26·20	0·24	−0·52	0·56	−0·58	0·58	94·1
50	52·60	50·36	24·79	−0·46	−0·25	−0·21	−0·20	−0·20	97·1
100	49·43	51·20	25·31	0·21	0·16	0·15	0·15	0·15	99·0
1 000	50·01	49·95	25·00	−0·01	−0·01	−0·02	−0·02	−0·02	99·8

Table 2: Column reactions for various truncations

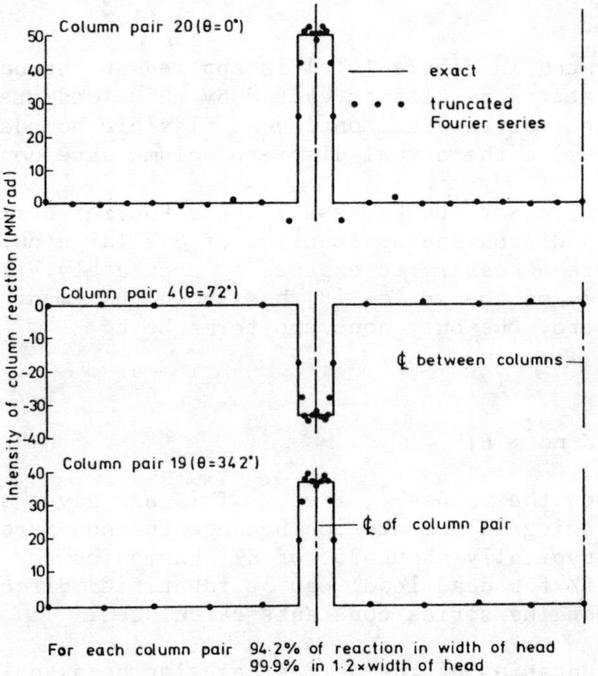

Figure 11: Column reactions, wind loading

resultant N_ϕ is compressive, is very large over the columns and drops to zero between the columns. In order to have a reasonable scale to graph, part of the distribution has been omitted. In fact N_ϕ at the base of the shell over the columns was equal, subject to truncation errors, to the column reaction. Effectively the unsupported part of the shell is shedding load to the supported part. Figure 12(b) shows the circumferential stress resultant N_θ. The changes caused by the discrete columns are complex. Such violent changes in membrane stress are accompanied by bending in the shell and the analysis gives those bending stress resultants as shown in Figure 13.

Figures 14 and 15 show results from the wind load analysis. The figures show N_ϕ and N_θ near to the windward meridia

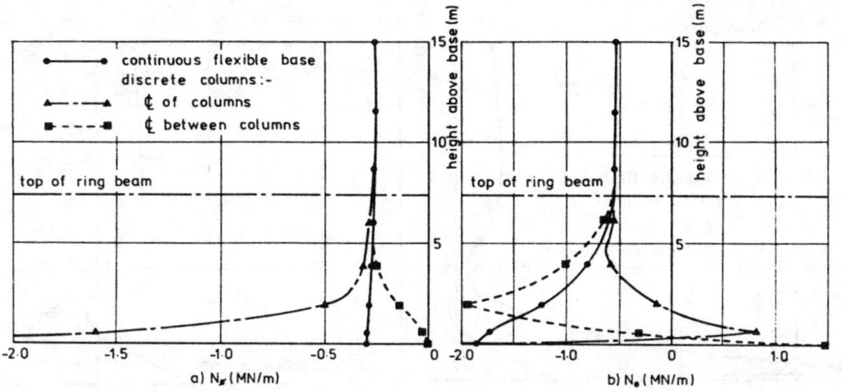

Figure 12: Membrane stress resultants - dead load

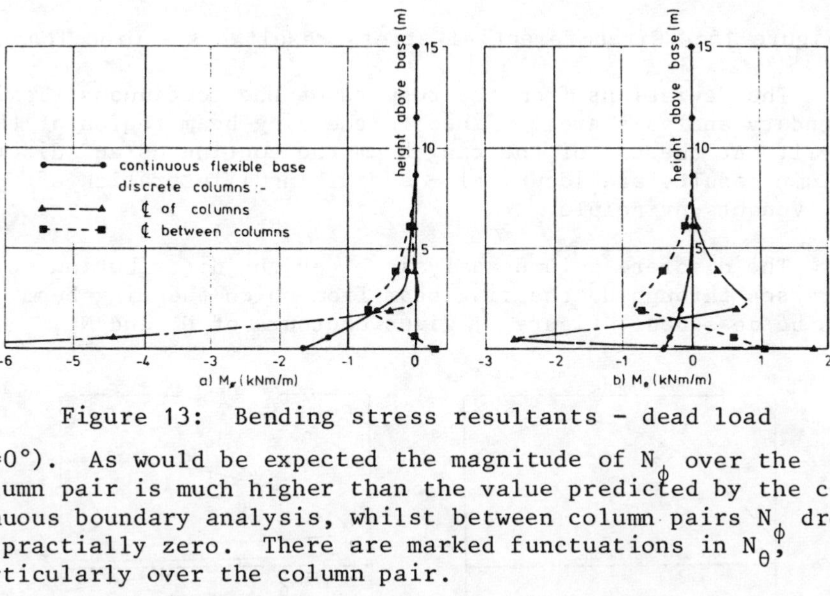

Figure 13: Bending stress resultants - dead load

(θ=0°). As would be expected the magnitude of N_ϕ over the column pair is much higher than the value predicted by the continuous boundary analysis, whilst between column pairs N_ϕ drops to practially zero. There are marked functuations in N_θ, particularly over the column pair.

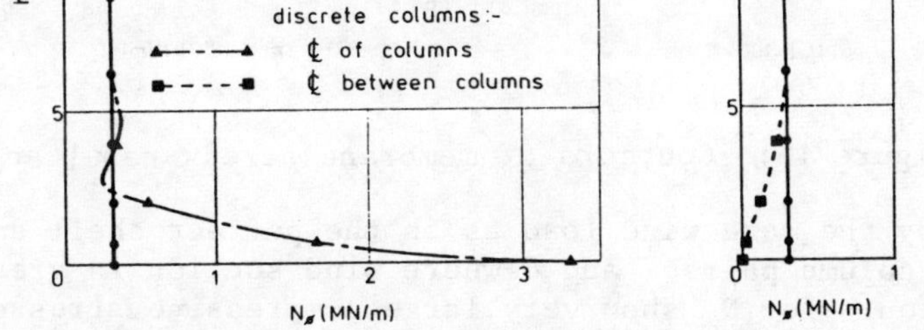

Figure 14: Meridional membrane stress resultants - wind load

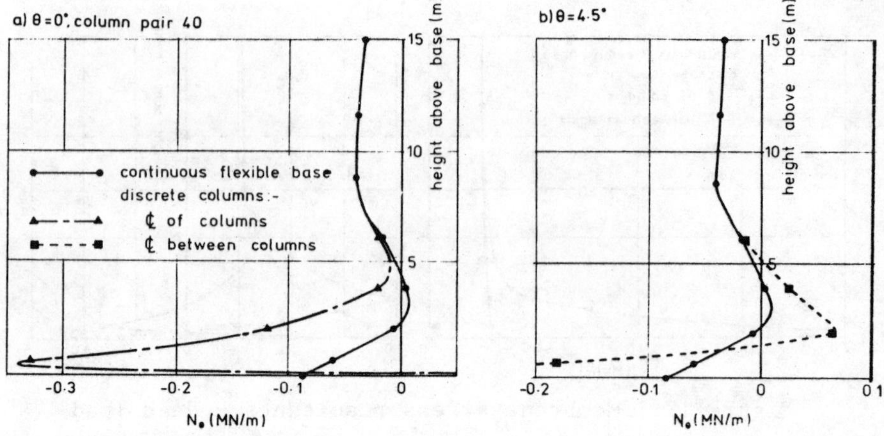

Figure 15: Circumferential stress resultants - wind load

The deviations from the results of the continuous flexible boundary analysis are confined to the ring beam region of the shell. At the top of the ring beam the continuous and discrete column results are identical - a striking illustration of St. Venants principle.

The discrete column analysis gives the distribution of stresses throughout the ring beam from which the ring beam can be designed. Figure 16 gives contours of N_ϕ and N_θ,

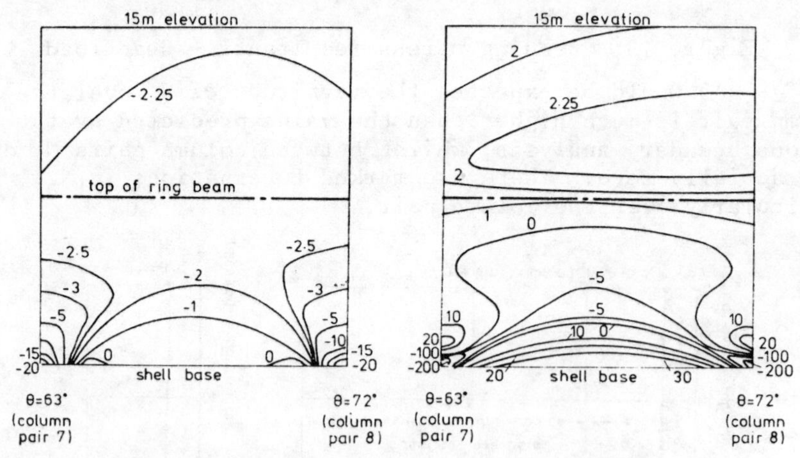

Figure 16: Contours of membrane stress resultants

caused by the same wind load as in the perfect shell analysis between column pairs 7 and 8 where wind suction is greatest. The contours for N_ϕ show very large compressive stresses over the columns, with practically zero stress between column pairs.

The build-up of stress local to the columns is obvious. The circumferential stresses, N_θ, are more complex. At the base of the shell there are large compressive values over the columns with tension between. At about 1.5m above the base the signs of the stresses are reversed. This indicates that the ring beam is behaving as a continuous beam over the columns.

Further computing economy is revealed by the results of the discrete column analysis. Since the significant changes in stress occur in the ring beam the element mesh can be redefined automatically for the discrete column correction, using very few elements for the main shell.

Nevertheless, the discrete column analysis, particularly with wind loading, uses significant amounts of computer time as shown in Table 3. However, it is only an option in the program. Generally the very economic continuous boundary analysis would be used, with the option only being exercised

Loading	*Base conditions*	*Number of significant harmonics (including zero harmonic)*	*CDC 7600 time (s)*
Dead load	Continuous flexible	1	9
	Discrete columns	26	75
'Most adverse' wind load	Continuous flexible	8	28
	Discrete columns	383	1 050

Table 3: Comparison of computer times

for the ring beam design. This analysis is still a considerable improvement from having to use a completely separate analysis for the ring beam.

5.3 Shells with Imperfections

5.3.2 Geometric approach

Geometric imperfections can have a marked effect on the distribution of stresses in cooling tower shells. Basically there are two types of imperfection, the axisymmetric and the localised 'patch' imperfections.

When using finite elements the obvious approach is to let the elements follow the geometry of the imperfection. A typical axisymmetric imperfection, as in Figure 17, could be

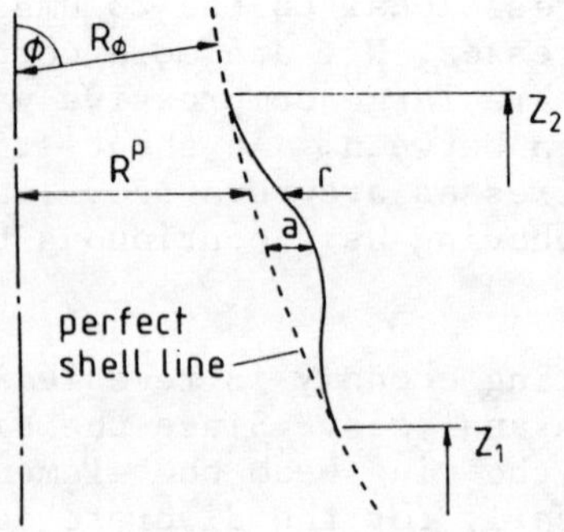

Figure 17: Details of axisymmetric imperfections

modelled using doubly curved ring elements. The shape of the imperfection would be given by

$$r(Z) = \frac{a}{2}\left[1 - \cos\frac{2\pi(Z-Z_1)}{(Z_2-Z_1)}\right] \tag{14}$$

and that of the imperfect shell by

$$R(Z) = R^P(Z) + r(Z) \tag{15}$$

where R^P is the horizontal radius of the perfect shell. Figure 17 also shows the parameters ϕ and R_ϕ which are required to define the element geometry. These are found from

$$\left.\begin{aligned} \cos\phi &= \frac{dR}{dZ} = \frac{dR^P}{dZ} + \frac{dr}{dZ} \\ \frac{1}{R_\phi} &= -\frac{d^2R}{dZ^2}\Big/\left\{1 + \left(\frac{dR}{dZ}\right)^2\right\}^{3/2} \end{aligned}\right\} \tag{16}$$

where $\dfrac{d^2R}{dZ^2} = \dfrac{d^2R^P}{dZ^2} + \dfrac{d^2r}{dZ^2}$.

The various derivatives in equation (16) can be determined from the separate geometries of the perfect shell and the imperfection. Similarly a localised 'patch' imperfection, as in Figure 18, would be defined by

$$r = \frac{a}{4}\left[1 - \cos\frac{2\pi(Z-Z_1)}{(Z_2-Z_1)}\right]\left[1 - \cos\frac{2\pi(\theta-\theta_1)}{(\theta_2-\theta_1)}\right] \tag{17}$$

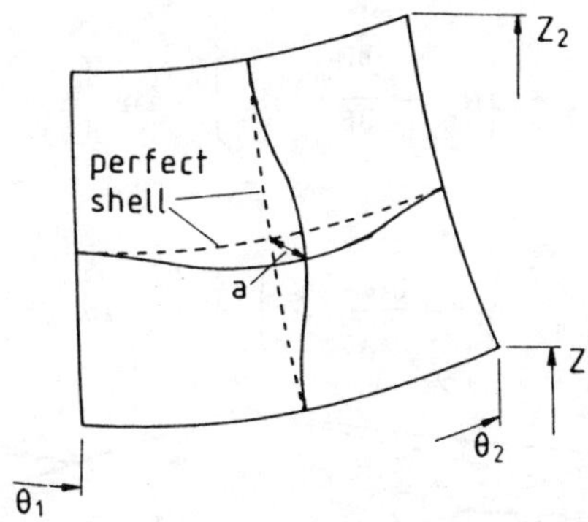

Figure 18: Details of patch imperfection

Either axisymmetric or general shell elements can be used to follow an axisymmetric imperfection. However, the general shell elements must be used for the patch imperfection. When using flat shell elements only the triangular elements can be made to follow the patch geometry. Gould et al [10] have described a mixed formulation in which the perfect part of the shell is described by ring elements while the imperfection area is modelled by general shell elements.

5.3.3 Equivalent load approach

Calladine [11] first showed that it is possible to model the effect of the imperfection by an equivalent additional load applied to the perfect shell. This has the advantage in cooling tower analysis that even patch imperfections can be modelled with the more economic semi-analytical technique.

The applied loads cause a system of stress resultants in the imperfect shell. Since the imperfect shell is inclined to the perfect shell, the membrane stress resultants of the imperfect shell have components tangential and normal to the perfect shell. The tangential components are membrane stress resultants in the perfect shell. The normal components can be considered as additional normal forces on the perfect shell. Applying these forces to the perfect shell will change the stress resultants and is equivalent to analysing the imperfect shell.

Figure 19 shows the meridional membrane stress resultants in an element of shell. Resolving normal to the perfect shell gives

$$\Delta F_{n\phi} = \left[- N_\phi \sin \alpha + \left(N_\phi + \frac{\partial N_\phi}{\partial S_\phi} \delta S_\phi\right) \sin \left(\alpha + \frac{\partial \alpha}{\partial S_\phi} \delta S_\phi\right)\right] \delta S_\theta$$

$$= \left[- N_\phi \alpha + N_\phi \left(\alpha + \frac{\partial \alpha}{\partial S_\phi} \delta S_\phi\right) + \frac{\partial N_\phi}{\partial S_\phi} \alpha \, \delta S_\phi\right] \delta S_\theta$$

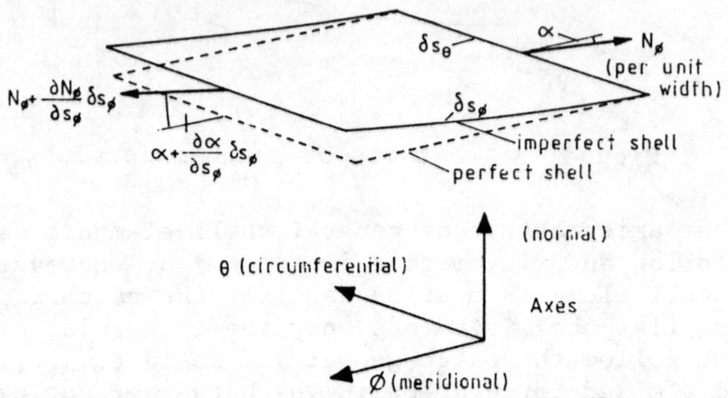

Figure 19: Meridional membrane stress resultants in the imperfect shell

assuming α is a small angle and ignoring higher order terms. Thus

$$\Delta F_{n\phi} = \left| N_\phi \frac{\partial \alpha}{\partial S_\phi} + \frac{\partial N_\phi}{\partial S_\phi} \alpha \right| \delta S_\phi \, \delta S_\theta$$

Since $\alpha \cong \tan \alpha = \dfrac{\partial r}{\partial S_\phi}$

$$\Delta F_{n\phi} = \left| N_\phi \frac{\partial^2 r}{\partial S_\phi^2} + \frac{\partial N_\phi}{\partial S_\phi} \frac{\partial r}{\partial S_\phi} \right| \delta S_\phi \, \delta S_\theta$$

$$q_{n\phi} = \frac{\Delta F_{n\phi}}{\delta S_\phi \delta S_\theta} = N_\phi \frac{\partial^2 r}{\partial S_\phi^2} + \frac{\partial N_\phi}{\partial S_\phi} \frac{\partial r}{\partial S_\phi} \qquad (18)$$

$q_{n\phi}$ is the normal pressure on the perfect shell due to the meridional stresses in the imperfect shell. Similar approaches for the circumferential and in-plane shear stress resultants (remembering the equality of cross-shears in thin shells) gives two further normal pressure terms. Adding all these terms gives the total normal pressure due to the imperfection.

$$q_n = N_\phi \frac{\partial^2 r}{\partial S_\phi{}^2} \quad {}_{\phi\theta} \frac{\partial^2 r}{\partial S_\phi \partial S_\theta} + N_\theta \frac{\partial^2 r}{\partial S_\theta{}^2}$$

$$+ \left(\frac{\partial N_\phi}{\partial S_\phi} + \frac{\partial N_{\phi\theta}}{\partial S_\theta}\right) \frac{\partial r}{\partial S_\phi} + \left(\frac{\partial N_\theta}{\partial S_\theta} + \frac{\partial N_{\phi\theta}}{\partial S_\phi}\right) \frac{\partial r}{\partial S_\theta} \tag{19}$$

The last two terms are small compared to the others and are usually ignored. The derivatives $\partial^2 r/\partial S_\phi^2$, etc., can be found directly from equation (14) or equation (17), but an iterative approach is required to find the stress resultants.

Initially the stress resultants in the perfect shell are found, ignoring the imperfection. These are then used to give a first approximation to q_n.

$$q_n^1 = N_\phi^P \frac{\partial^2 r}{\partial S_\phi^2} + 2N_{\phi\theta}^P \frac{\partial^2 r}{\partial S_\phi \partial S_\theta} + N_\theta^P \frac{\partial^2 r}{\partial S_\theta^2} \tag{20}$$

where the superscript P refers to the perfect shell.

When this load is applied to the shell it causes changes in the stress resultants, n_ϕ^1 etc., and m_ϕ^1 etc., so that

$$\begin{aligned} N_\phi^1 &= N_\phi^P + n_\phi^1 & M_\phi^1 &= M_\phi^P + m_\phi^1 \\ N_{\phi\theta}^1 &= N_{\phi\theta}^P + n_{\phi\theta}^1 & M_{\phi\theta}^1 &= M_{\phi\theta}^P + m_{\phi\theta}^1 \\ N_\theta^1 &= N_\theta^P + n_\theta^1 & M_\theta^1 &= M_\theta^P + m_\theta^1 \end{aligned} \tag{21}$$

The changes in the stress resultants can be used to find a second order load

$$q_n^2 = n_\phi^1 \frac{\partial^2 r}{\partial S_\phi^2} + 2n_{\phi\theta}^1 \frac{\partial^2 r}{\partial S_\phi \partial S_\theta} + n_\theta^1 \frac{\partial^2 r}{\partial S_\theta^2} \tag{22}$$

When this is applied to the shell it gives further changes in the stress resultants n_ϕ^2 etc. The process can be continued until convergence occurs.

For an axisymmetric imperfection

$$\frac{\partial^2 r}{\partial S_\phi \partial S_\theta} = \frac{\partial^2 r}{\partial S_\theta^2} = 0$$

so that

$$q_n = N_\phi \frac{\partial^2 r}{\partial S_\phi^2} \tag{23}$$

The only significant changes in the membrane stress resultants caused by this force are in N_θ, the circumferential component. Since n_ϕ^1 is insignificant only one cycle of iteration is required.

It has been shown [12] that the only limitation on the equivalent load method is that the amplitude of the imperfection must be small compared to the meridional and circumferential extent of the imperfection.

Standard techniques give the Fourier series constants for the equivalent loads in equations (29) and (22) [6]. For an axisymmetric imperfection the same number of harmonics as for the original applied load are required but at least 40 harmonics are required for the patch imperfection.

5.3.4 Results of imperfection analyses

Figure 20 shows the changes in stress resultants caused by an axisymmetric imperfection (amplitude 300mm, from elevation 47.9m to elevation 60.7m) in the shell of Figure 1. It can be seen that the various methods show good agreement for the membrane stress resultants but less good agreement for the lower order bending stress resultants. The differences between the elements themselves will contribute to these discrepancies.

Figure 21 shows results from a patch imperfection (amplitude and elevation as for the axisymmetric imperfection, angular extent 54° centred about the 72° meridian). The changes in stress resultants presented are at the 67.5° meridian. It can be seen that the changes from the second order load are small compared to those from the first order load so that it would be sufficiently accurate to use the first order load only. The figure also shows that using a pair of symmetrically placed imperfections had no effect on the results, whilst the amount of computation was halved.

Tables 4 and 5 give comparisons of the computer times required for the various methods. Clearly the semi-analytical method uses less computer time than the general shell elements. With axisymmetric imperfections it is better to follow the geometry of the imperfections.

5.4 Concrete Cracking

Since many cooling tower shells are cracked it is reasonable to try to predict crack formation. Milford and Schnobrich [7] have used a 9-node shell element to attempt this. They used a layered material model with smeared layers of reinforcement. Concrete cracking was predicted by a principal stress criterion. Cracks were allowed to rotate after formation as defined by the direction of the principal

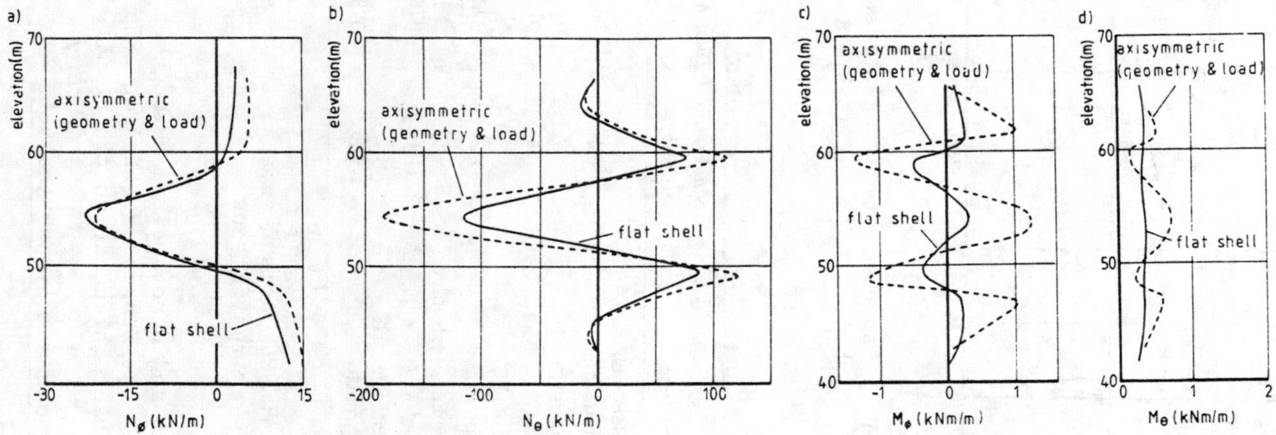

Figure 20: Stress resultants in shell with axisymmetric imperfection

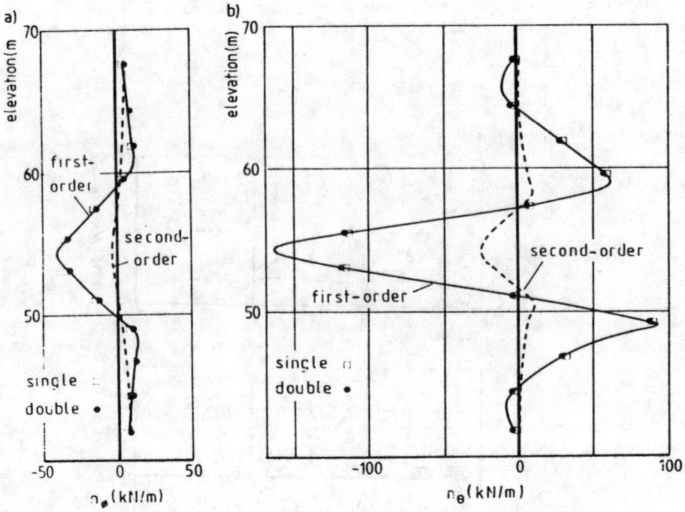

Figure 21: Changes in membrane stress resultants, shell with local imperfection

Type of analysis	*Number of elements*	*Total degrees of freedom*	*Computer time (ICL 2970 units)*
Axisymmetric elements, geometric approach	34	140	220
Axisymmetric elements, equivalent load	34	140	320
Flat triangle elements, geometric approach	1156	3780	1600
Flat triangle elements, first-order equivalent load	1156	3780	2300
Flat quadrilateral elements, geometric approach	578	3780	3200

Table 4: Comparison of analyses - axisymmetric imperfection

tensile strain and were smeared throughout the element. It was shown that the tension stiffening effect of the concrete had a significant influence on the results of the analysis.

5.5 Buckling and Non-linear Analysis

There has been a considerable amount of work on the application of finite elements to the buckling and non-linear analysis of cooling towers. In non-linear analysis by the semi-analytical method the individual harmonics do not

Type of analysis	Number of elements	Total degrees of freedom	Computer time (ICL 2970 units)
Axisymmetric elements, single imperfection, first-order equivalent load	34	140	1200
Axisymmetric elements, double imperfections, first-order equivalent load	34	140	700
Axisymmetric elements, double imperfections, second-order equivalent load	34	140	120
Flat triangle elements, geometric approach	1156	3780	1600
Flat triangle elements, first-order equivalent load	1156	3780	2300

Table 5: Comparison of analysis and patch imperfection

decouple. Thus the economy of the method is seriously eroded. Due to the amount of material on this aspect of cooling tower analysis the reader is referred to the Proceedings of the 2nd International Symposium on Natural Draught Cooling Towers[13].

6. CONCLUSIONS

The finite element method has been applied in a variety of ways to the analysis of cooling tower shells. This paper has outlined the possibilities, some in detail. A hyperbolic cooling tower shell is, of course, a shell of revolution. Consequently the methods described can be applied to any shell of revolution, such as a cylinder or a cone-toroid water tower.

7. REFERENCES

1. Martin, D. W. and Scriven, W. E. "The calculation of membrane stresses in hyperbolic cooling towers", Proc. Institution of Civil Engineers, 503-514, 1961.

2. Kemp, K. O. and Croll, K. O. "The role of geometric imperfections in the collapse of a cooling tower", The Structural Engineer, 54, 33-37, 1976.

3. Percy, J. H., Pian, T. H. H., Klein, S. and Navaratna, D. R. "Application of matrix displacement method to linear elastic analysis of shells of revolution", AIAA Journal, 3, 2138-45, 1965.

4. Niku, S. M. "Finite Element analysis of linear and geometrically non-linear hyperbolic cooling tower shells with various boundary conditions and geometric imperfections". Ph.D. Thesis, University of Southampton, 1984.

5. Moy, S. S. J. and Niku, S. M. "A new curved axisymmetric element for thin shell analysis", To be published.

6. Moy, S. S. J. and Niku, S. M. "Finite element techniques for the analysis of cooling tower shells with geometric imperfections", Thin-Walled Structures, 1, 239-263, 1983.

7. Milford, R. V. and Schnobrich, W. C. "The effect of cracking on the ultimate load of reinforced concrete cooling towers", in Natural Draught Cooling Towers, Proceedings of the 2nd International Symposium, (eds. P. L. Gould, W. B. Krätzig, I. Mungan and U. Wittek), Springer-Verlag, 319-332, 1984.

8. Moy, S. S. J. and Niku, S. M. "A direct method for the stress analysis of the ring beam and shell of a cooling tower". Thin-Walled Structures, 2, 129-151, 1984.

9. Merrylees, S. H. "Bending analysis of cooling towers - BACT02", Central Electricity Generating Board Report RD/C/N364, 1969.

10. Gould, P. L., Han, K. J. and Tong, G. S. "Analysis of hyperbolic cooling towers with local imperfections". in Natural Draught Cooling Towers, Proceedings of the 2nd International Symposium, (eds. P. L. Gould, W. B. Krätzig, I. Mungan and U. Wittek), Springer-Verlag, 397-411, 1984.

11. Calladine, C. R. "Structural consequences of small imperfections in elastic thin shells of revolution", Int. J. Solids and Structures, 8, 679-697, 1972.

12. Moy, S. S. J. and Niku, S. M. "The equivalent load method for the analysis of cooling towers with geometric imperfections", To be published.

13. "Natural Draught Cooling Towers, Proceedings of the 2nd International Symposium, Ruhr-Universität Bochum, Germany, September 5-7, 1984", (eds. P. L. Gould, W. B. Krätzig, I. Mungan and U. Wittek).

CHAPTER 10

COMPUTER-AIDED DESIGN FOR SERVICEABILITY OF PRESTRESSED AND NON-PRESTRESSED REINFORCED CONCRETE FRAMES

A. Ghali and M.M. Elbadry

Professor of Civil Engineering and Ph.D. Candidate
The University of Calgary, Alberta, Canada T2N 1N4

SUMMARY

Equations are presented for the analysis of stress and strain in a reinforced concrete cross section with or without prestressing accounting for the effects of creep and shrinkage of concrete and relaxation of prestressed steel. The prestressing may be of any magnitude varying from zero, allowing cracking to full prestressing eliminating cracks. The equations are employed in the analysis of continuous concrete frames for which the prestressing and the external loads are introduced in stages. The effect of cracking on the reactions and internal forces is analyzed by an iterative procedure.

Reference is made to an available computer program which performs the analysis. The equations and the program are simple and are easy to apply in practice. It is suggested that such a program can be routinely employed in the check of design of reinforced concrete structures for serviceability requirements using a micro-computer.

NOTATION

A	area
B	first moment of area of transformed section
b and b_w	width of flange and web of a T-section, respectively.
c	depth of compression zone in a fully-cracked section
{D} and {D*}	displacement vectors
d	distance below the top fibre of a section
E, $\bar{E}$	modulus of elasticity and age-adjusted elasticity modulus

$\{F\}$ and $\{F^*\}$	vectors of fixed-end forces
$[f]$	flexibility matrix
f_{ct}	strength of concrete in tension
$[H]$	transformation matrix
h_f and h	depth of flange and total depth of section, respectively
I	moment of inertia of transformed section
M	bending moment
N	normal force
$[S]$	stiffness matrix
s	mean crack spacing
t	time
w	mean crack width
α	modular ratio of steel
β_1 and β_2	coefficients 0.5 or 1 as specified below Eq. 30
γ	slope of stress diagram
Δ	increment or decrement
ε	normal strain
χ	aging coefficient
ϕ	creep coefficient
σ	stress
$\Delta\sigma_{pr}$, $\Delta\bar{\sigma}_{pr}$	intrinsic and reduced relaxations of prestressed steel
ζ	interpolation coefficient

Subscripts

c, ps, ns	concrete, prestressed and non-prestressed steel
cs	shrinkage of concrete
O	reference point
o	initial time
s	steel

1. INTRODUCTION

The stresses and the deformations of a reinforced concrete structure are affected by creep and shrinkage of concrete and relaxation of prestressed steel. Furthermore, when the strength of concrete in tension is exceeded, cracking occurs resulting in redistribution of stresses and important changes in deformations. Also, if the structure is statically indeterminate, changes in the reactions and in the internal forces result.

The present chapter is concerned with calculation of the instantaneous and time-dependent changes in stress and strain distributions in any cross section of a reinforced concrete frame with or without prestressing. The analysis gives also the displacements and an estimate of the average crack width. These are the analyses commonly required in a check of design for serviceability. Checks for ultimate strength requirements are beyond the scope of this chapter.

In current practice, the forces due to prestressing are treated as external forces applied on a plain concrete structure. The time-dependent change in prestressing force, commonly referred to as the prestress loss is estimated and its effect is treated in the same way as the initial prestressing force. The changes in prestress loss from section to section are commonly ignored. In most prestressed structures, a large amount of non-prestressed steel is present. The time-dependent changes in the prestressed and the non-prestressed steels can be of the same order of magnitude and thus both changes must be accounted for.

Cracking substantially reduces the flexural rigidity of a reinforced concrete section and hence results in change in the internal forces in a statically indeterminate structure. However, this change is usually ignored in practice because of complexity of the analysis.

With the aid of a computer, it is possible to perform with much more accuracy the computations involved in the check of design of serviceability. The approximate estimate of the time-dependent prestress loss is avoided. Instead, the conditions of equilibrium of forces and compatibility of strains in the concrete and steel in any section are employed to determine the changes in strain distribution and the corresponding changes in forces in each of these components. The input data for prestressing is simply the initial prestressing force magnitudes and locations at various sections. When post-tensioning is employed, the value of the initial prestressing force required in the data is equal to the jacking force less the loss due to friction and anchor set. It is not difficult to have this loss calculated by the computer, but the equations for this are not given here. The so-called balancing forces exerted on concrete wherever a prestressing tendon changes direction need not be calculated by the designer, as these are accounted for automatically by the analysis. Also, the effects of cracking on the reactions and the internal forces in statically indeterminate structures are accounted for by iterative computations.

For practical considerations, the analysis must be simple to code and use with a small computer. The assumptions adopted to achieve this goal are stated below.

2. ASSUMPTIONS

The presentation here is limited to plane frames for which the dimensions and the reinforcements and the parameters concerning the material properties are given. The loads including the prestressing are introduced in stages. The analysis is intended to give the reactions, the displacements and the internal forces and the strain and stress distributions at any section.

Shear deformations are ignored and cracking is assumed to occur due to bending only. Plane cross sections are assumed to remain plane during deformation. The reinforcements are assumed to remain within the elastic range.

The instantaneous strain and creep of concrete are assumed proportional to the applied stress; thus,

$$\Delta\varepsilon_c(t) = \frac{\Delta\sigma_c(t_o)}{E_c(t_o)}\,[1 + \phi(t,t_o)] \tag{1}$$

where $\Delta\varepsilon_c(t)$ is the total strain, instantaneous plus creep, due to a stress increment $\Delta\sigma_c(t_o)$ introduced at time t_o and sustained without change to time t; E_c is the modulus of elasticity of concrete at time t_o; $\phi(t,t_o)$ is the creep coefficient, which is the ratio of creep to the instantaneous strain.

A stress increment $\Delta\sigma_c(t,t_o)$ introduced gradually from zero at t_o to its full value at time t produces at time t a strain given by:

$$\Delta\varepsilon_c(t,t_o) = \frac{\Delta\sigma_c(t,t_o)}{E_c(t_o)}\,[1 + \chi\phi(t,t_o)] \tag{2}$$

This equation may be rewritten:

$$\Delta\varepsilon_c(t,t_o) = \frac{\Delta\sigma_c(t,t_o)}{\bar{E}_c(t,t_o)} \tag{3}$$

where $\bar{E}_c(t,t_o)$ is the age-adjusted modulus of elasticity of concrete

$$\bar{E}_c(t,t_o) = \frac{E_c(t_o)}{1 + \chi\phi(t,t_o)} \tag{4}$$

$\chi = \chi(t,t_o)$ is the aging coefficient, usually between 0.6 and 0.9; its value can be read from graphs or tables [1, 6].

The analysis is linear when cracking does not occur. Non-linearity occurs only when the tensile strength of concrete is exceeded producing cracking. After cracking, concrete in tension is ignored.

The concrete in tension contributes to some extent to the stiffness of a member. To account for this contribution, which is referred to as the tension stiffening, the strain distribution in a section is determined with the concrete in tension ignored and once more with cracking ignored. This gives an upper and a lower bound to the strain. The mean strain, which is the value used to calculate the displacements is determined by interpolation between these two bounds.

The value $\Delta\varepsilon_{cs}(t,t_o)$ is assumed constant parameter for each member; it represents the free (unrestrained) shrinkage of concrete during the period t_o to t.

The intrinsic relaxation value, $\Delta\sigma_{pr}$, usually given by suppliers of prestressed steel, represents the loss in tension in a tendon stretched between two fixed points. When the initial stress in prestressed steel is 0.5 its ultimate strength, the value of intrinsic relaxation is very small but increases rapidly as the initial tension approaches the ultimate strength. In concrete, the loss in tension due to creep and shrinkage of concrete has a similar effect on the amount of relaxation as a reduction in the initial tension. Thus, a reduced relaxation value $\Delta\bar{\sigma}_{pr}$ should be used in design of concrete structures. The value $\Delta\bar{\sigma}_{pr}$ equals the intrinsic relaxation multiplied by a reduction factor (close to 0.7; see Ref. 8).

3. SIGN CONVENTION

A normal tensile force, N, a tensile stress, σ and the corresponding strain, ε are positive. A bending moment, M producing tension at the bottom fibre and the corresponding curvature, ψ and slope, γ of the stress distribution diagram are positive. The symbol Δ indicates a change in value; a positive Δ represents an increase. Thus, the symbols $\Delta\varepsilon_{cs}$ and $\Delta\sigma_{pr}$ representing the changes in the strain in concrete due to shrinkage and in the tension in prestressed steel due to relaxation are always negative quantities.

4. INSTANTANEOUS STRAIN

Figure 1a shows a typical cross section of a reinforced concrete member. Introduce at any instant a normal force ΔN at an arbitrary reference point O combined with a bending moment ΔM. The change in strain due to these forces is given

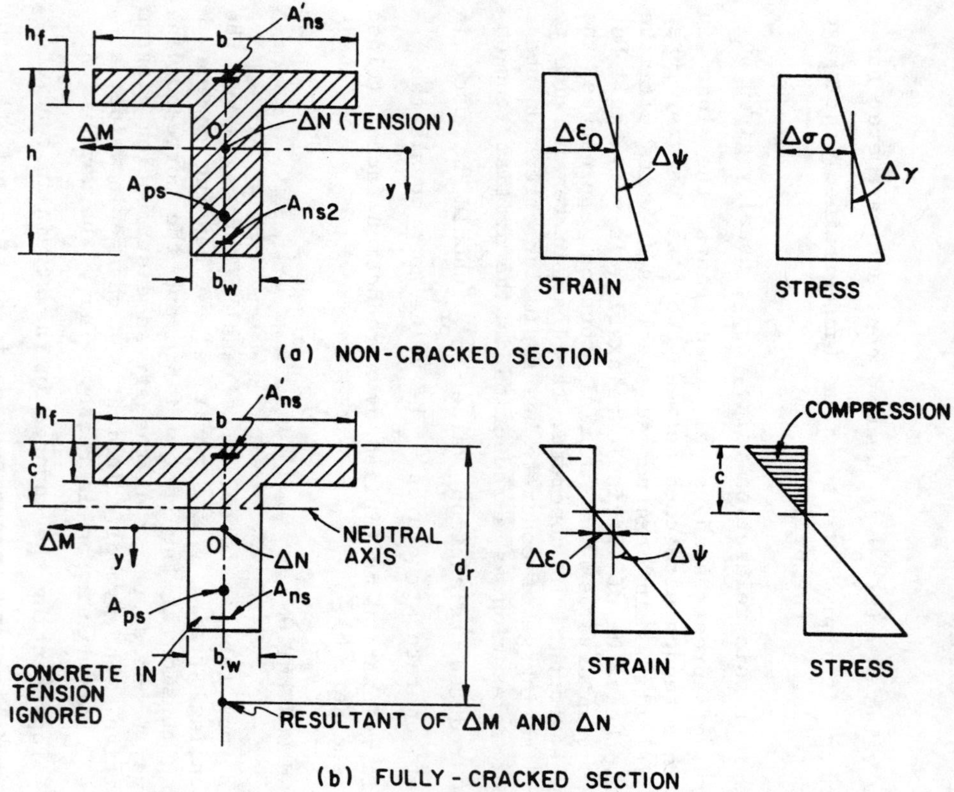

Fig. 1 Symbols and their positive sign convention

by:

$$\begin{Bmatrix} \Delta\varepsilon_0 \\ \Delta\psi \end{Bmatrix} = \frac{1}{E_c(A\,I - B^2)} \begin{bmatrix} I & -B \\ -B & A \end{bmatrix} \begin{Bmatrix} \Delta N \\ \Delta M \end{Bmatrix} \tag{5}$$

where $\Delta\varepsilon_0$ is the change in strain at 0 and $\Delta\psi$ is the slope of strain diagram, or the curvature; A, B and I are the area, its first and second moments about an axis through 0 of a transformed section composed of the area of concrete plus the area of the reinforcements, each multiplied by its modulus of elasticity divided by E_c, with E_c being the modulus of elasticity of concrete at the instant considered.

If 0 is chosen at the centroid, B = 0 and Eq. 5 will take the simpler form:

$$\begin{Bmatrix} \Delta\varepsilon_0 \\ \Delta\psi \end{Bmatrix} = \frac{1}{E_c} \begin{Bmatrix} \frac{\Delta N}{A} \\ \frac{\Delta M}{I} \end{Bmatrix} \tag{6}$$

Equation 5 will be preferred here, because the properties of the transformed section and the position of its centroid, change with the change of E_c and when cracking occurs. Thus, the reference point 0 for any section and a normal axis through 0 are maintained constant through all steps of the analysis.

At any fibre of coordinate y (Fig. 1a), the strain and stress are:

$$\Delta\varepsilon = \Delta\varepsilon_0 + y\,\Delta\psi \tag{7}$$

$$\Delta\sigma = E\,\Delta\varepsilon \tag{8}$$

Subscripts c, ps or ns may be employed with the symbols E, σ and ε to refer to concrete, prestressed or non-prestressed steel, respectively.

Equations 5 to 8 are applicable to cracked or non-cracked sections. In the former case, $E_c = 0$ for the concrete in the tension zone. Thus, the use of the equations should be preceded by determination of the depth c of the compression zone (Fig. 1b).

5. DEPTH OF COMPRESSION ZONE

Figure 1b shows the strain and stress distribution due

to forces {ΔN, ΔM} which produce cracking at the bottom fibre of a T-section. Here the concrete in tension is ignored and the section in this state is referred to as fully-cracked. As above, the force ΔN is assumed at the reference point 0. The depth c is determined by solution of the following cubic equation [6, 9]:

$$c^3 + a_1c^2 + a_2c + a_3 = 0 \tag{9}$$

where

$$a_1 = -3d_r \tag{10}$$

$$a_2 = -\frac{6}{b_w}\{h_f(b - b_w)(d_r - \frac{2}{3}h_f) + \Sigma[A'_s(\alpha_s - 1)(d_r - d'_s)] + \Sigma[\alpha_s A_s(d_r - d_s)]\} \tag{11}$$

$$a_3 = \frac{6}{b_w}\{\frac{h_f^2}{2}(b - b_w)(d_r - \frac{h_f}{2}) + \Sigma[A'_s d'_s(\alpha_s - 1)(d_r - d'_s)] + \Sigma[\alpha_s A_s d_s(d_r - d_s)]\} \tag{12}$$

with

$$\alpha_s = \frac{E_s}{E_c} \tag{13}$$

and the summations performed over all steel layers; b and h_f are width and depth of the flange; b_w is the width of the web; d_r is the distance from the top fibre to the location of the resultant of ΔN and ΔM; A'_s and d'_s are cross section area and distance from top to a steel layer in the compression zone and A_s and d_s have the same meaning but for a layer in the tension zone. An additional subscript p or n may be used with s to indicate prestressed or non-prestressed steel.

When ΔN = 0, the depth c is determined by solving the quadratic equation:

$$c^2 + a_1c + a_2 = 0 \tag{14}$$

where

$$a_1 = \frac{2}{b_w}\{h_f(b - b_w) + \Sigma[A'_s(\alpha_s - 1)] + \Sigma(\alpha_s A_s)\} \tag{15}$$

$$a_2 = -\frac{2}{b_w}\{\frac{h_f^2}{2}(b - b_w) + \Sigma[A'_s d'_s(\alpha_s - 1)] + \Sigma(\alpha_s A_s d_s)\} \quad (16)$$

The equations presented in this section apply only when the top fibre is in compression and when the stress changes sign within the height of the section. This occurs when the resultant of ΔN and ΔM is compressive situated above the core of the transformed non-cracked section, or when the resultant is tensile situated below the core.

In the above equations it is assumed that the stress in concrete is zero before the introduction of ΔN and ΔM. In partially prestressed structures, the sections are often designed to have no cracking due to the combination of prestressing and sustained dead load and cracking is allowed only due to an additional transient live load. Let ΔN and ΔM be forces due to live load introduced at any instant on a section for which the stress distribution is known. This stress distribution may be defined by two values: σ_0 and γ representing the value of stress at the reference point 0 and the slope of the stress diagram.

The pair of forces $\{\Delta N, \Delta M\}$ must be partitioned into a decompression part to be applied on a non-cracked section and a remaining part on a fully-cracked section. The decompression part is given by:

$$\begin{Bmatrix} \Delta N \\ \Delta M \end{Bmatrix}_{\text{decompression}} = \begin{bmatrix} A & B \\ B & I \end{bmatrix} \begin{Bmatrix} -\sigma_0 \\ -\gamma \end{Bmatrix} \quad (17)$$

where A, B and I are properties of the transformed non-cracked section at the time of application of the live load. The strain due to the decompression forces

$$\begin{Bmatrix} \Delta\varepsilon_0 \\ \Delta\psi \end{Bmatrix}_{\text{decompression}} = -\frac{1}{E_c} \begin{Bmatrix} \sigma_0 \\ \gamma \end{Bmatrix} \quad (18)$$

The forces to be applied on a fully-cracked section are

$$\begin{Bmatrix} \Delta N \\ \Delta M \end{Bmatrix}_{\text{fully-cracked}} = \begin{Bmatrix} \Delta N \\ \Delta M \end{Bmatrix} - \begin{Bmatrix} \Delta N \\ \Delta M \end{Bmatrix}_{\text{decompression}} \quad (19)$$

The depth c depends upon the eccentricity $(\Delta M/\Delta N)_{\text{fully-cracked}}$ and may now be determined by solving Eq. 9.

The parameters $\varepsilon_O(t_o)$ and $\psi(t_o)$ are assumed known; they represent the instantaneous strain distribution due to prestressing and dead load introduced at time t_o. The change in strain due to creep and shrinkage of concrete and relaxation of prestressed steel during a period t_o to t is given by (derived from Eq. 5):

$$\begin{Bmatrix} \Delta\varepsilon_O(t,t_o) \\ \Delta\psi(t,t_o) \end{Bmatrix} = \frac{1}{\bar{E}_c(\bar{A}\,\bar{I} - \bar{B}^2)} \begin{bmatrix} \bar{I} & -\bar{B} \\ -\bar{B} & \bar{A} \end{bmatrix} \begin{Bmatrix} -\Delta N \\ -\Delta M \end{Bmatrix} \quad (20)$$

where $\bar{A}$, $\bar{B}$ and $\bar{I}$ are properties of the age-adjusted transformed section composed of the area of concrete plus the areas of the reinforcements, each multiplied by its modulus of elasticity and divided by $\bar{E}_c$, the age-adjusted elasticity modulus of concrete (Eq. 4). $\{\Delta N, \Delta M\}$ are forces required to restrain artificially the time-dependent deformations and may be calculated as the sum of the three components:

$$\begin{Bmatrix} \Delta N \\ \Delta M \end{Bmatrix} = \begin{Bmatrix} \Delta N \\ \Delta M \end{Bmatrix}_{creep} + \begin{Bmatrix} \Delta N \\ \Delta M \end{Bmatrix}_{shrinkage} + \begin{Bmatrix} \Delta N \\ \Delta M \end{Bmatrix}_{relaxation} \quad (21)$$

where

$$\begin{Bmatrix} \Delta N \\ \Delta M \end{Bmatrix}_{creep} = -\bar{E}_c\,\phi \begin{bmatrix} A_c & B_c \\ B_c & I_c \end{bmatrix} \begin{Bmatrix} \varepsilon_O(t_o) \\ \psi(t_o) \end{Bmatrix} \quad (22)$$

$$\begin{Bmatrix} \Delta N \\ \Delta M \end{Bmatrix}_{shrinkage} = -\bar{E}_c\,\Delta\varepsilon_{cs} \begin{Bmatrix} A_c \\ B_c \end{Bmatrix} \quad (23)$$

where A_c, B_c and I_c being properties of concrete section without reinforcement, and

$$\begin{Bmatrix} \Delta N \\ \Delta M \end{Bmatrix}_{relaxation} = \Sigma \begin{Bmatrix} A_{ps} & \Delta\bar{\sigma}_{pr} \\ A_{ps}\,y_{ps} & \Delta\bar{\sigma}_{pr} \end{Bmatrix} \quad (24)$$

With the deformation restrained, a change in stress must develop given by:

$$\Delta\sigma_{restraint} = -\bar{E}_c[\phi\,\varepsilon_c(t_o) + \Delta\varepsilon_{cs}] \quad (25)$$

The time-dependent changes in stress in concrete and the

two types of reinforcement are:

$$\Delta\sigma(t,t_o) = \Delta\sigma_{restraint} + \bar{E}_c\ \Delta\varepsilon_c(t,t_o) \tag{26}$$

$$\Delta\sigma_{ns}(t,t_o) = E_{ns}\ \Delta\varepsilon_{ns}(t,t_o) \tag{27}$$

$$\Delta\sigma_{ps}(t,t_o) = \Delta\bar{\sigma}_{pr} + E_{ps}\ \Delta\varepsilon_{ps}(t,t_o) \tag{28}$$

For a reinforced concrete section without prestressing Eqs. 20-27 apply, with $A_{ps} = 0$.

The equations presented in this section may also be used to determine the time-dependent effects in a section subjected to forces which produce cracking at t_o. This involves the assumption that the depth c of the compression zone does not vary with time and the values $\bar{A}$, $\bar{B}$ and $\bar{I}$ are those of a fully-cracked section.

7. TENSION STIFFENING

The concrete in the tension zone contributes to some extent to the axial and flexural stiffnesses of a cracked member. For calculation of displacements, it is more accurate to use mean values of the axial strain and curvature rather than the values calculated for a fully-cracked section. The mean values are determined by interpolation between lower and upper bounds. The former is determined with cracking neglected and the latter is calculated ignoring the concrete in tension. The following empirical interpolation equation [2, 4] is adopted:

$$\begin{Bmatrix}\Delta\varepsilon_O\\ \Delta\psi\end{Bmatrix}_{mean} = (1-\zeta)\begin{Bmatrix}\Delta\varepsilon_O\\ \Delta\psi\end{Bmatrix}_{non\text{-}cracked} + \zeta\begin{Bmatrix}\Delta\varepsilon_O\\ \Delta\psi\end{Bmatrix}_{fully\text{-}cracked} \tag{29}$$

The interpolation coefficient ζ is given by:

$$\zeta = 1 - \beta_1\ \beta_2\ (\frac{f_{ct}}{\Delta\sigma_{max}})^2 \tag{30}$$

where $\Delta\sigma_{max}$ is the change in stress at the extreme tension fibre in a non-cracked section due to the forces $\{\Delta N,\ \Delta M\}_{fully\text{-}cracked}$ (Eq. 19); f_{ct} is the tensile strength of concrete; β_1 = 1 or 0.5 for high bond or plain reinforcing bars, respectively; β_2 = 1 or 0.5; the value 1 is to be used for first loading and 0.5 for the case when the load is applied in a sustained manner or in a large number of cycles.

The mean value of crack width at the level of a steel layer

$$w = \zeta \, s \, \Delta\varepsilon_{s \text{ fully-cracked}} \qquad (31)$$

where $\Delta\varepsilon_{s \text{ fully-cracked}}$ is the change in strain in steel calculated for a fully-cracked section; s is the mean spacing between cracks. Prediction of s can be made by an empirical equation [2, 4] which includes parameters related to bar spacing and diameter, concrete cover, bond quality and the reinforcement ratio.

The equations presented in this and in the earlier sections are concerned with the strains and stresses in individual sections due to instantaneous or time-dependent effects. The computer program CRACK [5] performs the analysis involved.

8. MEMBER STIFFNESS

Figure 2a shows a typical member of a plane frame with six degrees of freedom located at reference points 0_1 and 0_2 at the two end sections. Prevent rigid body motion by fixing the right end and generate a flexibility matrix [f] corresponding to the three coordinates at the left end. The elements on any column j of the matrix [f] are:

$$f_{1j} = -\int_o^{\ell} \varepsilon_{0j} \, dx \; ; \; f_{2j} = -\int_o^{\ell} \psi_j \, x \, dx \; ; \; f_{3j} = \int_o^{\ell} \psi_j \, dx \qquad (32)$$

where ε_{0j} and ψ_j are the strain at the reference axis $0_1 \, 0_2$ and the curvature evaluated at any section at distance x from 0_1 due to a unit force at coordinate j. The values of ε_0 and ψ needed for the numerical evaluation of the integrals are to be determined using Eq. 5 at a number of sections for which the geometry and cross section areas of reinforcements are given as data. For analysis of instantaneous displacements, use the modulus of elasticity of concrete and the properties of the transformed section at the time of application of the load. When the analysis is for the time-dependent changes during a period t_o to t, use the age-adjusted elasticity modulus $\bar{E}_c(t,t_o)$ and the properties of the age-adjusted transformed section; this gives the age-adjusted flexibility $[\bar{f}]$.

After cracking, the same Eq. 5 can be used to calculate

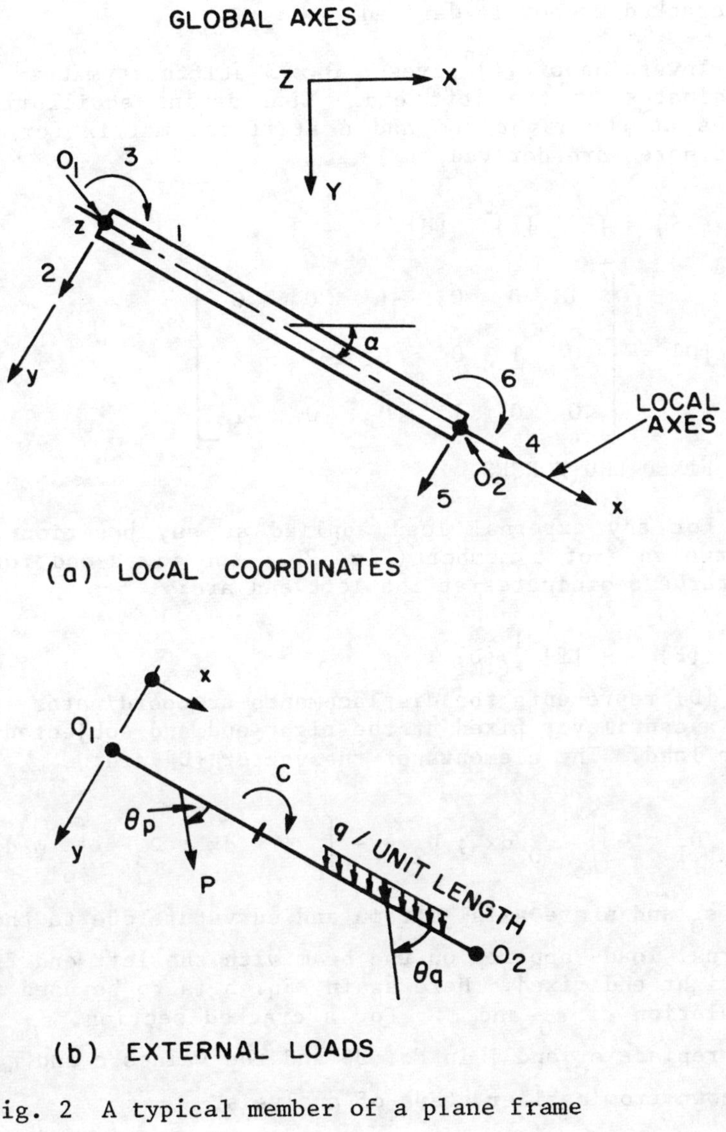

Fig. 2 A typical member of a plane frame

mean values ε_{Om} and ψ_m according to Eq. 29. This requires that the depth c of the compression zone and the interpolation coefficient ζ be known in advance; for this reason, an iterative procedure will be necessary. When ε_{Om} and ψ_m are used in Eq. 32, instead of ε_O and ψ, the flexibility of the cracked member is derived.

Inversion of [f] gives a 3 x 3 stiffness matrix for the coordinates at the left end. Considering equilibrium, the forces at the right end and a stiffness matrix for the six coordinates are derived:

$$[S] = [H]^T \; [f]^{-1} \; [H] \tag{33}$$

where

$$[H] = \begin{bmatrix} 1 & 0 & 0 & -1 & 0 & 0 \\ 0 & 1 & 0 & 0 & -1 & \ell \\ 0 & 0 & 1 & 0 & 0 & -1 \end{bmatrix}$$

9. FIXED-END FORCES

For any external load applied at any position between the two ends of a member (Fig. 2b), the fixed-end forces at the three coordinates at the left end are:

$$\{F\} = - \; [f]^{-1} \; \{D\} \tag{35}$$

Here {D} represents the displacements at coordinates 1, 2 and 3 of a cantilever fixed at the right end and subjected to the given load. The elements of the vector {D} are:

$$D_1 = - \int_o^{\ell} \varepsilon_O \, dx \; ; \; D_2 = - \int_o^{\ell} \psi \, x \, dx \; ; \; D_3 = \int_o^{\ell} \psi \, d\ell \tag{36}$$

Here ε_O and ψ are axial strain and curvature due to the external loads applied on the beam with the left end free and the right end fixed. Here again, Eq. 5 is to be used for calculation of ε_O and ψ. For a cracked section, ε_{Om} and ψ_m must replace ε_O and ψ in Eq. 36 and the values c and ζ must be known from earlier steps of analysis.

From equilibrium, the forces at the right end can be determined; the six fixed end forces may be expressed as:

$$\{F^*\} = [H]^T \; \{F\} + \left\{ \begin{array}{c} 0 \\ \hline R \end{array} \right\} \tag{37}$$

where {R} is a vector of the three reactions due to external loads applied on a cantilever fixed at the right end.

Equations 35-37 may be employed to calculate the changes in fixed-end forces for a period t_o to t due to creep, shrinkage and relaxation by substituting in Eq. 36 the values $\Delta\varepsilon_O(t,t_o)$ and $\Delta\psi(t,t_o)$ determined by Eq. 20.

10. STEPS OF ANALYSIS

The conventional displacement method of analysis [7] is applied to determine the instantaneous changes of displacements and internal forces due to an increment of external loads. The nodal displacements and internal forces at various sections existing before application of the load increment are assumed known. If these internal forces produce cracking at any section, the values c and ζ are also known. At the beginning of the analysis, before application of any load, the displacements and internal forces are zero; thus, c equals the full height of the section and $\zeta = 0$.

Increments of displacements, reactions or internal forces occur either instantaneously at the time of application of forces or prestressing or develop gradually during a specified period due to creep, shrinkage and relaxation. All increments, instantaneous or gradual, are determined by the same steps of analysis:

1. Calculate for each member the relative end displacements {D} and the fixed-end forces {F*} by Eqs. 35-37. The fixed-end forces for the members are assembled with the forces applied directly at the nodes (if any) and a conventional linear analysis is performed to determine increments of nodal displacements, reactions and internal forces. The values of c and ζ, needed here at any section, are based on the internal forces existing prior to the application of the new load increment.

2. Add the increment in internal forces to the existing values. When cracking occurs, c and ζ should be changed accordingly. Apply Eq. 5 (or Eqs. 5 and 29 for cracked sections) to determine the change in axial strain and curvature due to the increment in internal forces. Add this change to the existing values and use Eq. 36 to determine the displacements {D} of the left end of each member treated as a cantilever.

3. Add the increment in nodal displacements to the existing values and calculate, for each member, the relative displacement of the left end with respect to the right end:

$$\{D\} = [H]\{D^*\} \qquad (38)$$

where $\{D^*\}$ are the displacements, after addition of the increment, at the six coordinates at the ends (Fig. 2a).

4. Without cracking, the three displacements calculated by Eq. 38 for each member will be identical to the values calculated in step 2 and no further calculations are required. When this is not the case, calculate the difference in the three values for each member and use Eq. 35 and 37 to obtain a vector of residual fixed-end forces. When the residual forces are smaller than specified values the analysis may stop. Alternatively, the difference in $\{D\}$ may also be used in the convergence test.

5. The residual forces calculated for the individual members in step 4 are assembled and applied with sign reversed to the assembled structure. The stiffness matrix to be employed is based on the values c and ζ updated in step 2. This analysis gives increments in nodal displacements and internal forces.

6. Go to step 2.

The computer program CPF (Cracked Plane Frames, Ref. 3) includes the program CRACK as a subroutine and performs the analysis in the above steps. The program gives the instantaneous values and the time-dependent changes in the displacements, the reactions and the internal forces, the stress and strain in concrete and steel and the crack width at chosen sections. The program is simple and requires a small core storage and can be fitted to a micro-computer.

11. EXAMPLE

The computer program CPF is employed for the analysis of the continuous beam in Fig. 3. The beam has one post-tensioned parabolic tendon and the prestressing force introduced at time t_o is 2900 kN (assumed constant at all sections). The beam self-weight, 16kN/m comes into effect at the same time. A live load, 14 kN/m is introduced at time t. The beam dimensions and the cross section areas of prestressed and non-prestressed steels are indicated in the figure. Other data are:

$E_c(t_o) = 25$ GPa; $E_c(t) = 30$ GPa; $E_{ps} = 186$ GPa; $E_{ns} = 200$ GPa;

$\phi(t,t_o) = 3$; $\chi = 0.8$, $\Delta\varepsilon_{cs}(t,t_o) = -300\times10^{-6}$; $\Delta\bar{\sigma}_{pr} = -90$ MPa;

$\beta_1 = 1.0$; $\beta_2 = 0.5$; $f_{ct} = 2.5$ MPa; mean crack spacing,

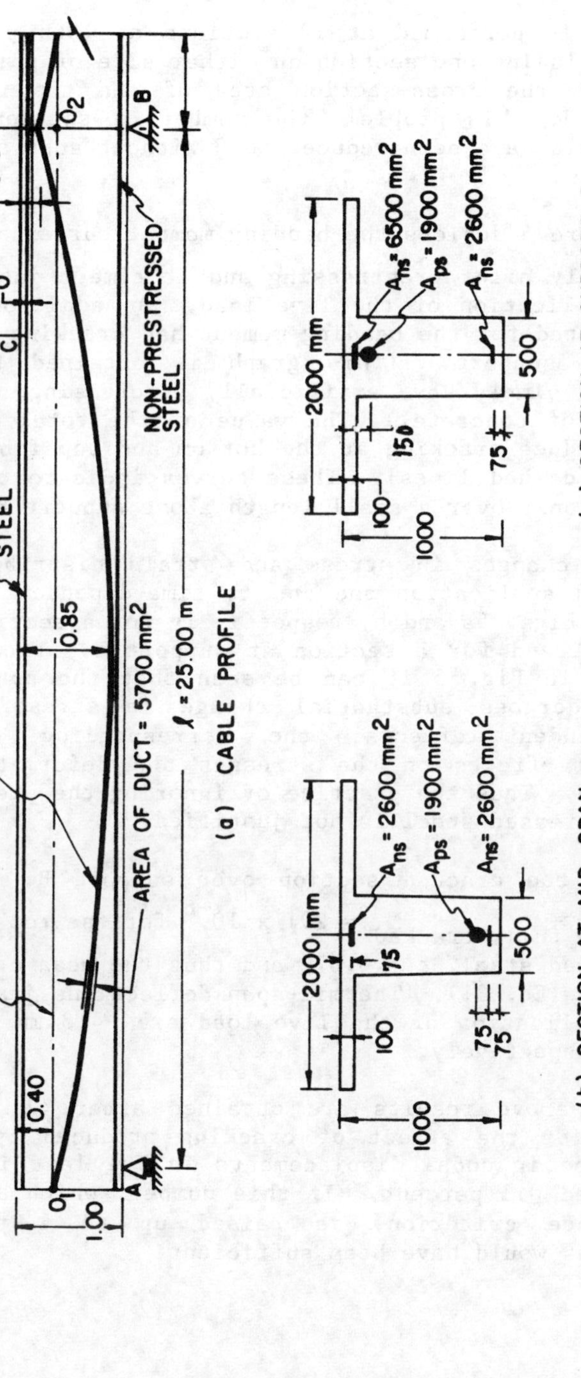

Fig. 3 Two-span continuous prestressed beam example

s = 300 mm.

Because of symmetry, the structure to be analyzed has only two nodes O_1 and O_2 and one member. The reference axis O_1O_2 is chosen 0.40 m below the top edge of the member. The analysis is performed at 12 sections at 10th points of the span, including one section on either side of point C where a change in the cross-section area of the top reinforcement occurs. In this problem, the number of sections between A and C could have been reduced to 3 without substantial change in results.

Figure 4 depicts the bending moment curves at time t_o immediately after prestressing and at time t just before and after application of the live load. An additional curve is also plotted for the bending moment had cracking not occurred over the support. (This graph is obtained by the same program, simply by artificially increasing the tensile strength of concrete.) The value of the total moment which would produce cracking at the bottom and top fibres are also plotted (dashed lines). These curves indicate that cracking occurred only over a small length above support B.

The changes in stress and strain distributions after each load application and due to time-dependent effects are shown in Figs. 5a and b, respectively for a section 10 m from support A and for a section at support B. From the stress diagrams in Fig. 5 it can be seen that the non-prestressed steel undergoes substantial change in stress due to the time-dependent effects; the corresponding forces have important effects on the stresses and deformations of the structure. Thus the practice of ignoring the presence of the non-prestressed steel is not justified.

For the cracked section over support B, the analysis gives $\Delta\varepsilon_{s\ \text{fully-cracked}} = 291 \times 10^{-6}$ for the top non-prestressed steel, $\zeta = 0.67$, and thus the mean crack width, w = 0.06 mm (Eq. 31). The mid-span deflections just before and after application of the live load are -4.3 mm (upward) and 10 mm, respectively.

The above results are obtained after 4 iterations to account for the effect of cracking produced by live load. The change in nodal displacements in the last iteration did not exceed 0.1 percent. If this number, which specifies the convergence criterion, is raised up to 1 percent, two iterations would have been sufficient.

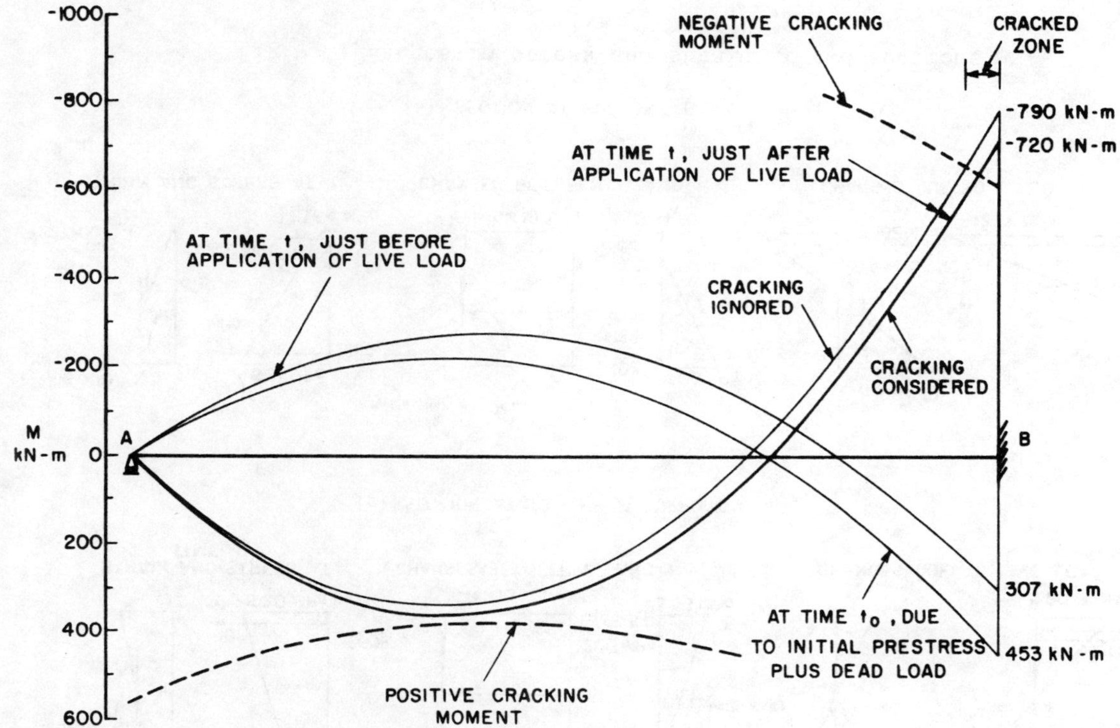

Fig. 4 Changes in bending moment due to the time-dependent effects and due to cracking.

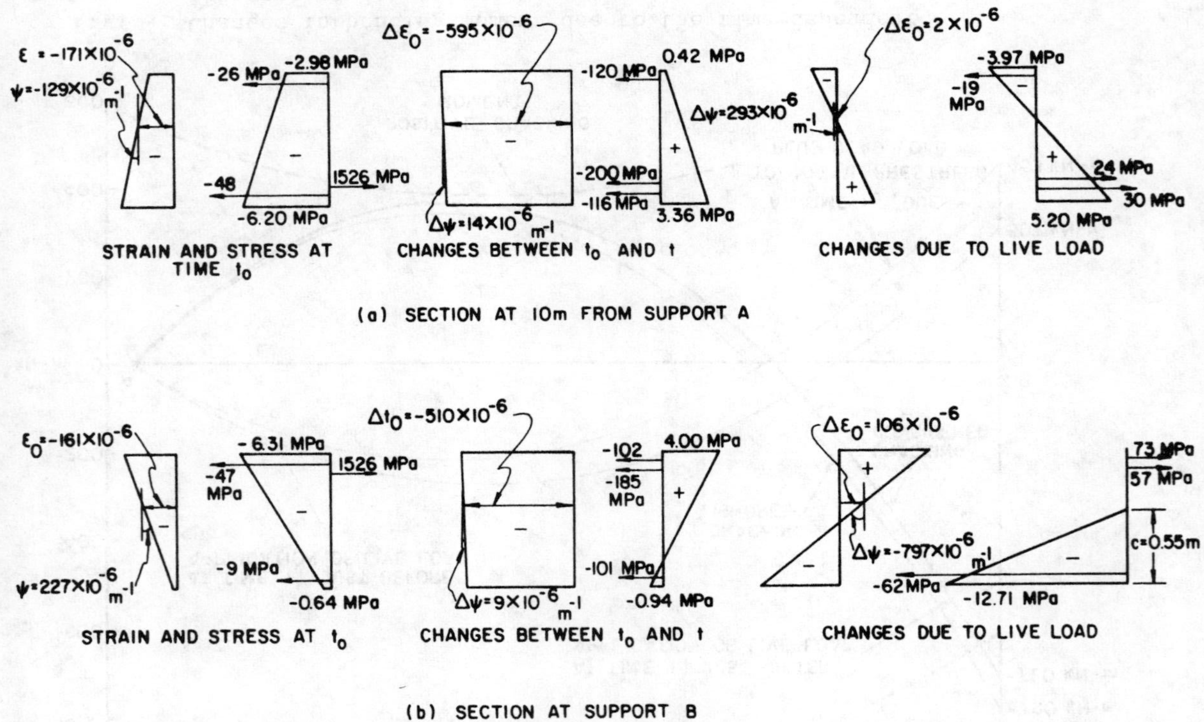

Fig. 5 Changes in stress and strain at two sections

12. CONCLUSION

For accurate prediction of stresses and deformations in reinforced concrete structures with or without prestressing, the effects of creep shrinkage, relaxation and cracking must be taken into account. In prestressed frames, the presence of non-prestressed reinforcement should not be ignored. The equations required for the analysis are presented. They are relatively simple but require repetitive calculations. Reference is made to a computer program which can perform the analysis for continuous plane frames, including the non-linear effects of cracking on the statically indeterminate forces. The program requires limited core storage and thus can be fitted to a micro-computer.

13. REFERENCES

1. Bazant, Z.P., "Prediction of concrete creep effects using age-adjusted effective modulus method", American Concrete Institute Journal, Proceedings Vol. 69, No. 4, pp. 212-217,April 1972.

2. Comité Euro-International du Béton (CEB) - Fédération Internationale de la Précontrainte (FIP), Model Code for Concrete Structures, (MC-78), CEB, 6 rue Lauriston, F-75116, Paris, France, 1978.

3. Elbadry, M.M. and Ghali, A., User's Manual and Computer Program CPF: Cracked Plane Frames, Research Report No. CE85-2, Department of Civil Engineering, The University of Calgary, Alberta, Canada, 1985.

4. Favre, R., Beeby, A.W., Falkner, H., Koprna, M. and Schiessel, P., Cracking and Deformations, CEB Design Manual, Swiss Federal Institute of Technology, Lausanne, Switzerland.

5. Ghali, A. and Elbadry, M.M., Manual for Computer Program CRACK, Research Report No. CE85-1, Department of Civil Engineering, The University of Calgary, Alberta, Canada, 1985.

6. Ghali, A. and Favre, R., Concrete Structures: Stresses and Deformations, Chapman and Hall, London and New York, approx. 400 pp. (in print).

7. Ghali, A. and Neville, A.M., Structural Analysis - A Unified Classical and Matrix Approach, 2nd Edition, Chapman and Hall, London, 1978.

8. Ghali, A. and Trevino, J., "Relaxation of Steel in Prestressed Concrete", Prestressed Concrete Institute

Journal, Proceedings Vol. 30, NO. 6, 16 pp., Nov.-Dec. 1985.

9. Korn, G.A. and Korn, T.M., Mathematical Handbook for Scientists and Engineers, McGraw-Hill Co., New York, 2nd Edition, 1968.

CHAPTER II

DYNAMIC ELASTO-PLASTIC ANALYSIS OF REINFORCED CONCRETE SLABS AND APPLICATION TO PLASTIC DESIGN OF SOME BUILDING STRUCTURES

M. Aufaure*, D. Chauvel**, Y. L'Huby**
Electricité De France :
* 6, quai Watier - 78400 Chatou (France)
** 12-14, avenue Dutrievoz - 69628 Villeurbanne Cedex (France)

SUMMARY

Some reinforced concrete structures in nuclear power plants must be dimensioned under accidental dynamic loadings. In order to take advantage of a possible plastic behaviour of materials, we have developed a specific program, Dalton, to compute slabs and walls.

In Dalton, the slabs, subjected to bending, are modelled with 3-node triangular plate elements. Dynamic elasto-plastic behaviour is computed by an explicit time-integration procedure and the Johansen's criterion is used, in combination with the associated normality law and a kinematic hardening. The boundary conditions are spring supports in deflection and rotation.

We have validated this program by testing a series of slabs under dynamic loading. By computing, we have found a good agreement with the tests.

Then we have used Dalton to compute some R.C. slabs submitted to an aircraft impact and other accidental shocks and we have inferred a simplified rule to dimension R.C. structures, under that kind of loading, by an easy static elastic calculation.

I - INTRODUCTION

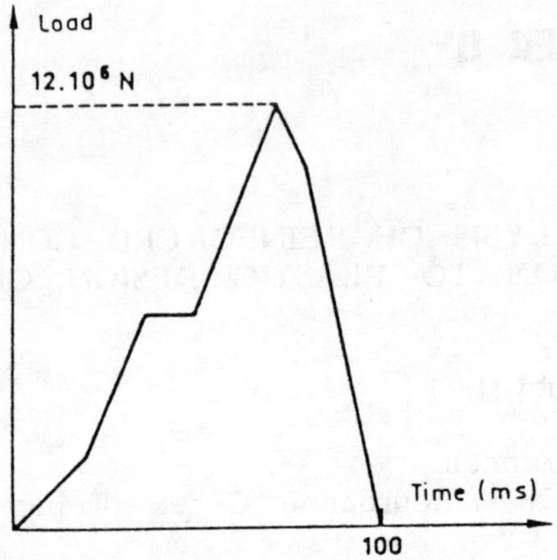

Fig. 1. Time history of the impact-load of a military aircraft (lear-jet)

Some R.C. structures in nuclear power plants must be dimensioned to support accidental actions such as : aircraft crashes (Fig. 1), pipe whipping, explosions or earthquakes.

We can reduce the construction costs significantly by taking advantage of the possible plastic behaviour of the structure under such loadings [1], [2].

We introduce subsequently : the computation program Dalton, the experimental tests which allowed us to validate it, and the simplified calculation rules derived from it.

2 - THE DALTON-PROGRAM

2.1 - Computation algorithm

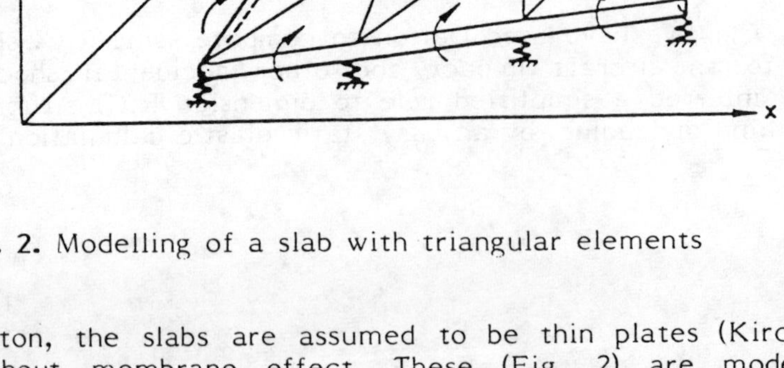

Fig. 2. Modelling of a slab with triangular elements

In Dalton, the slabs are assumed to be thin plates (Kirchoff theory) without membrane effect. These (Fig. 2) are modelled with nonconforming 3-node triangular finite elements [3, p. 244]. The integration over the element uses three Gauss points. The edge of the slab may have any shape. It bears upon spring supports

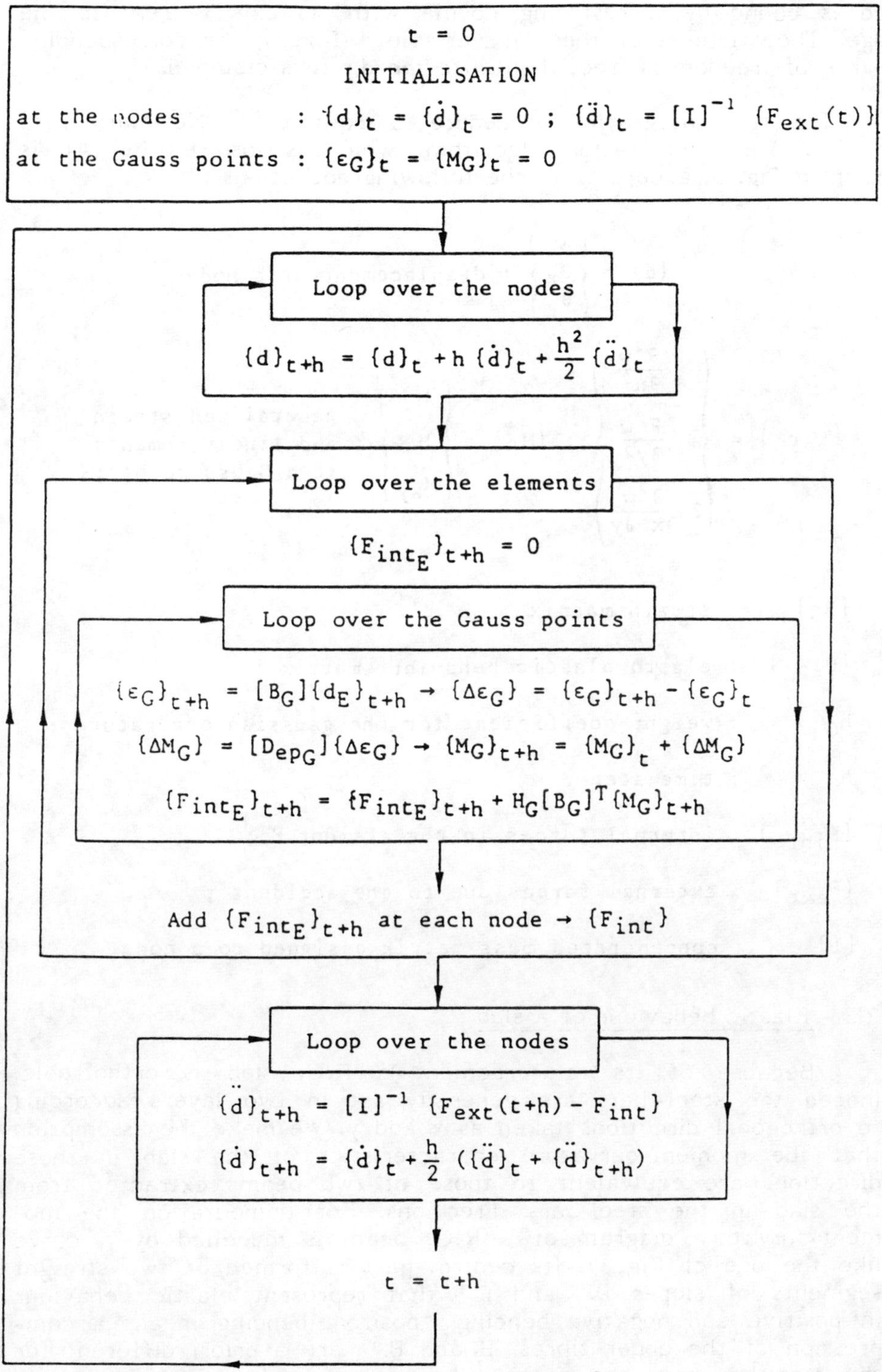

Fig. 3. Flow chart of the program.

and is bound by a restoring couple with its axis tangent to the edge. The stiffness of these is variable : if zero, the corresponding degree of freedom is free, if infinite the d.o.f. is clamped.

The time-integration procedure is explicit (β -Newmark with $\beta = 0$). The computation algorithm, which is inspired by [4], is shown in Fig. 3, according to the following definitions :

$$\{d\} = \begin{Bmatrix} w \\ \theta_x \\ \theta_y \end{Bmatrix} \text{ : displacement of a node ;}$$

$$\{\varepsilon_G\} = \begin{pmatrix} \dfrac{\partial^2 w}{\partial x^2} \\ \dfrac{\partial^2 w}{\partial y^2} \\ 2\,\dfrac{\partial^2 w}{\partial x\,\partial y} \end{pmatrix} \; ; \; \{M_G\} = \begin{Bmatrix} M_x \\ M_y \\ M_{xy} \end{Bmatrix} \text{ : generalized strain and linear moments at a Gauss point ;}$$

$[B_G]$: strain matrix ;

$[D_{ep_G}]$: elasto-plastic behaviour matrix ;

H_G : weight coefficient for the gaussian quadrature ;

h : time-step ;

$\{F_{int_E}\}$: internal forces in the element E ;

$\{F_{ext}\}$: external forces due to the accident ;

$[I]$: concentrated mass matrix assigned to a node.

2.2 - Elastic behaviour of a slab

Because of its reinforcement a R.C. slab is orthotropic. Indeed the steel-bars are generally set in two layers according to orthogonal directions taken as x and y. We make the assumption that the moment-curvature characteristics of the slab in these directions are equivalent to those of two beams extracted from the slab in the steel-bars directions. For computation the moment-curvature diagram of a R.C. beam is modelled by a curve like the one of Fig. 4. Its central part is formed of two straight segments of slopes P_e^+ and P_e^- that represent elastic behaviour in positive and negative bending : positive bending means a compression of the upper fibres. P_e^+ and P_e^- are a priori different for the steel-bars are not necessarily symmetrically disposed on both sides of the mid-plane and the cracks (Fig. 15) do not initiate on the upper and lower faces at the same absolute value of the moment.

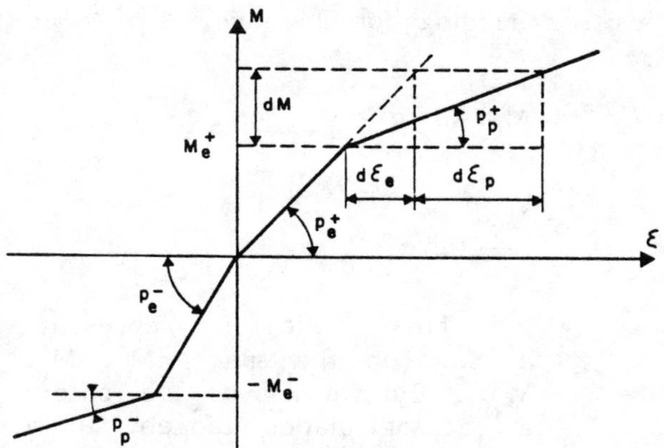

Fig. 4. Idealization of the moment-curvature diagram of a R.C. beam

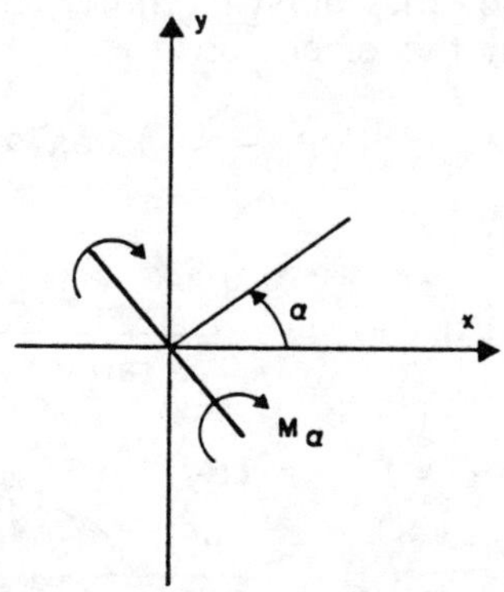

Fig. 5. Linear moment in the direction of angle α

The elastic incremental behaviour of the slab is given by :

$$\{dM\} = [D]\,\{d\mathcal{E}\}$$

with [5, Chap. 11 and 6, Chap. 1] :

$$[D] = \begin{bmatrix} P_{ex} & \nu\sqrt{P_{ex}\,P_{ey}} & 0 \\ \nu\sqrt{P_{ex}\,P_{ey}} & P_{ey} & 0 \\ 0 & 0 & \frac{1-\nu}{2}\sqrt{P_{ex}\,P_{ey}} \end{bmatrix} .$$

We shall see at the end of (2.4) what values are to be taken for P_{ex} and P_{ey} ; ν is the Poisson's ratio of concrete.

2.3 - Yield criterion

For a beam, plasticity appears when the bending moment reaches the value M_e^+ or $-M_e^-$ respectively in positive and negative bending (Fig. 4). The two extreme half-lines of slopes P_p^+ and P_p^- represent hardening.

For a plate, we have adopted the following Johansen's yield criterion [1, Chap. 7]. One admits that, in the direction α (Fig. 5), the elastic limit moment in positive or (/) negative bending is :

$$M_{e\alpha}^{+/-} = M_{ex}^{+/-}\cos^2\alpha + M_{ey}^{+/-}\sin^2\alpha \quad .$$

Now, in the direction α, the bending moment is, because of its tensorial nature :

$$M_\alpha = M_x\cos^2\alpha + M_y\sin^2\alpha + M_{xy}\sin 2\alpha .$$

So the elastic limit, in positive bending for instance, is reached in the direction α if :

$$(M^{+}_{ex} - M_x)\cos^2\alpha + (M^{+}_{ey} - M_y)\sin^2\alpha - M_{xy}\sin 2\alpha = 0$$

or :

$$\frac{M^{+}_{ex} - M_x}{2\tan\alpha} + \frac{M^{+}_{ey} - M_y}{2}\tan\alpha - M_{xy} = 0.$$

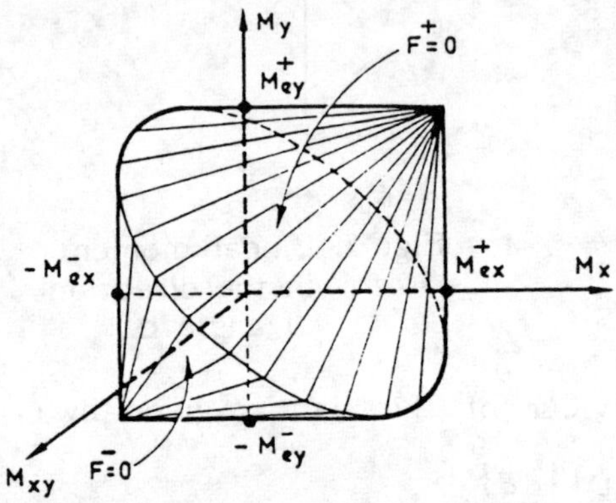

Fig. 6. Yield cones in the three-dimensional space of linear moments

This equation represents a plane in the space (M_x, M_y, M_{xy}). By varying α we obtain a set of planes tangent to a surface - the yield surface - the equation of which is obtained by eliminating α between the previous equation and its derivative with respect to α :

$$-\frac{M^{+}_{ex} - M_x}{\sin^2\alpha} + \frac{M^{+}_{ey} - M_y}{\cos^2\alpha} = 0.$$

Finally the equation of the yield surface in positive bending is :

(1) $$F^{+}(M_x, M_y, M_{xy}) = -(M_x - M^{+}_{ex})(M_y - M^{+}_{ey}) + M^2_{xy} = 0$$

and in the same way in negative bending :

(2) $$F^{-}(M_x, M_y, M_{xy}) = -(M_x + M^{-}_{ex})(M_y + M^{-}_{ey}) + M^2_{xy} = 0.$$

These equations represent two cones in the M-space (Fig. 6).

2.4 - Hardening law

Some tests on reinforced concrete beams have made us assume that their hardening was purely kinematic.

In positive bending, beyond the elastic limit M^{+}_{e}, the moment increment dM gives the strain increment $d\varepsilon$ so that (Fig. 4) :

$$dM = P^{+}_{p}\, d\varepsilon .$$

$d\varepsilon$ is separated into elastic and plastic parts :

(3) $$d\varepsilon = d\varepsilon_e + d\varepsilon_p$$

with :

(4) $$dM = P^{+}_{e}\, d\varepsilon_e .$$

Thus :

$$dM = \frac{P_e^+ \, P_p^+}{P_e^+ - P_p^+} \, d\varepsilon_p .$$

After the increment dM, the elastic limit moment in positive bending becomes :

$$M_p^+ = M_e^+ + dM = M_e^+ + \frac{P_e^+ \, P_p^+}{P_e^+ - P_p^+} \, d\varepsilon_p :$$

Owing to the purely kinematic hardening the elastic limit moment in negative bending becomes at the same time :

$$M_p^- = M_e^- - dM,$$

so that $(M_p^+ + M_p^-)$ is always equal to $(M_e^+ + M_e^-)$.

Thus, with the following definitions :

$$h_x^+ = \frac{P_{ex}^+ \, P_{px}^+}{P_{ex}^+ - P_{px}^+} ; \; h_y^+ = \frac{P_{ey}^+ \, P_{py}^+}{P_{ey}^+ - P_{py}^+} ; \; h_x^- = \frac{P_{ex}^- \, P_{px}^-}{P_{ex}^- - P_{px}^-} ; \; h_y^- = \frac{P_{ey}^- \, P_{py}^-}{P_{ey}^- - P_{py}^-}$$

and Q being the representative point of the linear moments in the M-space, we have :

if Q is on the surface $F^+ = 0$ (positive bending) : $\left\{ \begin{array}{l} dM_{px}^+ = h_x^+ \, d\varepsilon_{px} \\ dM_{py}^+ = h_y^+ \, d\varepsilon_{py} \end{array} \right.$,

if Q is on the surface $F^- = 0$ (negative bending) : $\left\{ \begin{array}{l} dM_{px}^+ = h_x^- \, d\varepsilon_{px} \\ dM_{py}^+ = h_y^- \, d\varepsilon_{py} \end{array} \right.$

and, at any time :

$$M_{px}^+ + M_{px}^- = M_{ex}^+ + M_{ex}^-$$

$$M_{py}^+ + M_{py}^- = M_{ey}^+ + M_{ey}^- .$$

Naturally as soon as a point of the plate has entered into plasticity one must replace $M_{ex}^{+/-}$ and $M_{ey}^{+/-}$ by $M_{px}^{+/-}$ and $M_{py}^{+/-}$ in the formulae (1) and (2) for this point.

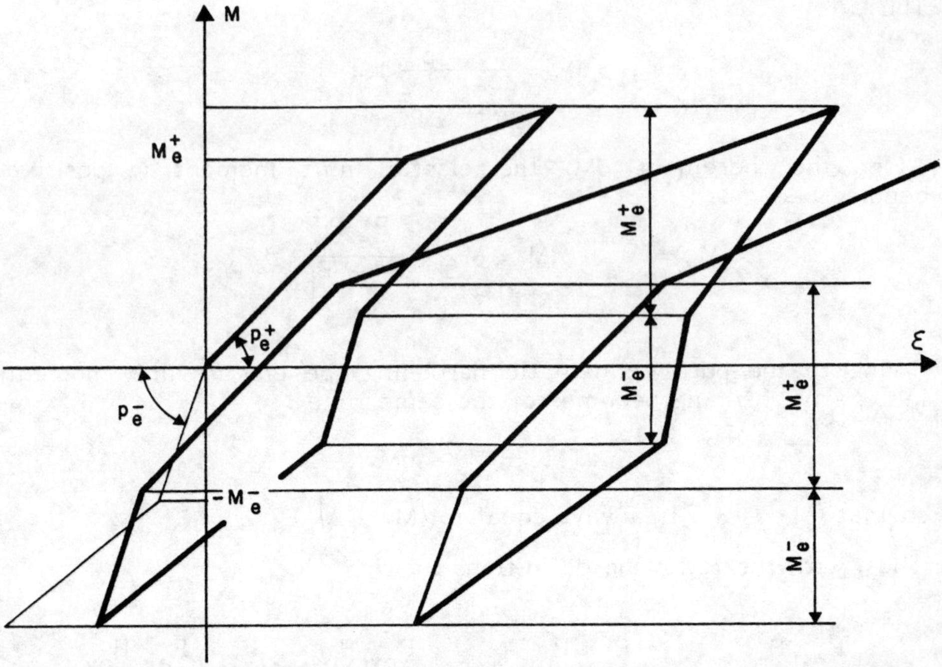

Fig. 7. Cycles of loading-unloading

During an elastic loading or unloading, the elastic slope P_e depends on the size of $(M_p^+ - M)$ (Fig. 7) :

$$\text{if } M_p^+ - M < M_e^+, \; P_e = P_e^+ ;$$

$$\text{if } M_p^+ - M > M_e^+, \; P_e = P_e^- .$$

2.5 - Flow rule

In plastic situation (F = 0 and dF = 0), the formulae equivalent to (3) and (4) for a plate are :

$$\left\{ dM \right\} = [D] \left\{ d\,\varepsilon_e \right\} = [D] \left(\left\{ d\varepsilon \right\} - \left\{ d\,\varepsilon_p \right\} \right)$$

We have taken up the "normality law" associated with the F^+ and F^- functions, i.e. :

$$\left\{ d\,\varepsilon_p \right\} = \lambda \, \frac{\partial F}{\partial \{M\}} .$$

Now F must remain equal to zero during the strain increment, so :

$$F \left(\left\{ M + dM \right\}, \left\{ \varepsilon_p + d\,\varepsilon_p \right\} \right) = 0$$

or :

$$F(\{M\},\{\varepsilon_p\}) + \left\{\frac{\partial F}{\partial\{M\}}\right\}^T \{dM\} + \left\{\frac{\partial F}{\partial\{\varepsilon_p\}}\right\}^T \{d\varepsilon_p\} = 0.$$

$F(\{M\}, \{\varepsilon_p\})$ should be zero but it is not exactly so because of numerical approximations and the previous relation forces the representative point of M+dM to be closer to the actual yield surface than the simple relation :

$$dF = 0.$$

It follows that :

$$\lambda = \frac{F(\{M\},\{\varepsilon_p\}) + \left\{\frac{\partial F}{\partial\{M\}}\right\}^T [D]\{d\varepsilon\}}{\left\{\frac{\partial F}{\partial\{M\}}\right\}^T [D] \left\{\frac{\partial F}{\partial\{M\}}\right\} - \left(\frac{\partial F}{\partial\varepsilon_{px}}\frac{\partial F}{\partial M_x} + \frac{\partial F}{\partial\varepsilon_{py}}\frac{\partial F}{\partial M_y}\right)}.$$

About the previous right-hand side expression one notices that :

i) the numerator is the value taken by F if the strain increment were purely elastic ;

ii)

$$\frac{\partial F^{+/-}}{\partial M_x} = +/- M_{py}^{+/-} - M_y$$

$$\frac{\partial F^{+/-}}{\partial M_y} = +/- M_{px}^{+/-} - M_x$$

$$\frac{\partial F^{+/-}}{\partial M_{xy}} = 2\, M_{xy}$$

$$\frac{\partial F^{+/-}}{\partial \varepsilon_{px}} = -\, h_x^{+/-}\, \frac{\partial F^{+/-}}{\partial M_x}$$

$$\frac{\partial F^{+/-}}{\partial \varepsilon_{py}} = -\, h_y^{+/-}\, \frac{\partial F^{+/-}}{\partial M_y} \;;$$

iii) the values of the slopes to be taken in [D] are P_{ex}^{+} and P_{ey}^{+} or P_{ex}^{-} and P_{ey}^{-} depending on whether the representative point of the linear moments is on the surface $F^{+} = 0$ or $F^{-} = 0$.

3 - VALIDATION TESTS

3.1 - The tests

These tests [7] consisted in submitting square slabs to a soft impact meant to simulate an aircraft impact (low shearing, prevailing bending) at about 1/10 scale. The experimental device is shown on Fig. 8.

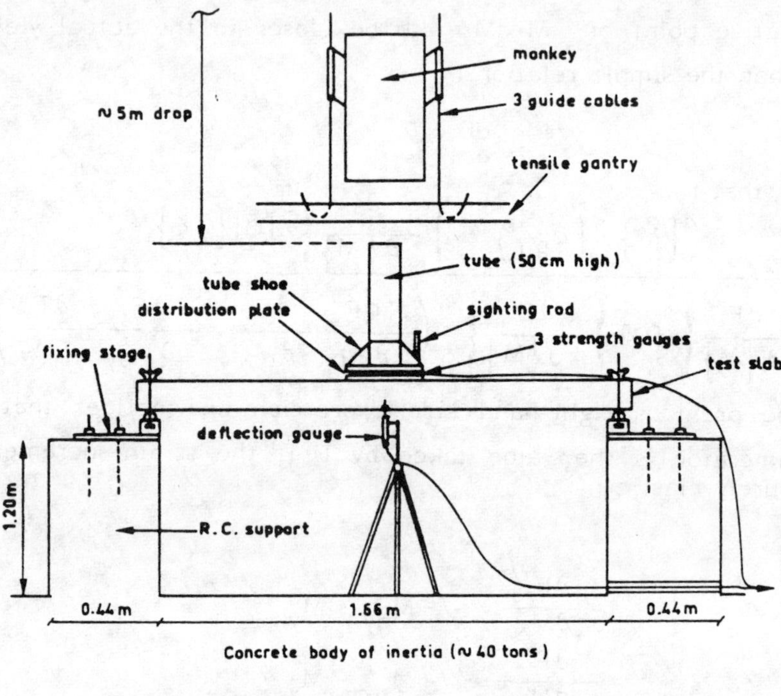

Fig. 8 : General view of the testing device

There were two types of slabs, with different bearing conditions (Fig. 9) :

- Slabs simply supported at their edges. These slabs were 1.90 m long, 0.12 m thick (1.80 m between supports).

- Slabs embedded at their edges. These slabs were 2.50 m long, 0.12 m thick (1.90 m between embedments).

Each series of slabs included slabs with various reinforcement rates, so that we could have behaviours between elasticity and bending close to the failure. The slabs were symmetrically reinforced with high-bond steel.

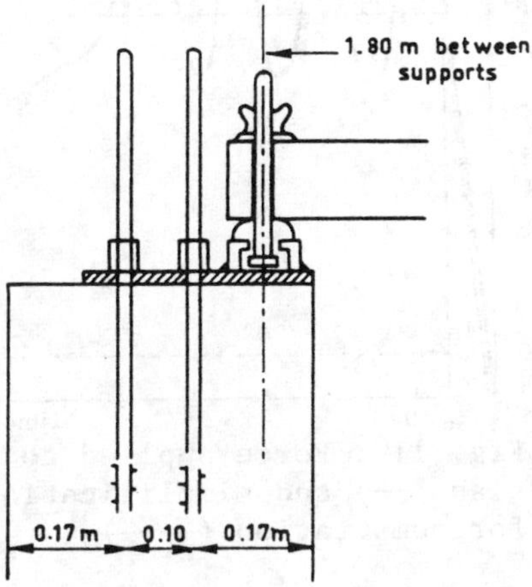

Fig. 9a : realization of the simple support condition

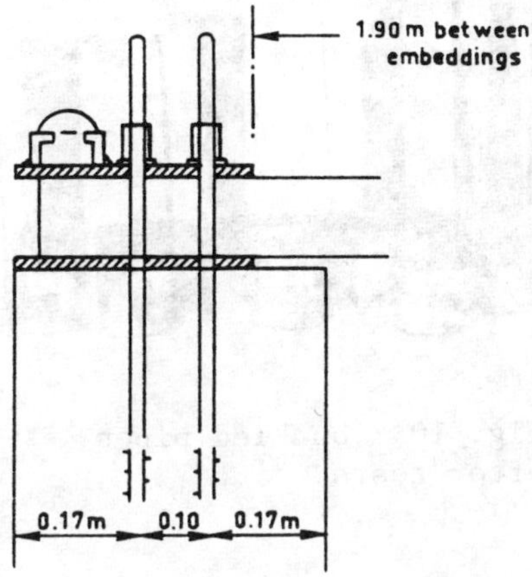

Fig. 9b : realization of the embedding condition

The shock was not exactly the 1/10 reproduction of the signal F(t) (Fig. 1) ; the aim of the tests was simply to reproduce the same types of phenomena as in real slabs under the impact of a plane, with formation of plastic hinges and potential failure under bending. Nevertheless, the peak level had been kept (reduced at the scale) to derive the same type of behaviour in bending, and the loading area is also kept (still at 1/10 scale) to simulate the same shear levels around the contact area.

The loading force was generated by the dynamic crushing of one or two vertical steel pipe(s) by a falling mass. The pipe buckles locally on itself under the impact, starting at the impacted end, with successive formation of regular folds (Fig. 10). The signal F(t) delivered to the slab is the dynamic buckling force of the pipe (Fig. 11) ; each ripple of the signal represents the formation of an additional fold. In order to have a series of regular folds and to prevent from a greater initial peak force, we previously initiated the first fold by a triangular deformation shape.

The crushing force F(t) is so characterized by a climbing part and a series of regular oscillations around a mean level F_o which can be adjusted by selecting the proper pipe (diameter, thickness, steel grade) and the loading speed (i.e. the dropping height of the tup).

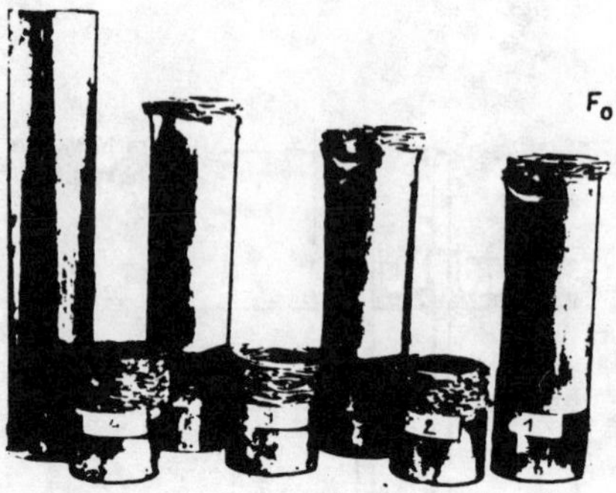

Fig. 10 : buckled pipes after tests.

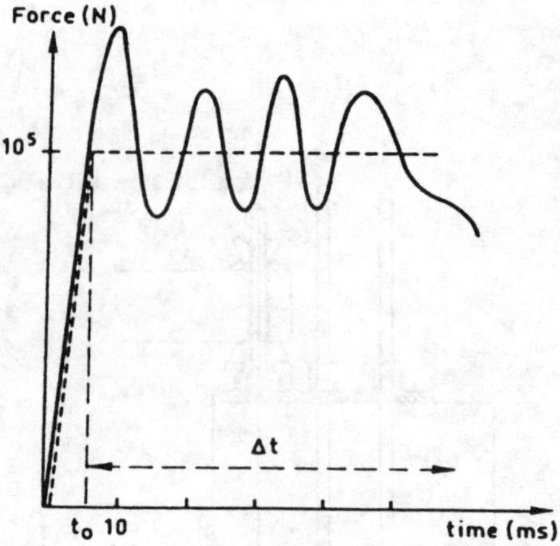

Fig. 11 : Force applied to the slab (—) and simplification for computation (----)

In these tests, we used 72-76 mm rolled-welded steel of 500 MPa ultimate simple tensile stress ; the dropping height was 5 m in case of one tube and 7.30 m in case of two tubes ; the mean force F_o was about $10.\ 10^4$ N (1 tube) or $22.\ 10^4$ N (2 tubes). The total time of the oscillations Δt was about 30 ms (1 tube) or 45 ms (2 tubes) so that it was long compared to the eigen-period T of the slabs (about 10 ms).

To ensure the stability of the pipes and avoid punching problems, the pipes were held in place by a steel shoe and a steel distribution plate (30 cm diameter).

The impact force F(t) was measured, during the test, with three quartz cells integrated between the shoe and the distribution plate ; the displacement at midspan was measured during the impact with a deflection transducer, set under the slab and also with a sighting-rod and a fast camera turning at 3000 or 5000 pictures per second ; strains at a few points of the slabs were measured with strain gauges (3 gauges at the middle of the underside, 2 gauges on the diagonals of the upperside of the slab), but these measures were generally unusable because of the detachment of the gauges.

The response of the slabs under the impact results in a damped oscillating deflection. The testing of a slab with a given reinforcement is characterized by one single number : the maximum value reached by that deflection :w_{max}. The characteristics of the slabs and the results of the main tests are mentioned in the Tables 1 and 2, with the following symbols :

μ : ratio of the steel area to the total area for a cross-section of the slab (Fig. 13) ;

M_e : initial yield bending moment per meter ;

T : longest free oscillation period of the slab ;

w_{max} : largest central deflection during the impact ;

t_{maxl} : time when w_{max} is reached ;

w_{res} : residual central deflection.

Some slabs were submitted to several successive impacts. That's why the tables sometimes give several results for the same slab. The maximum value of the deflection as a function of the yield-moment M_e is plotted on Fig. 12.a and Fig. 12.b (the numbers plotted on the figures correspond to the numbers of the slabs in the tables).

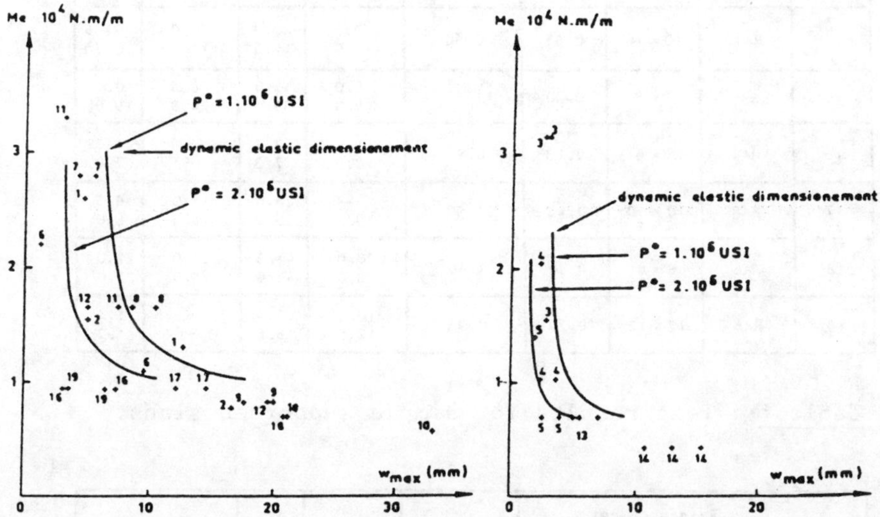

a. Simple supported slabs b. Embedded slabs

Fig. 12 : Maximum deflection as a function of reinforcement rate.

We can see that, until the reduction of reinforcement doesn't exceed 30 % the elastic-limit reinforcement, the maximum deflection varies very little. But it increases very significantly above a 60 % reduction.

The results shown in the tables and plotted on the figures take into account only the tests of the slabs with one steel tube (i.e. with a mean load of $10.\ 10^4$ N). The results of the tests with 2 tubes (mean load of $22.\ 10^4$ N) are similar.

Slab number	Reinforcement			M_e (N.m/m)	T (10^{-3} s)	w_{max} (mm)	t_{max1} (10^{-3} s)	w_{res} (mm)
	Steel Ø (mm)	Spacing (cm × cm)	μ (%)					
1	HA8	10 × 10	0.42	2.60	11.0 19.0	4.5 25.1	8.0 14.0	- 4.9
2	HA6	8 × 8	0.30	1.55	14.0 -	5.0 33.1	8.0 24.0	- 17.1
6	HA8	10 × 10	0.42	2.20	11.0 -	1.0 19.0	7.5 17.0	0.19 5.8
7	HA14	10 × 10	1.28	5.60	- -	8.0 10.8	11.0 10.0	1.0 0.1
8	HA10	10 × 10	0.65	3.30	15.0 14.0	17.0 20.5	15.0 14.0	4.5 3.3
9	HA6	8 × 8	0.30	1.65	16.0 16.0	35.0 40.0	24.0 21.0	17.5 18.0
10	HA6	12 × 12	0.20	1.15	20.0	66.0	32.0	45.0
11	HA10	10 × 10	0.65	3.30	8.5 14.0	2.9 14.5	7.0 11.0	0.25 0.75
12	HA6	8 × 8	0.30	1.65	9.0 16.0	5.0 39.0	8.5 24.5	0.7 19.0
16	HA4	8 × 8	0.13	0.95	11.0 12.5	3.2 7.3	- -	1.3 2.0
17	HA8	10 × 10	0.42	1.90	- 19.0	24.0 29.0	19.0 16.5	8.5 6.0
18	HA6	8 × 8	0.30	1.40	20.0 18.0	41.5 42.0	28.0 25.0	22.0 22.0
19	HA6	12 × 12	0.20	0.95	10.0 12.5	3.4 6.6	- -	1.4 1.2

Table 1 : Test results for simple supported slabs.

Slab number	Reinforcement			M_e (10^4 N.m/m)	T (10^{-3} s)	w_{max} (mm)	t_{max1} (10^{-3} s)	w_{res} (mm)
	Steel Ø (mm)	Spacing (cm × cm)	μ (%)					
3	HA10	10 × 10	0.65	3.15	12.0 9.0 12.0	2.6 2.4 5.3	9.5 6.5 6.5	0.7 0.3 0.4
4	HA8	10 × 10	0.42	2.05	10.0 11.0 13.0	2.1 4.6 7.2	7.0 7.5 6.0	0.6 0.8 0.6
5	HA6	8 × 8	0.30	1.40	10.0 12.0 13.0	1.9 5.0 7.8	7.0 8.0 9.0	0.1 1.4 0.9
13	HA6	8 × 8	0.30	1.40	11.5 14.0 14.0	10.0 11.0 14.0	12.0 11.0 12.0	3.0 1.0 1.0
14	HA6	12 × 12	0.20	0.95	18.0 20.0 20.0	21.5 26.0 30.5	19.0 18.0 17.0	8.0 6.5 5.0

Table 2 : Test results for embedded slabs.

3.2 - Moment-curvature relations and equivalent bilinear modelization

If we want to compute R.C. slabs with Dalton, we have to give bilinear moment-curvature relations for each axis-direction x and y and each sense of bending (positive or negative). For a given slab, we have then four relations ; a given relation (i.e. a given sense and a given direction) is characterized in the program by three data (Fig.4) :

- the elastic slope p_e,
- the yield-moment M_e,
- the plastic slope p_p.

When we know the features of a cross-section of the slab (or the beams corresponding to the directions x and y of the slab (Fig.13)), it is easy to determine the positive and negative moment-curvature diagrams of the slab. A computer program (MOCO) calculates these relations.

We give it the dimensions of the concrete section, the position and areas of the steel bars, the real stress strain relations of concrete and steel (Fig.14) and the program produces the moment-curvature relation (Fig.15) which takes into account concrete cracking in the calculated section.

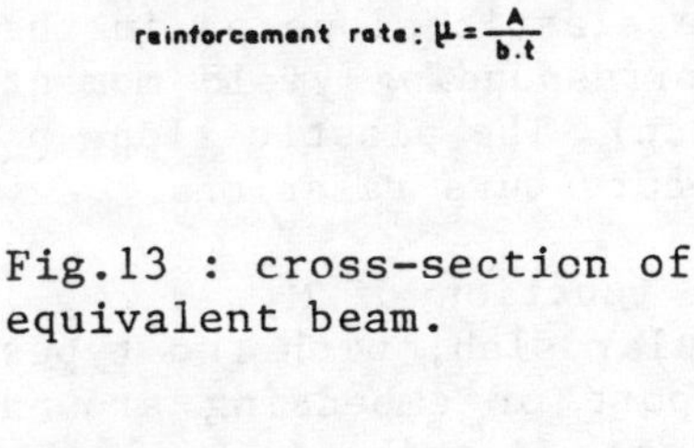

Fig.13 : cross-section of the equivalent beam.

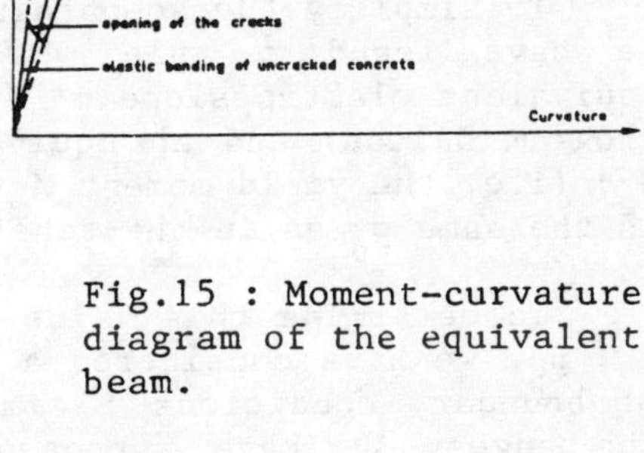

Fig.15 : Moment-curvature diagram of the equivalent beam.

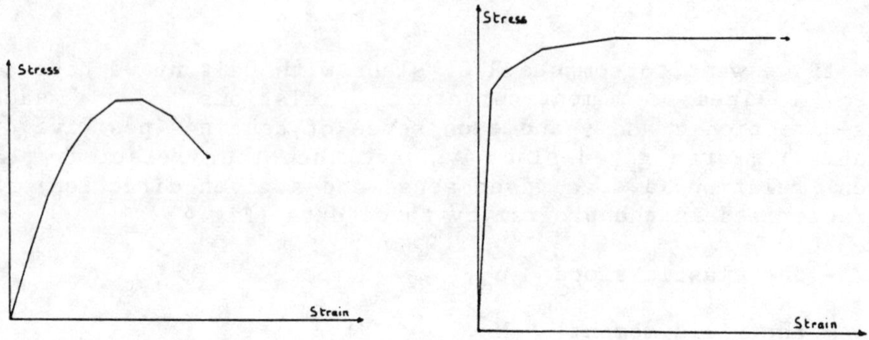

a. Concrete b. Steel

Fig.14 : Stress-strain relations of the materials

On the moment-curvature relation, we can observe an elastic part with two phasis : first, when concrete remains uncracked, we have elastic behaviour with the slope p_1 ; then we have formation and opening of a crack until we join a new elastic slope p_2, which corresponds to the elastic strain of steel. The real plastic phasis is not linear but can be easily schematized by a half-line of slope p_p.

So one real moment-curvature relation is characterized by five parameters :

- the cracking moment M_f,
- the yield moment M_e,
- the uncracked elastic slope p_1,
- the cracked elastic slope p_2,
- the plastic slope p_p.

To simplify the so-obtained relations in bilinear curves, we have found a rule which makes possible to find an equivalent elastic slope p* (i.e. the elastic slope p_e in the program Dalton) and an equivalent corresponding yield moment M_e* (i.e. the yield moment M_e in Dalton). The plastic slope p_p is the same p_p as in the real moment-curvature relation.

To determine this value p* as a function of M_f, M_e, p_1, p_2, p_p, we have considered a rectangular slab, with two types of boundary conditions : simple support or embedding around the edges. We have supposed a linear distribution of the components M_x and M_y of the bending moment with respect to the coordinates x and y at a given time t and the torsion

ORDER NUMBER

PO NYAATP08

BOOK NUMBER

9965199

BAY NUMBER

2271-11

AUTHOR	Hinton, E.
TITLE	Computational Modelling of Reinforced

component M_{xy} is assumed to be neglectible ; this distribution corresponds approximatively to such a slab subjected to a central load (as we have verified on computed similar slabs). We have supposed such a load function that the maximum moment in the slab is a linear function of time and that the yield moment M_e (M_{ex} or M_{ey}) is reached at the time t_o. p* is then such that we obtain the same mean bending energy during the space time $[o, t_o]$ with the real idealized relation (Fig.16a) and with the simplified bilinear relation (Fig.16b).

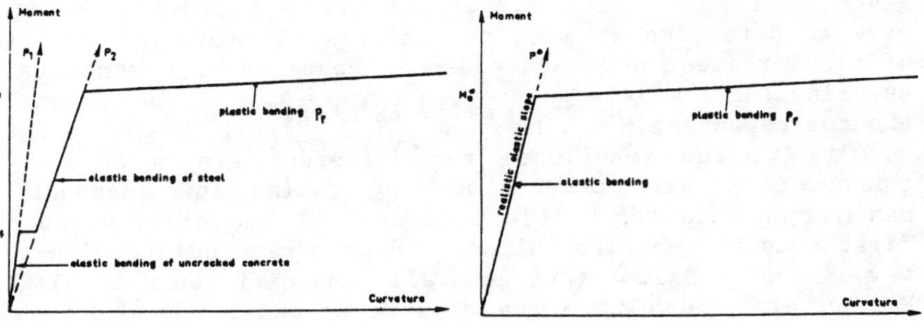

a. Real law. b. Simplified law.

Fig.16 : Moment-curvature relations.

With the next notations :

$$K_M = \frac{M_f}{M_e} \qquad K_p = \frac{P_2}{P_1} \qquad K'_p = \frac{P_p}{P_2}$$

We come to :

$$(5) \qquad p^* = \frac{p_p\,(1 - x)}{K'_p - x} \qquad \text{and :} \qquad M_e{}^* = M_e\,(1 - x)$$

Where :

$$x = -\frac{1}{2} + \sqrt{\frac{1}{4} + \frac{K_M{}^3\,K'_p}{1 - K'_p}\,(1 - K_p)(1 - 3\,\mathrm{Ln}K_M + \frac{9}{2}\,\mathrm{Ln}^2\,K_M)}$$

We can note that with the simplified relation we have a plasticized zone at the middle of the slab or along the edges (in the case of the embedded slab) and not with the real law because we have supposed that the maximum moment is the yield moment M_e ; but this plasticized zone is very small because M_e* is very close to M_e.

In every cases of slabs that reach or exceed the yield moment, we can use that determination rule for p* without fearing a too large error : we have calculated that the difference of méan bending energy between the p*-rule (5) and an expression of p* adapted to the accurate maximum value M_{max} (>Me) reached in the slab is lower than 5 % for the usual values of parameters. In case of a completely elastic behaviour, the rule should be adapted (in main cases, it is sufficient to replace M_e in the formula (5) by the maximum moment M_{max} reached or expected) because otherwise the difference with the real law could become too large. For a given real slab, with different zones of reinforcement, we have to determine p* with the formula (5) for each direction of reinforcement and each sense as concerns each zone, before entering these values as $\underline{p}^+_{ex}$, $\underline{p}^-_{ex}$, $\underline{p}^+_{ey}$, p^-_{ey} in Dalton (and the corresponding M^+_{ex}, M^-_{ex}, M^+_{ey}, M^-_{ey}). If we want a better result, we can reevaluate the different slopes p* and the components of the moments M_e* by taking into account the maximum moments reached in each zone of the slab, given by a first computation with Dalton, and then recalculate with these new values in Dalton (but generally the difference is slight). We can also reevaluate the plastic slopes p_p in order to be nearer the real curves up to the values M_{max}.

In the case of the tests studied here, we have only two moment curvature relations for each slab (in x and y directions, corresponding to the directions of reinforcement) because the steel layers are symmetrical with respect to the mid-plane, and are the same for the whole slab (only one zone of reinforcement). Fig.17 gives an example of these relations and the corresponding bilinear simplified laws with the above mentionned rule (p_p is chosen by taking into account the maximum value of the expected curvature, corresponding to the loading).

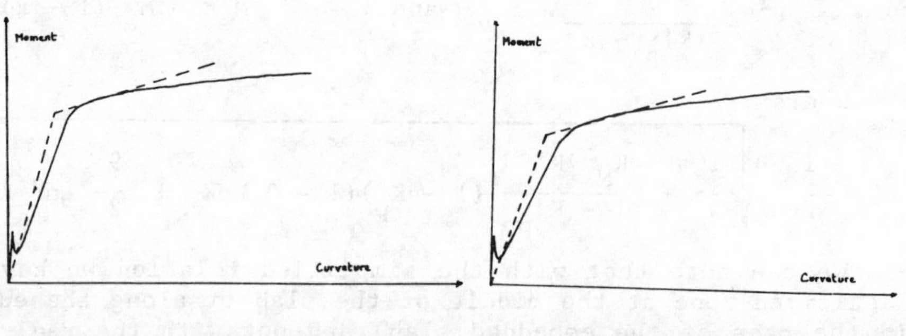

Fig.17 : Moment-curvature relations of a tested slab (n°3)

3.3 Computation of the tested slabs

In order to validate the program (Dalton), we have computed the tested slabs. As we would have had too many computations if we wanted to calculate every test, we have first tried to recover the experimental results as a whole with a limited amount of computed cases. We have first calculated the two moment-curvature relations of every slab with Moco and then we have determined the mean value of the main parameters of the bilinear schematized laws (p^*, p_p). To represent suitably all the tests, we have computed two series of sample slabs, with two values of the elastic slope p^* that encompass the different values p^* calculated with Moco and the rule (5) ; to simplify, we have considered that $p_x^* = p_y^* = p^*$ and that p_p is the same in all cases. In each series, we have varied the amount of reinforcement, i.e. the yield moment Me*(Mex = Mey = Me*) from a level corresponding to elastic behaviour, down to a level corresponding to the failure. We have submitted these sample slabs to the simplified loading (climbing part till to and then a constant load Fo : see Fig. 11). The elasto-plastic computations with Dalton give the maximum value of deflection for every level of reinforcement (proportional to Me*). The curves expressing that deflection with respect to the yield-moment for the two values p^* can be compared with the experimental measures (Fig. 12). We can also compare the plasticized area, according to the program, with the crack-growth on the tested slabs(Fig. 18).

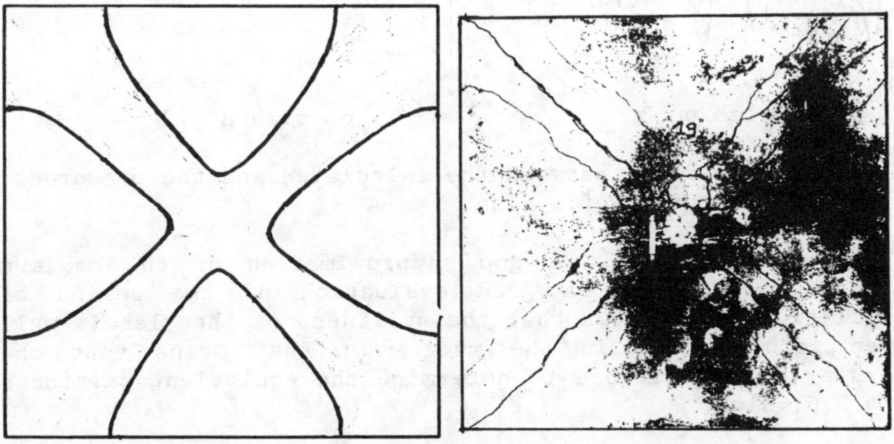

Fig. 18 : Comparison between growth of plasticized area (Dalton) and a picture of crack-growth (tested slab).

Taking into account the dispersion of the test results and the simplifications of the computations, we consider that we have a good agreement, at least if the reinforcement rate is not too low (it is the case in the real R.C. slabs in which we are interested).

Moreover, we have recalculated a few representative tested slabs with the accurate paramaters p_x^*, p_y^*, p_{px}, p_{py}, M_{ex}^*, M_{ey}^*, determined with the p*-rule (5). The loading was also schematized by a multi-linear curve (a climbing part till to, a constant part Fo during Δt, a decreasing part till F = 0) according to the recordings at the gauges (a few computations with an oscillating load F(t) have shown that it is suitable to simplify by taking the mean load Fo) corresponding to the tests of the given slab. Considering the imprecision of the measures and the dispersion between the tests on a same slab, we have used in fact two loading functions, which give a minimum and a maximum evaluations of the deflection. This second set of computations ought to validate more especially the p*-rule.

The different slabs computed with this method show a very good agreement between the tests and the computations (ex. Fig. 19).

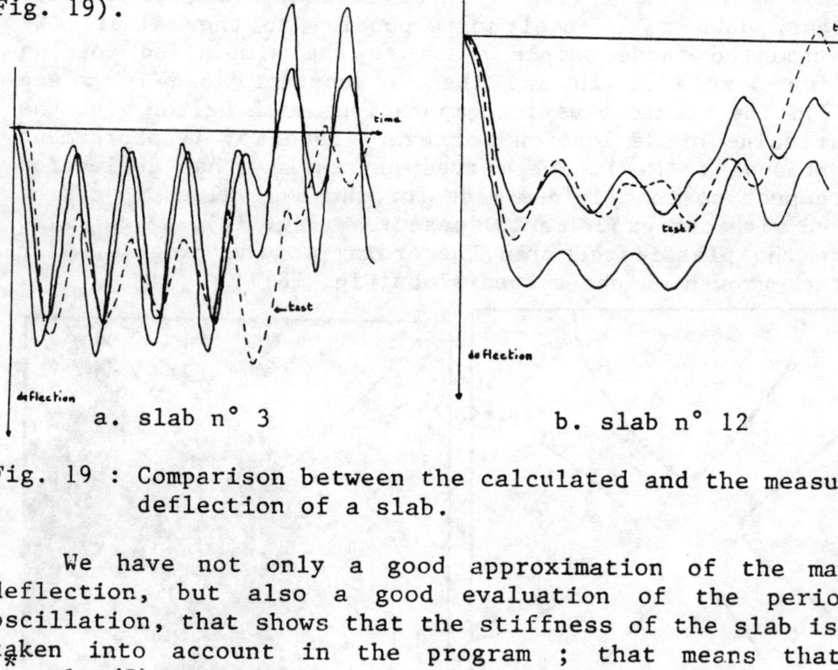

a. slab n° 3 b. slab n° 12

Fig. 19 : Comparison between the calculated and the measured deflection of a slab.

We have not only a good approximation of the maximum deflection, but also a good evaluation of the period of oscillation, that shows that the stiffness of the slab is well taken into account in the program ; that means that the p^*-rule (5) is suitable to determine the equivalent stiffness of a R.C. slab.

These various comparisons between tests and computations allow us to consider that Dalton (and the associated p*-rule) suitably models reinforced concrete slabs subjected to soft shocks.

4 - APPLICATION TO PLASTIC DESIGN IN R.C. STRUCTURES OF NUCLEAR POWER PLANTS.

As a safety provision, some reinforced concrete structures in nuclear power stations must be able to support highly unlikely accidental dynamic loadings such as the impact of a military aircraft. Obviously it is not necessary to dimension these structures so that, in any case, they should have only elastic deformations. We may accept that, in case of an aggression of the preceding type, they could undergo plastic deformations, provided that there is no breaking.

To design these structures, it is not possible to use dynamic elastic calculations as a general rule, because such calculations are very expansive and the design offices have not always the suitable computer programs. So we were lead to find simplified rules to design R.C. slabs or walls under accidental dynamic loadings by using a usual elastic static method.

Since these accidental loads are dynamic, strictly elastic desigh ought to take into account a dynamic load factor : for instance, the maximum deflection reached at a given point is greater under dynamic loading than under the static load equal to the maximum value. Many computations with Dalton on sample slabs submitted to aircraft loadings have shown that a slab with reinforcement reduced by 40 % with regard to the elastic dynamic design can easily support this loading without fearing failure (maximum deformations in the slab are far from the failure). So we have taken up this simplified rule (for aircraft-crash, explosions, etc...) : the design offices use elastic static design rules with the load $F_{eq.}$ which is the maximum of F(t) multiplied by the dynamic load factor and by 0.6 (the load factor is in fact the maximum of the different dynamic load factors evaluated for the studied kind of loading and the different kinds of slabs encountered in the structure we have to study). In a few particular cases of loading (grouping of pipe rupture with earthquake for example) with this loading depending on the location of the structure or with very complicated structures for which it is not easy to determine the dynamic load factor, the method is a bit different, but always lead to the definition of an equivalent elastic static load associated to elastic static design.

In all cases, the extra-loading would be easily absorbed by the plastic resources of the structure.

We have both to comply with safety requirements and to justify a high capacity of resistance above the elastic limit thanks to the ductility of materials. We have determined an ultimate state considered as allowable, based on the local deformations achieved through the elastic plastic computation :

- 0.35 % deformation for compressed concrete,

- 0.80 % deformation for the most strained steel bars.

We have shown, by computations with Dalton, that slabs of the above-mentionned standard design we have in our nuclear power plants, subjected to the real accidental loadings, would never reach the ultimate state defined above and would have a sufficient margin before failure. Fig. 20 and Fig. 21 give an example concerning the anti-aircraft slabs.

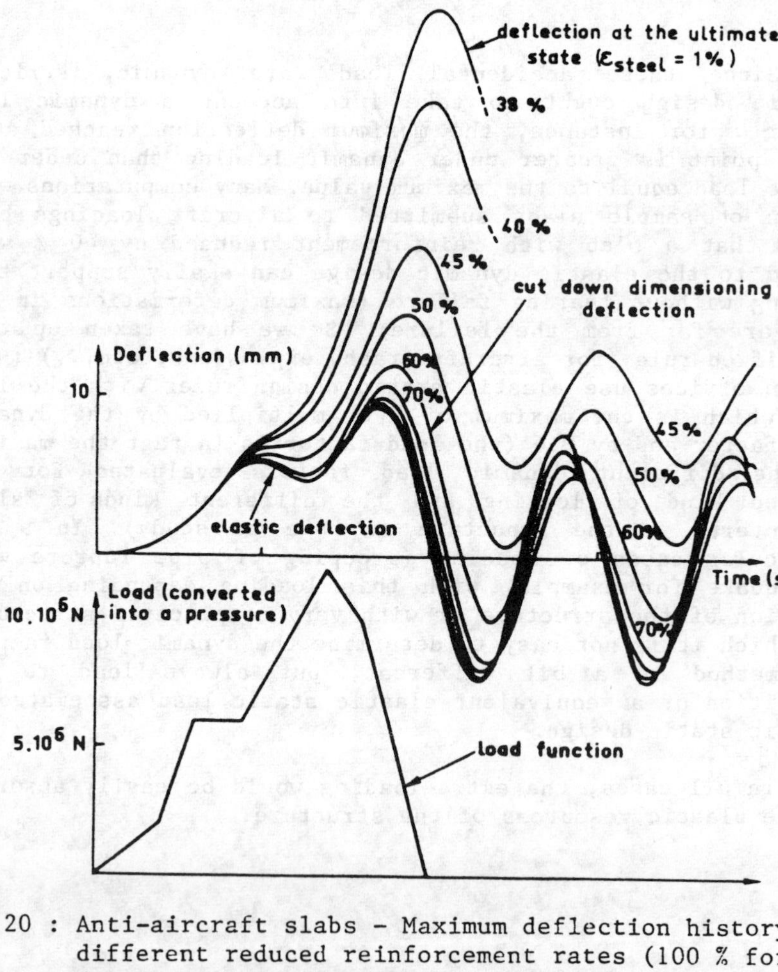

Fig. 20 : Anti-aircraft slabs - Maximum deflection history at different reduced reinforcement rates (100 % for dynamic elastic design)

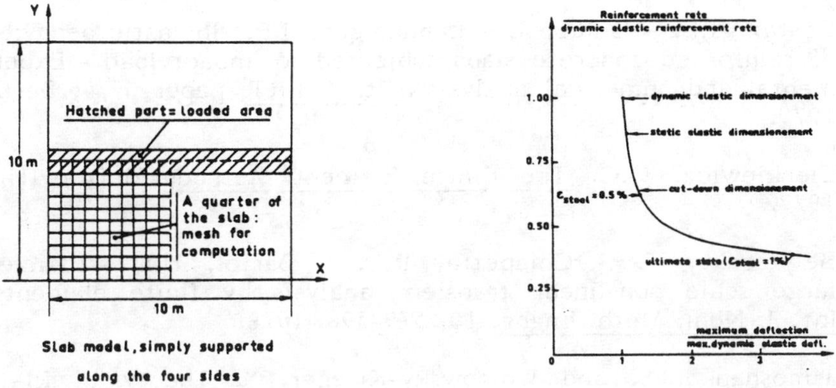

Fig. 21 : Maximum reduced deflection as a function of reduced reinforcement.

5 - CONCLUSION

Our Dalton-program being validated, we have used it to solve many specific dynamic problems, which we often find in nuclear R.C. structures (pipe ruptures, earthquakes, and grouping of these phenomena in casemates or swimming-pools for example).

This program is very suitable to study the behaviour of plane structures under different types of dynamic loadings. As concerns nuclear power plants, in which some structures must be designed under many various dynamic phenomena, Dalton is very useful to define simplified design rules and to validate these rules, by taking into account plastic behaviour of materials.

6 - REFERENCES

1. Save, M. and Massonnet, C. Calcul plastique des constructions, vol. II, Centre belgo-luxembourgeois d'information de l'acier, Bruxelles, 1972.

2. Costaz, J.L. - Dulac, J. - Laboudigue, J.F. "Inelastic behaviour of reinforced concrete slabs subjected to impact load - Experimental and numerical analysis", 5th SMIRT, paper J7/4, Berlin, 1979.

3. Zienkiewicz, O.C. The Finite Element Method, McGraw-Hill, 1977.

4. Belytschko, T. - Chiapetta, R.L. - Bartel, H.D. "Efficient large scale non-linear transient analysis by finite elements", Int. J. Num. Meth. Engng, 10, 579-596, 1976.

5. Timoshenko, S. and Woinowsky-Krieger, S. Theory of Plates and Shells, McGraw-Hill, 2nd ed., 1959.

6. Troitsky, M.S. Stiffened Plates, Bending Stability and Vibrations, Elsevier, 1976.

7. Dulac J. and Giraud J.P. "Impact testing of reinforced concrete slabs", 6th SMIRT, paper J7/1, Paris, 1981.

CHAPTER 12

NONLINEAR ANALYSIS OF REINFORCED CONCRETE PLATES AND SHELLS USING A THREE DIMENSIONAL MODEL

M. Cervera and E. Hinton
Dept. of Civil Engng., University College of Swansea, U.K.

SUMMARY

A fully three dimensional finite element computational model is presented which may be used to analyse reinforced concrete structures in an efficient and accurate manner. Emphasis is placed on the analysis of plate and shell structures and several examples are presented to highlight various features of the model.

1. INTRODUCTION

Much effort has been dedicated over recent years to the development of computational models for the nonlinear analysis of reinforced concrete plate and shell systems. Most of this research work has been carried out using plate and shell finite element models. The resulting formulations are quite sophisticated and are able to account for both material and geometric non-linearities.

The usual models for reinforced concrete plate and shell structures incorporate so-called 'layered elements' which consist of a series of through-thickness 'layers', each assumed to be in a state of plane stress. This approach has several disadvantages:

(i) The mid-ordinate rule used to integrate the stiffness matrices and residual forces through the thickness is not very accurate, and as many as eight to twelve layers must be used.

(ii) The large number of sampling points used per element makes the evaluation of the stiffness matrix and internal forces vector quite expensive.

(iii) The 'plane stress' assumption for the layers implies that the normal stress perpendicular to the mid-surface is neglected. Furthermore, in some nonlinear models in-plane stresses are treated separately from transverse shear stresses, usually in an inconsistent way.

(iv) Cracks are usually assumed to be formed as a result of in-plane stresses only, the effect of the transverse shear stresses being disregarded when checking for cracking. As a result, all cracks are normal to the mid-surface, which is an unrealistic constraint which prevents the model from being able to represent non-flexural failures.

(v) The elements used are usually of the Mindlin-type. This results in an 'a priori' imposed constant distribution of transverse shear strains through the thickness.

(vi) Concrete slabs are usually moderately thick so consideration of transverse shear strains and stresses may be be significant, especially near the supports.

All these considerations highlight the necessity of developing fully three-dimensional computational models which may be used to analyse reinforced concrete structures more efficiently and accurately than the existing models which are based on plate and shell formulations. This paper describes a first step in that direction.

2. CONCRETE MATERIAL MODEL

2.2 Generalities

The analysis and design of reinforced concrete structures is fraught with difficulties, mostly due to the complex material behaviour involved. This can be appreciated by considering the following facts:

(a) the structural system is composed of two very different materials, concrete and steel,

(b) bond forces and bond slip between the reinforcement and the surrounding concrete are difficult to model with complete confidence,

(c) concrete is a composite of aggregate, cement, paste, water and unfilled voids - this makes the stress-strain relationship for concrete nonlinear even for low stress levels,

(d) concrete cracks under relatively low tensile stresses, since its tensile strength is much lower than its compressive strength,

(e) concrete deformations are affected by phenomena such as creep and shrinkage and depend on load and environmental history,

(f) relatively little data is available for multiaxial stress states in concrete.

These complexities have led to the development of many models for the analysis of plain and reinforced concrete. Material models based on linear-elasticity, nonlinear-elasticity, elasto-plasticity, elasto-viscoplasticity, endochronic theory, plasticity-damage (fracturing) models, as well as fracture mechanics have been extensively used in the present research area [1],[2]. However, no definitive model seems to have been achieved, and many can be employed in different applications.

2.2 Aims of the present model

One aim of the present work is to present a material model suitable for nonlinear static analysis of three-dimensional reinforced structures under monotonically increasing load, without regard to geometric nonlinearities.

The proposed model is kept as simple as possible so that it can be easily implemented in finite element systems. Its relative simplicity allows it to be used in the analysis of real engineering structures without involving prohibitively expensive calculations. Nevertheless, the model is shown to be capable of idealising the nonlinear behaviour of reinforced concrete structures. Thus, it provides a realistic computational tool for tracing the overall behaviour of these structures up to ultimate load, within an engineering degree of accuracy.

The material properties needed to define the model have a clear physical meaning and they can be estimated from the compressive strength of the concrete, usually the only material data available.

2.3 Behaviour in triaxial compression

As stated in Section 2.1 experimental evidence indicates that the stress-strain relationship for concrete is nonlinear even for low stress levels. The inelastic deformation may be separated into recoverable and irrecoverable components. In the present model elasticity theory is used for the recoverable strain components, and a strain-hardening plasticity approach is employed to model the irrecoverable parts of the deformation.

Four conditions have to be established to be able to evaluate the stress state for a given strain state:

(a) the yield condition,
(b) the flow rule,
(c) the hardening rule, and
(d) the crushing condition.

2.3.1 Yield condition

Under a triaxial stress state the yield criterion for concrete is generally assumed to be dependent on the three stress invariants. Nevertheless, a yield function depending only on the mean normal stress, I_1, and the shear stress invariant, J_2, has proved to be adequate for most practical situations. The yield criterion adopted here is of such a type [3], and it can be written as

$$f(I_1,J_2) = [\beta(3J_2) + \alpha I_1]^{\frac{1}{2}} = \sigma_0 \tag{1}$$

where α and β are two material parameters and σ_0 is the equivalent effective stress.

It should be noted that the Huber-Mises yield condition can be recovered by assuming $\alpha = 0.0$ and $\beta = 1.0$.

As our main concern will be for situations which are not very different from biaxial stress states, it is reasonable to obtain the material parameters by fitting biaxial test results:

for the uniaxial compression test:

$$\sigma_1 = f_c' \qquad \sigma_2 = \sigma_3 = 0 \tag{2}$$

for the biaxial compression test:

$$\sigma_1 = \sigma_2 = \gamma f_c' \qquad \sigma_3 = 0 \tag{3}$$

If Kupfer's results [4] are to be fitted then γ equals 1.16 and the use of (2) and (3) gives as

$$\alpha = 0.355\sigma_0 \qquad \beta = 1.355$$

The resulting yield criterion is compared in Figure 1 with the experimental results of Kupfer et al in the biaxial principal stress space.

This yield criterion can be used as the basis for an elastic-perfectly plastic model if σ_0 is taken to be equal to the ultimate compressive stress, f_c', obtained from an uniaxial test. In such a model the response is assumed to be elastic until the effective stress reaches the value of f_c'. After that, the concrete is assumed to flow until the crushing surface is reached.

It is more realistic to use a work-hardening model in which σ_0 is a function of a hardening parameter (see Section 2.3.3). A value $\sigma_0 = cf'_c$ defines a surface limiting the elastic behaviour. N.B. A typical value for c is 0.3. When this surface is reached inelastic deformation begins and a hardening rule monitors the expansion of the yield surface under further loading. In this way, a whole family of 'loading surfaces' is defined, as shown in Figure 2.

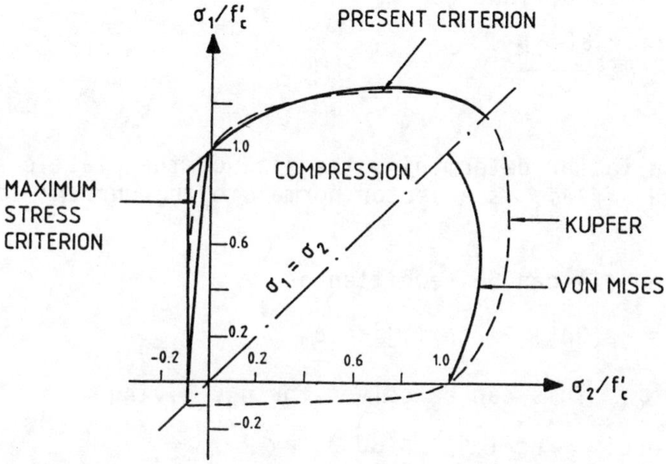

Figure 1 Compressive yield and tensile cracking criterion for concrete - biaxial stress space

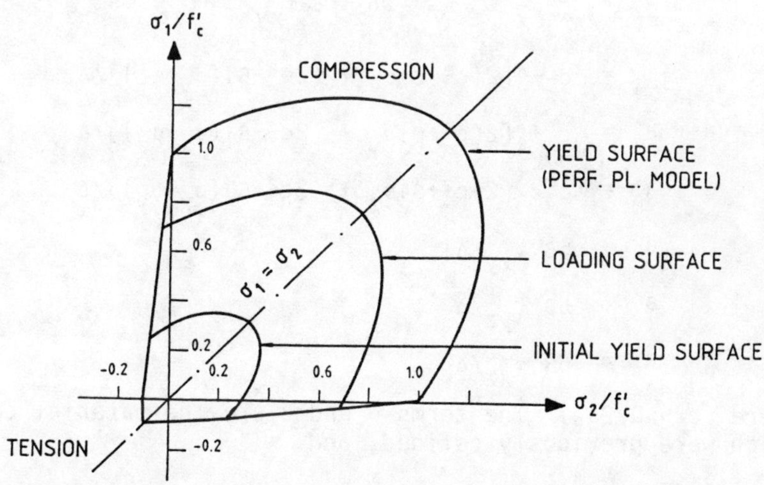

Figure 2 Two-dimensional stress space representation of the concrete constitutive model

2.3.2 The Flow Rule

In plasticity theory a flow rule must be defined so that during the inelastic response the plastic strain increments can be evaluated for a given stress increment.

Although very little supportive experimental evidence is availible an associative flow rule will be employed here. This means that the plastic deformation rate vector will be assumed to be normal to the yield surface. The plastic strain increment is then defined [5] as

$$d\varepsilon_{ij}^{p} = d\lambda \frac{\partial f(\underline{\sigma})}{\partial \sigma_{ij}} \tag{4}$$

where $d\lambda$ is a factor determining the size of the plastic strain increment and $\partial f/\partial \sigma_{ij}$ is a vector normal to the current loading surface.

Equation (1) can be rewritten as

$$f(\underline{\sigma}) = [\beta(3J_2) + 2c\sigma_0 I_1]^{\frac{1}{2}} = \sigma_0 \tag{5}$$

with $c = \alpha/2\sigma_0$. This can be solved for σ_0, giving

$$f(\underline{\sigma}) = cI_1 + [c^2 I_1{}^2 + \beta(3J_2)]^{\frac{1}{2}} \tag{6}$$

The yield function derivatives which define the flow vector $\underline{a}$ can be now evaluated as

$$\underline{a}^T = [\frac{\partial f}{\partial \sigma_x}, \frac{\partial f}{\partial \sigma_y}, \frac{\partial f}{\partial \sigma_z}, \frac{\partial f}{\partial \tau_{xy}}, \frac{\partial f}{\partial \tau_{yz}}, \frac{\partial f}{\partial \tau_{xz}}]$$

$$\begin{aligned}
a_1 &= c + [2(c^2+\beta)\sigma_x + (2c^2-\beta)(\sigma_y+\sigma_z)]/\Delta \\
a_2 &= c + [2(c^2+\beta)\sigma_y + (2c^2-\beta)(\sigma_x+\sigma_z)]/\Delta \\
a_3 &= c + [2(c^2+\beta)\sigma_z + (2c^2-\beta)(\sigma_x+\sigma_y)]/\Delta \\
a_4 &= 6\beta\, \tau_{xy}/\Delta \\
a_5 &= 6\beta\, \tau_{yz}/\Delta \\
a_6 &= 6\beta\, \tau_{xz}/\Delta
\end{aligned} \tag{7}$$

where $c = \alpha/2\sigma_0$. The terms α and β are the material constants which were previously defined, and

$$\Delta = 2[(c^2+\beta)(\sigma_x^2+\sigma_y^2+\sigma_z^2) + (2c^2-\beta)(\sigma_x\sigma_y+\sigma_y\sigma_z+\sigma_x\sigma_z) + 3\beta(\tau_{xy}^2+\tau_{yz}^2+\tau_{xz}^2)]^{\frac{1}{2}}$$

The plastic multiplier can be found to be

$$d\lambda = (H' + \underline{a}^T \underline{D}\, \underline{a})^{-1}\, \underline{a}^T \underline{D}\, d\underline{\varepsilon} \quad (8)$$

where H' is the hardening parameter to be defined,
$\underline{D}$ is the elastic constitutive matrix, and
$d\underline{\varepsilon}$ is the total strain increment vector.

The complete elasto-plastic incremental stress-strain relation is given by the expression

$$d\underline{\sigma} = \underline{D}_{ep}\, d\underline{\varepsilon} \quad (9)$$

with

$$\underline{D}_{ep} = \underline{D} - (\underline{D}\, \underline{a}\, \underline{a}^T \underline{D})(H' + \underline{a}^T \underline{D}\, \underline{a})^{-1} \quad (10)$$

2.3.3 The hardening rule

The hardening rule is necessary to define the change of position of the 'loading surfaces' during plastic deformation. A relationship between the accumulated plastic strain and the effective stress σ_0 is required to control the position of the current 'loading surface'. The concepts of effective plastic strain and effective stress are therefore used to allow for extrapolation from the results of an uniaxial test to the multiaxial situation.

In the present work such a relationship is assumed to be the conventional 'Madrid Parabola' which may be written as

$$\sigma_0 = E_0\varepsilon - E_0\varepsilon^2/2\varepsilon_0 \quad (11)$$

where σ_0 is the effective stress,
E_0 is the initial Young's modulus,
ε is the current total strain, and
ε_0 is the total strain at peak stress f_c' .

Upon substitution of the elastic strain, $\varepsilon_e = \sigma_0/E_0$, in (11) the desired relation is obtained as

$$\sigma_0 = - E_0\varepsilon_p + (2E_0{}^2\varepsilon_0\varepsilon_p)^{\frac{1}{2}}, \; cf_c' < \sigma_0 < f_c' \qquad (12)$$

From (12) it is also possible to obtain the hardening parameter

$$H' = \frac{d\sigma_0}{d\varepsilon_p} = E_0((\frac{\varepsilon_0}{2\varepsilon_p})^{\frac{1}{2}} - 1) \qquad (13)$$

As can be observed from (12) four material properties are needed to define the hardening rule: the ultimate compressive strength, f_c', the initial elasticity modulus, E_0, a constant defining the initial yield stress, c, and the strain at peak stress, ε_0. As can be seen in Figure 3, very good agreement with Kupfer's experimental results can be achieved by taking

$$c = 0.3 \quad \text{and} \quad \varepsilon_0 = 2\, f_c'/E_0 \qquad (14)$$

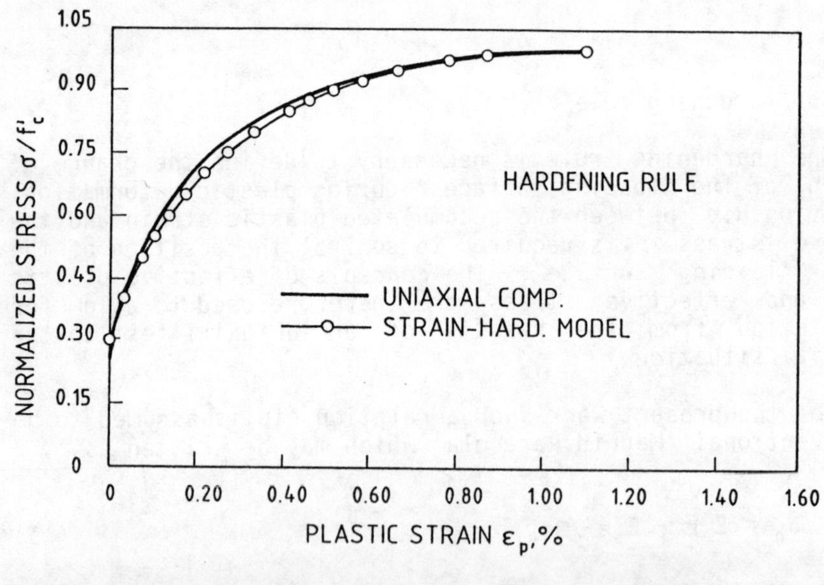

Figure 3 Comparison of hardening rule with the inelastic uniaxial compressive strains of Kupfer

2.3.4 The crushing condition

The yield criterion and the hardening rule define a family of expanding 'loading surfaces' which allow the evaluation of the stress increments and the plastic strain increment for a given total strain increment. Inelastic deformation proceeds until crushing occurs. The crushing type of fracture is a strain controlled phenomenon. A failure surface in the strain space must be defined so that this kind of fracture can be taken into account. A simple way of incorporating this feature into the model, despite the lack of experimental data on concrete ultimate deformation capacity under multiaxial loading, is to convert the yield criterion (described in terms of stresses) directly into strains. Thus,

$$\beta(3J_1') + \alpha\, I_1' = \varepsilon_u{}^2 \qquad (15)$$

where I_1' and J_2' are strain invariants and ε_u is an ultimate total strain value that can be extrapolated fron uniaxial tests.

When the strains reach the crushing surface, the concrete is assumed to lose all its stress bearing capacity and stiffness.

2.4 Behaviour in tension

The main feature of plain concrete material behaviour is its low tensile strength, which results in tensile cracking at very low stress compared with the failure stress in compression. Consequently, cracking is the major cause of nonlinearity in most reinforced concrete structures.

Sophisticated theoretical models have been developed to describe the cracking and post-cracking response of concrete [6-8]. However, the usefulness of these models must be questioned in the light of the uncertainties commonly associated with the properties of reinforced concrete members [9]. The value of the concrete tensile strength is difficult to measure, and considerable scatter is obtained even in laboratory specimens. In situ, uncontrolled environmental and loading conditions will cause higher statistical scatter. In the presence of reinforcing steel the prediction of crack initiation and cracking direction is far more difficult. Consequently, very simple criteria are commonly employed by most analysts to predict tensile fracture. Usually, a maximum tensile stress or maximum tensile strain criterion is used for this purpose. In the present work, concrete in tension is modelled as a linear elastic-brittle material and the maximum tensile stress criteria is employed to distinguish elastic behaviour from brittle fracture.

The smeared crack approach is adopted for crack modelling. Unlike the discrete crack approach this avoids adjustment of the finite element mesh and allows cracks to form in arbitrary directions, not just following the element boundaries. In this approach cracked concrete is treated as a continuum, that is, cracking is taken into account by modifying the material properties, but not the structural topology.

2.4.1 Cracking criterion

As explained above the maximum tensile stress criterion is adopted. For concrete at a previously-uncracked sampling point, the principal stresses and their directions are evaluated. If the maximum principal stress exceeds a limiting value a crack is formed in a plane orthogonal to the offending stress. Thereafter, the concrete behaviour is no longer isotropic - it becomes orthotropic as the local material axes coincide with the principal stress directions.

A maximum of two sets of cracks are allowed to form at each sampling point. For simplicity the crack directions are assumed to be orthogonal.

Once the first set of cracks is formed at a point, a search is performed in the orthogonal plane to determine the maximum stress in that plane and its direction. Again, if that stress exceeds a limiting value, a new set of cracks is assumed to form and the local material axes are fixed.

The limiting value required to define the onset of cracking is established as follows:

(i) in the <u>triaxial tension zone</u>

$$\sigma_{io} = f_t' \qquad (i=1,2,3) \tag{16}$$

as there is experimental evidence that the triaxial tensile strength of concrete is almost independent of the stress ratio.

(ii) for <u>tension-tension-compression</u> and <u>tension-compression-compression stress states</u>, linearly decreasing tensile strength equations of the following form are used

$$\sigma_{io} = f_t'(1 + \frac{\sigma_{i+1}}{f_c'}); \quad \sigma_{i+1} < 0 \tag{17}$$

$$\sigma_{io} = f_t'(1 + \frac{\sigma_{i+1}}{f_c'})(1 + \frac{\sigma_{i+2}}{f_c'}); \quad \sigma_{i+1}, \ \sigma_{i+2} < 0 \tag{18}$$

These equations incorporate the observation that compression in one direction favours microcracking in the others, thus reducing the tensile capacity. The resulting cracking surfaces are shown in Figure 4.

2.4.2 Postcracking behaviour

Postcracking behaviour is based on the experimental evidence showing that after cracking concrete is still able to play an active role in stiffening the member, this influence progressively becoming weaker.

This evidence justifies the introduction of the tension - stiffening concept, illustrated in Figure 5. The stress normal to the crack does not drop to zero immediately when the crack is formed. It decreases with increasing crack width, w, or with the nominal 'tensile strain' across the crack, calculated as w/l , where l is a characteristic length in the analysis - the gauge length in an experiment. Many curves have been suggested, with linear, multi-linear, parabolic, exponential forms, etc. The manner in which parameters defining the tension-stiffening curve are chosen is more critical than its actual shape.

In the analysis of reinforced concrete structures using finite elements it has been found that the overall stiffness and predicted ultimate load decrease with the degree of refinement of the element meshes. When the number of the elements increases the cracking model exhibits greater softening. To overcome this difficulty and to relate the constitutive model to objective measures an approach is applied which assumes fracture energy, G_f, to be a material property [10-12].

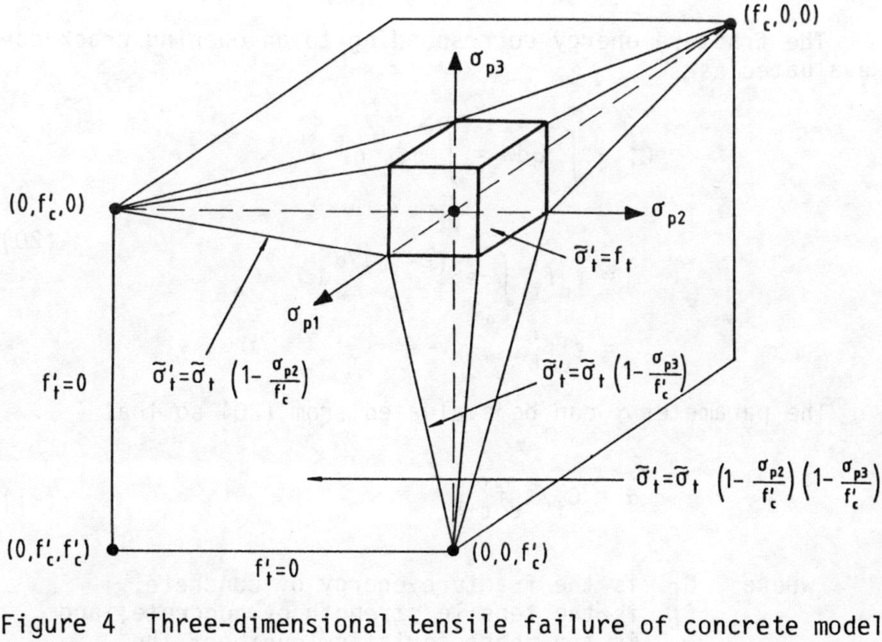

Figure 4 Three-dimensional tensile failure of concrete model

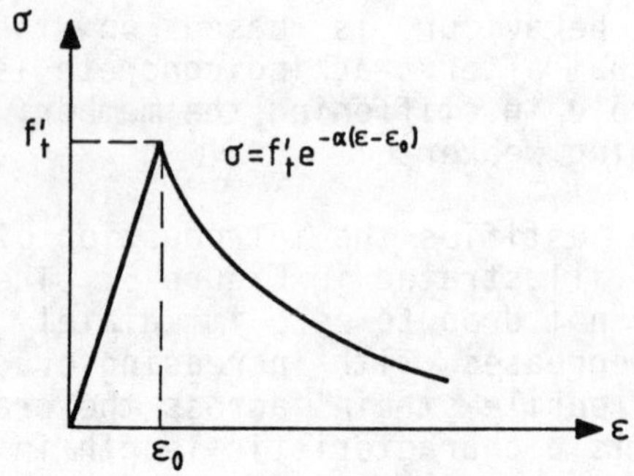

Figure 5 Tension stiffening curve

Assuming an exponential curve is used to simulate the tension - stiffening effect, then

$$\sigma = f_t' \, e^{-(\varepsilon-\varepsilon_0)/\alpha} \tag{19}$$

where f_t' is the tensile strength of concrete,
ε_0 is the strain at cracking,
α is the softening parameter, and
ε is the nominal tensile strain in the cracked zone.

The fracture energy corresponding to an opening crack can be evaluated as

$$\begin{aligned} G_f &= \int_0^\infty \sigma dw = \int_{\varepsilon_0}^\infty \sigma \ell_c \, d\varepsilon \\ &= \ell_c f_t' \int_{\varepsilon_0}^\infty e^{-(\varepsilon-\varepsilon_0)/\alpha} d\varepsilon \\ &= \ell_c f_t' \alpha \end{aligned} \tag{20}$$

The parameter α can be evaluated from (20) so that

$$\alpha = G_f/\ell_c f_t' \tag{21}$$

where G_f is the fracture energy of concrete,
f_t' is the tensile strength of concrete, and
ℓ_c is the characteristic length of the sampling point.

In the present work ℓ_c is taken for each sampling point to be

$$\ell_c = (d V)^{1/3} \tag{22}$$

where dV is the volume of concrete represented by the sampling point.

(i) The Cracked D-Matrix

Once cracked, the concrete is assumed to become orthotropic with the material axes oriented in the direction of the offending tensile stresses. The $\underline{D}$-matrix is constructed with reference to the material axes and it is then transformed into the global system.

At cracked sampling points, Poisson's ratio is assumed equal to zero, the $\underline{D}$-matrix thereby becoming a diagonal matrix.

In many previous works the elastic modulus is assumed to equal zero in the directions normal to cracked planes - this implies constant stress under further loading - see Figure 6a. Although this has proved successful in many situations it has been found that it can sometimes lead to non-positive-definite stiffness matrices and subsequent numerical difficulties. Such a problem can be easily overcome by assuming a progressively decreasing stiffness in the considered direction, that is, using a secant elastic modulus for $\underline{D}$-matrix evaluation rather than assuming the elastic modulus equals zero. The secant modulus, E, can be evaluated as

$$E_r = \alpha\ (\text{current}) / \ \varepsilon\ (\text{current}) \leq E_c \tag{23}$$

(See Figure 6b.)

The resulting $\underline{D}$-matrix for doubly-cracked concrete has the form

$$\begin{bmatrix} \Delta\sigma_1 \\ \Delta\sigma_2 \\ \Delta\sigma_3 \\ \Delta\tau_{12} \\ \Delta\tau_{23} \\ \Delta\tau_{13} \end{bmatrix} = \begin{bmatrix} E_{r_1} & & & & & \underline{0} \\ & E_{r_2} & & & & \\ & & E_c & & & \\ & & & \beta_1 G & & \\ & & & & \beta_2 G & \\ \underline{0} & & & & & \beta_1 G \end{bmatrix} \begin{bmatrix} \Delta\varepsilon_1 \\ \Delta\varepsilon_2 \\ \Delta\varepsilon_3 \\ \Delta\gamma_{12} \\ \Delta\gamma_{23} \\ \Delta\gamma_{13} \end{bmatrix} \tag{24}$$

$$\Delta\underline{\sigma} = \underline{D}_c\ \Delta\underline{\varepsilon}$$

For singly-cracked concrete $E_{r_2} = E_c$ and $\beta_2 = 1.0$.

As can be seen in (24) the shear modulus is also modified for cracked concrete, accounting for phenomena such as aggregate interlock, dowel action, etc. The shear retention factor is usually given a value of

$$\beta = 0.5 \tag{25}$$

During the loading process a previously-opened crack can begin to close and eventually close totally or re-open again. The present model allows for closing and re-opening of existing cracks along the secant path shown in Figure 6c. Once a crack is completely closed, the concrete is assumed to recover its initial strength normal to the crack.

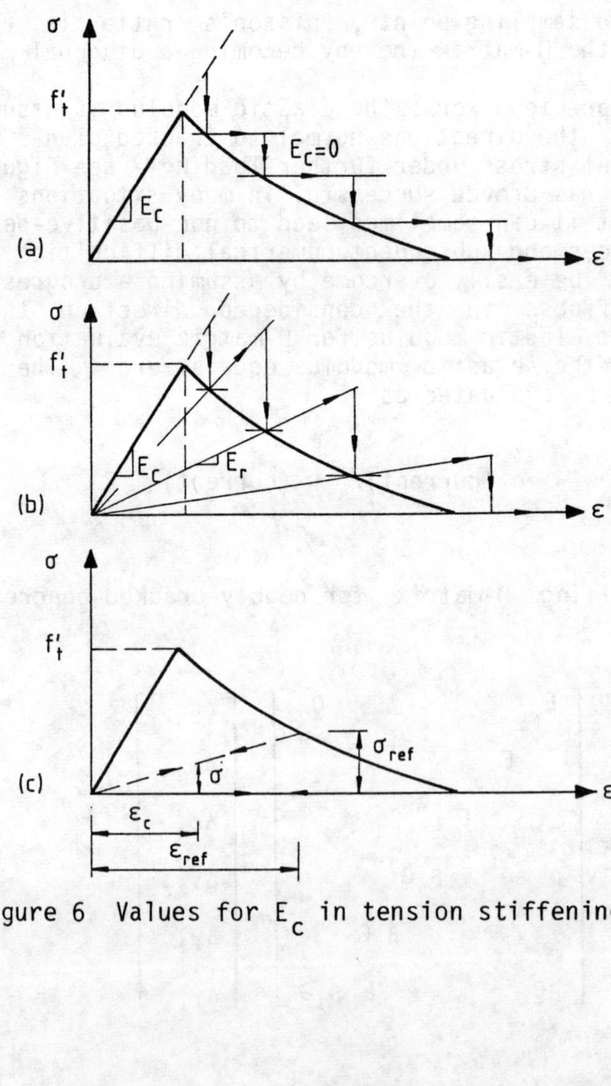

Figure 6 Values for E_c in tension stiffening curves

2.4.3 Compressive behaviour of cracked concrete

Secondary cracking can form normal to the initial cracking plane, or if compression stresses are present, plastic yielding of the concrete and, eventually, crushing may occur.

At a cracked point, the following algorithm is adopted:

(i) The stresses in the plane parallel to the initial cracking plane are evaluated and checked and secondary cracking accounted for if it is occurs. The tensile and shear stresses are modified if necessary.

(ii) This modified stress state is then employed to evaluate the effective stress. The stresses are then adjusted to satisfy the yield criterion if necessary.

(iii) The strains are checked to ascertain if crushing has occurred.

Figure 7 illustrates the process described in the biaxial principal stress space.

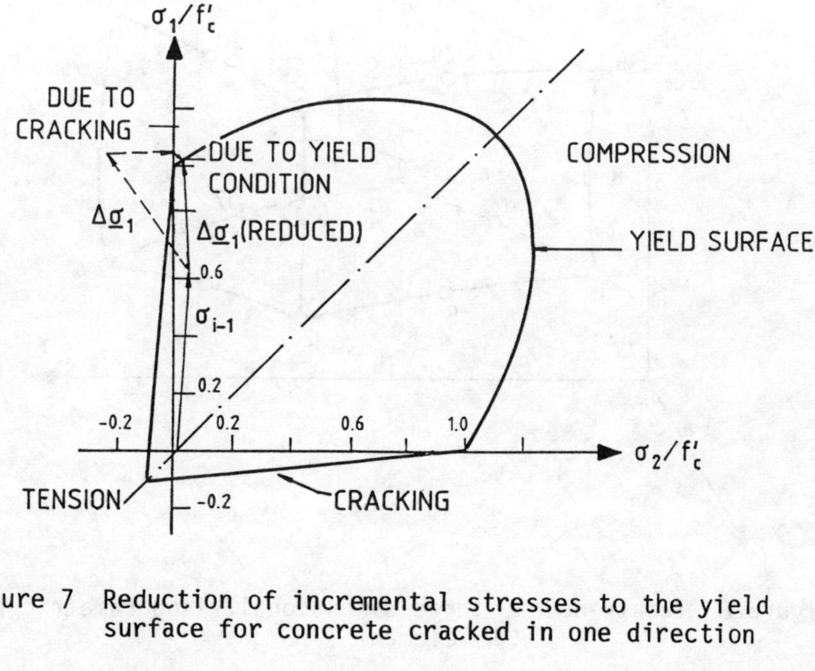

Figure 7 Reduction of incremental stresses to the yield surface for concrete cracked in one direction

It must be noted that when cracking has occurred, the normal and shear strain components directly affected by the cracking are not the real strains sustained by the concrete, but purely nominal strains used by the smeared crack model. Therefore, they must not be taken into account when the crushing condition is verified.

3. MATERIAL MODEL FOR THE REINFORCEMENT

In the present work the steel reinforcement is assumed to have uniaxial properties in the direction of the bars. An elasto-plastic formulation is adopted in which linear isotropic hardening is assumed after initial yielding. Unloading is assumed to occur elastically.

4. FINITE ELEMENT MODEL

4.1 3-D element

The 20-noded isoparametric element is used for this work and it is shown in Figure 8. The derivation of the element stiffness matrix can be found in many books [13,14].

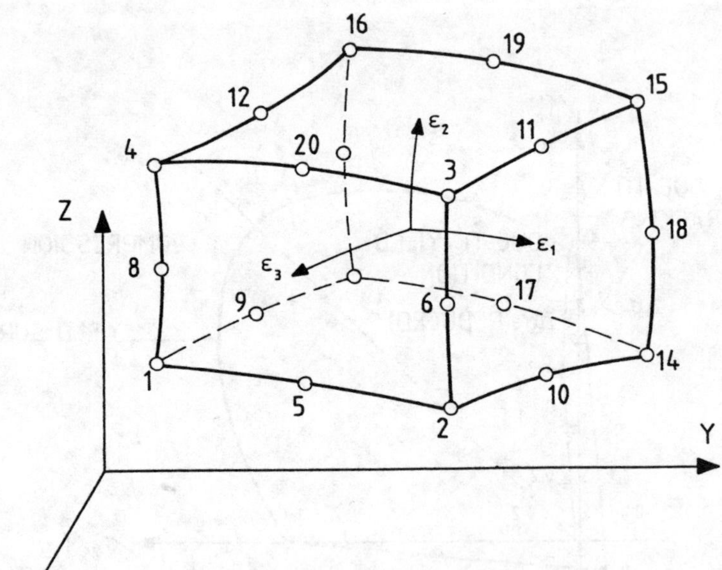

Figure 8 Twenty-noded, three dimensional, isoparametric brick element

The contribution to the stiffness matrix from the solid concrete is determined from the expression

$$\underline{K}_c = \int_V \underline{B}^T \underline{D}_c \underline{B} \, |\underline{J}| \, d\xi \, d\eta \, d\zeta \qquad (26)$$

in which $\underline{B}$ is the strain-displacement matrix and $|\underline{J}|$ is the Jacobian matrix determinant.

4.1.1 Integration rule

Twenty-noded, isoparametric, hexahedral, finite elements are highly competitive in practice. However, for linear problems, a considerable proportion of the total computation time is often absorbed in the numerical integration of the stiffness matrix coefficients. It is thus very important to choose a suitable integration scheme that is both accurate and computationally efficient.

For problems with material nonlinearities, the previous observation is even more critical since the evaluation of the stiffness matrix is frequent and, equally important, the integration points are also the sampling points for the stresses and material state which are used to evaluate the residual forces. On the one hand, few sampling points should be used to reduce the computation time. However, on the other hand, the use of a larger number of stress sampling points implies that a more complete picture of the material state is provided. Clearly, a compromise must be sought.

Usually, either Gauss-Legendre or Lobatto 3x3x3 product rules are used with 20 noded elements. Gauss quadrature is more accurate, while Lobatto rules possess the advantage of having sampling points on the element faces where the peak stresses usually occur.

In the present work we use a different set of rules, first presented by Irons [15], which have proved both accurate and efficient.

These rules are designed to integrate complete polynomials. They integrate correctly polynomials of the form $\Sigma C_{ijk} x^i y^j z^k$, $i+j+k \leq n$ whereas conventional Gauss-Legendre rules correctly integrate polynomials with more terms, i.e. $i,j,k \leq n$ - an unnecessary accuracy.

The rules can be expressed in the form

$$\int_{-1}^{+1}\int_{-1}^{+1}\int_{-1}^{+1} f(x,y,z)dxdydz = Af(0,0,0)+ \qquad \text{(1 term)}$$

$$B\{f(-b,0,0) + f(b,0,0)+....\} + \qquad \text{(6 terms)}$$

$$C\{f(-c,-c,-c) + f(-c,-c,c) + \ldots\} \quad \text{(8 terms)} \qquad (27)$$

The coefficients in (27) are for the following different rules:

	A	B	C	b	c
Rule 14	0.	0.886427	0.335180	0.795822	0.758787
Rule 15a	1.564444	0.355556	0.537778	1.	0.674100
Rule 15b	0.712137	0.686227	0.396312	0.848418	0.727662

The distribution of the sampling points over the volume of the element can be seen in Figure 9.

Of the three rules presented, rule 15a is perhaps the most attractive in the present study because

(i) it has sampling points at the centres of the element faces, and

(ii) the sum of the weights of the five points in the top and the bottom levels is slighter greater (31,33%) than in the two others.

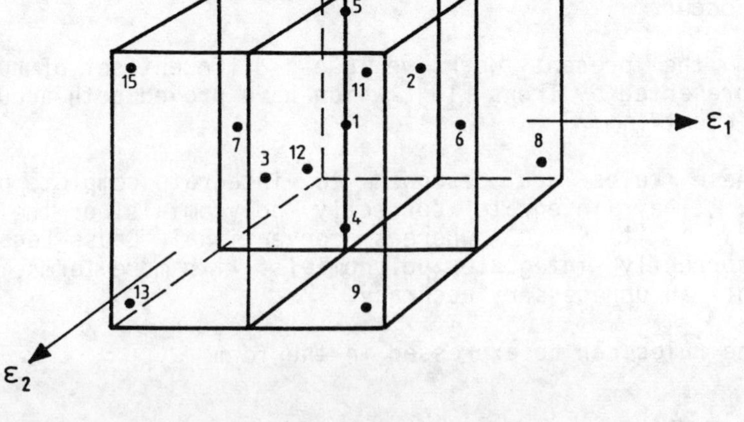

Figure 9 Three dimensional concrete element with 20 nodes and 15 integrating points

4.2 Reinforcement idealisation

Perfect bond is assumed between the steel reinforcement and the surrounding concrete. This assumption of compatibility of displacements and strains between the steel and concrete allows the reinforcement to be treated as an integral part of the 3D element. Its stiffness is added to that of the concrete to obtain the global stiffness of the element, so that

$$\underline{K} = \underline{K}_c + \underline{K}_s \tag{28}$$

Each set of reinforcement bars is smeared as a 2D membrane 'layer' of equivalent thickness, placed inside the concrete element in such a position that one of the isoparametric coordinates (ξ , η , ζ) is constant, as shown in Figure 10. The angle the bars make with the x-axis of a local Cartesian system tangent to the steel membrane at each sampling point is input as data.

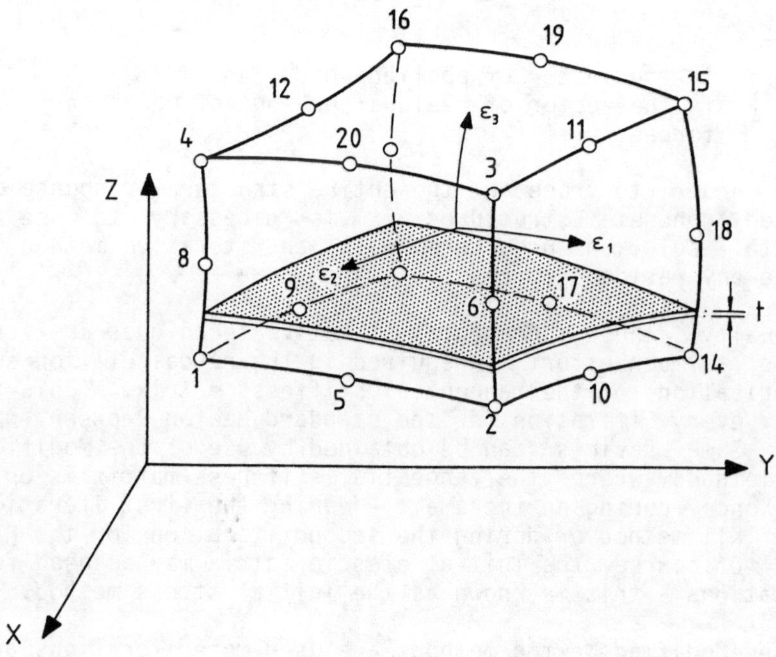

Figure 10 Three dimensional concrete element with reinforcement layer

This procedure is analogous to treatment of reinforcement in the 'layered elements' usually employed in the analysis of reinforced plates and shells.

The stiffness contribution of the steel can be evaluated as

$$\underline{K}_s = \iint_\Omega \underline{B}^T \underline{L}^T \underline{D}_s \; \underline{L} \; \underline{B} \; t \; d\Omega \tag{29}$$

where t is the equivalent thickness of the membrane and $\underline{L}$ is a transformation matrix from the local to the global system.

5. NONLINEAR SOLUTION TECHNIQUES

5.1 Incremental-iterative procedures

The equilibrium equations in the nonlinear analysis of reinforced concrete structures may be expressed as

$$\underline{\Psi} = \int \underline{B}^T \underline{\sigma} dv - \underline{f} = \underline{0} \tag{30}$$

in which $\underline{f}$ is the vector of applied loads, and
$\underline{\Psi}$ is the vector of residual or 'out-of-balance' forces.

In order to trace out the entire structural response of reinforced concrete structures it is necessary to use an incremental solution procedure with an iterative method to dissipate any residual forces.

Usually, these incremental-iterative techniques are very expensive as much effort is required in the re-calculation and re-factorisation of the tangential stiffness matrix. This is done at every iteration for the standard Newton-Raphson (NR) method. Some savings can be obtained by use of the Modified Newton methods where the tangential stiffness matrix is only updated once during an increment - during the first iteration for the KT1 method or during the second iteration for the KT2 method. Of course, the initial elastic matrix may be used for all iterations - this is known as the initial stress method.

When Modified Newton methods are used more iterations are required to achieve solution convergence for each load increment than with the standard NR method.

5.2 Line search procedures

Line search procedures have been reported to be quite effective in the solution of material and geometrically nonlinear problems. In particular, they have been succesfully employed in the nonlinear analysis of concrete structures and they are therefore used in the present work to improve the convergence of the Modified Newton method.

In problems with N-degrees of freedom, given an approximation $\underline{d}_i$ to the solution of a system of equations

$$\underline{\Psi}(\underline{d}_i) \simeq \underline{0} \tag{31}$$

any suitable iterative scheme will choose a search direction $\delta\underline{d}_i$ in N-dimensional space in which to look for a better approximation. The new guess will lie in the line $\underline{d}_{i+1} = \underline{d}_i + \eta_i \delta\underline{d}_i$. We cannot expect to find the exact solution with this approximation but we can choose the scalar multiplier step length η_i so that the component of Ψ_{i+1} in the search direction is very small, thus implying that equilibrium is achieved. This can be interpreted as

$$\delta\underline{d}_i^T \, \underline{\Psi}_{i+1} \, (\underline{d}_{i+1}(\eta_i)) = 0 \tag{32}$$

An exact line search can be prohibitively expensive since it requires frequent calculation of the residual forces in each trial. In the present work a less stringent requirement is adopted in which

$$|\delta d_i^T \, \underline{\Psi}_{i+1}(\eta_{i,j})| < \varepsilon \, |\delta \underline{d}_i^T \, \underline{\Psi}_i| \tag{33}$$

where j is the line search trial number, and
ε is an input tolerance.

5.3 Arc-length methods

Arc-length methods have gained much popularity due to their ability to deal with geometrically nonlinear problems involving snap-through and snap-back behaviour. In addition to overcoming the limit point problem, the arc-length methods are computationally beneficial through the entire loading range and therefore they have been implemented for the present study.

The basic idea in all arc-length methods is to modify the load level at each iteration, rather than holding the applied load level constant during a load step, so that the solution follows some specified path until convergence is attained [16]. The process is shown in Figure 11. The selected path may be in

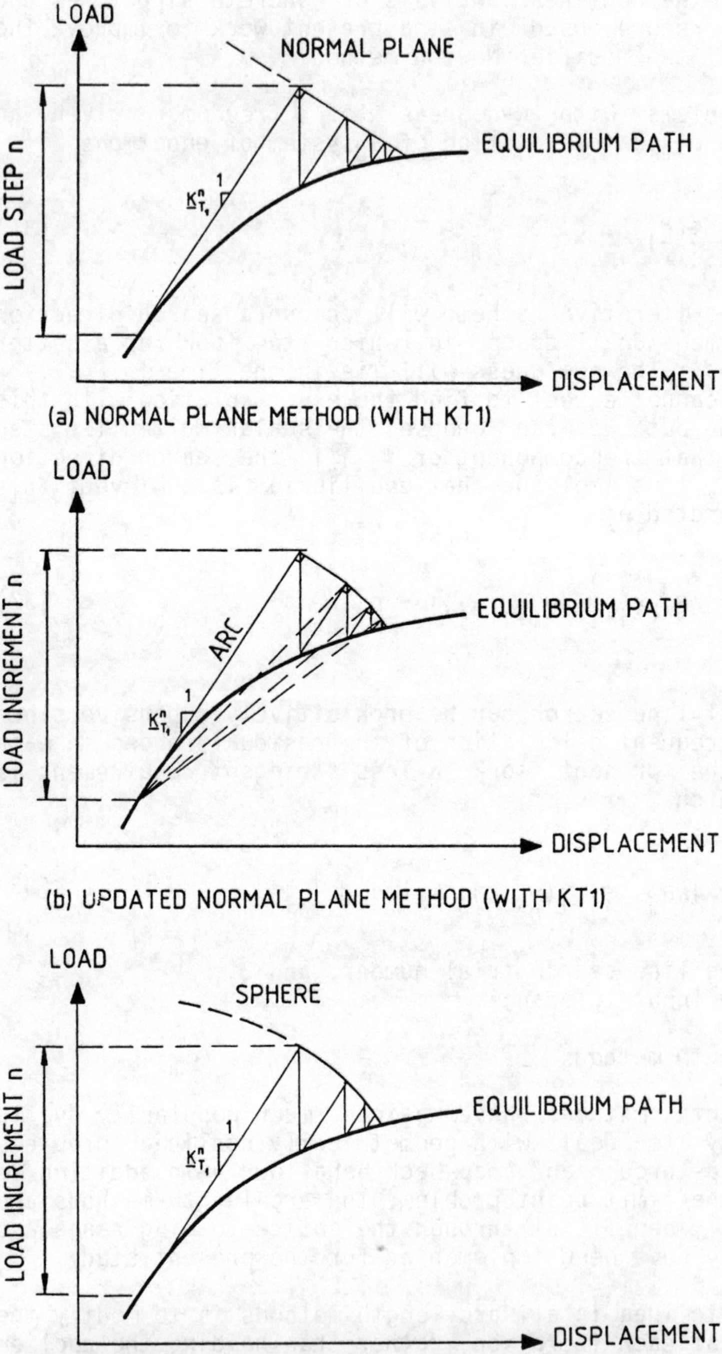

(a) NORMAL PLANE METHOD (WITH KT1)

(b) UPDATED NORMAL PLANE METHOD (WITH KT1)

(c) SPHERICAL PATH METHOD (WITH KT1)

Figure 11 Arc length methods

a normal plane (the normal-plane method, Figure 11a), or normal to the arc at the current iteration (the updated-normal-plane method, Figure 11b), or part of a sphere of a given radius (the spherical-path method, Figure 11c). In the arc-length methods the loading should be proportional, which means that an increment of load can be given as $\Delta\lambda_i^n \underline{f}$ where $\Delta\lambda_i^n$ is a scaling factor for increment n at iteration i and $\underline{f}$ is some reference applied load vector.

The effect of varying the load level during the iterative process can be explained as follows:

The out-of-balance force vector $\underline{\Psi}_i$ at load level λ_i is given by

$$\underline{\Psi}_i(\lambda_i) = \underline{r}_i - \lambda_i \underline{f} \tag{34}$$

where $\underline{r}_i$ is the vector of internal equivalent nodal forces. Similarly, for some adjusted load level $\lambda_i + \delta\lambda_i$

$$\underline{\Psi}_i(\lambda_i + \delta\lambda_i) = \underline{r}_i - (\lambda_i + \delta\lambda_i)\, \underline{f} \tag{35}$$

or, by using (34)

$$\underline{\Psi}_i(\lambda_i + \delta\lambda_i) = \underline{\Psi}_i(\lambda_i) - \delta\lambda_i \underline{f} \tag{36}$$

Upon substitution from the basic equations for the iterative process

$$\underline{\Psi}_i^{\,n} = -\underline{K}_{Ti}^{\,n}\ \delta\underline{d}_i^{\,n} \tag{37}$$

and premultipliying by $-[\underline{K}_{Ti}^{\,n}]^{-1}$, we obtain

$$\delta\underline{d}_i(\lambda_i + \delta\lambda_i) = \delta\underline{d}_i(\lambda_i) + \delta\lambda_i\ \underline{K}_{Ti}^{-1}\,\underline{f} \tag{38}$$

in which $\delta\underline{d}_i(\lambda_i)$ are the iterative displacements corresponding to the residual forces $\underline{\Psi}_i(\lambda_i)$.

Equation (38) describes the effect of varying the load level during an iteration i on the resulting iterative displacements $\delta\underline{d}_i$. This fundamental equation is the basis for all arc-length methods which differ only in the method for evaluating $\delta\lambda_i$ and the specified path. The detailed derivation of expressions for $\delta\lambda_i$ are given elsewhere [17].

5.4 Convergence criteria

When using an incremental-iterative procedure to solve a nonlinear problem a convergence criterion must be established to define when the iterative process can be stopped and a new load increment applied.

Displacement-based criteria can be used, in which iterations are stopped when the norm of the iterative displacement vector is lower than a specified percentage of the norm of the displacement vector for the first iteration of the increment. A drawback of this criterion is that it does not ensure that equilibrium has been accomplished.

Residual force convergence criteria are far more reliable. In such criteria the norm of the out-of-balance force vector is checked against the norm of the total applied load vector or the vector of calculated reactions. Convergence is achieved when

$$\frac{||\underline{\Psi}_i||}{\max\left(||\underline{r}_i||,\ \lambda_i||\underline{f}||\right)} \leq \varepsilon \tag{39}$$

where $||\underline{r}||$ is the norm of the reaction vector, and ε is a specified tolerance.

This criterion is used in the present work.

6. NUMERICAL EXAMPLES

6.1 General

In the previous sections theoretical and computational aspects of the nonlinear analysis of reinforced concrete structures have been described. In this section some practical applications of the proposed model are presented and comparisons with experimental and other numerical solutions are given.

6.2 The Bresler-Scordelis Beam

6.2.1 Description of the beam and its loading

The simply supported, reinforced concrete beam tested by Scordelis and Bresler [18], is shown in Figure 12. This beam has a span of 12 ft. with cross-sectional dimensions 21.75 in. x 9 in. It has no web reinforcement and the longitudinal reinforcement consists of four bars with a cross-sectional total area 4 in².

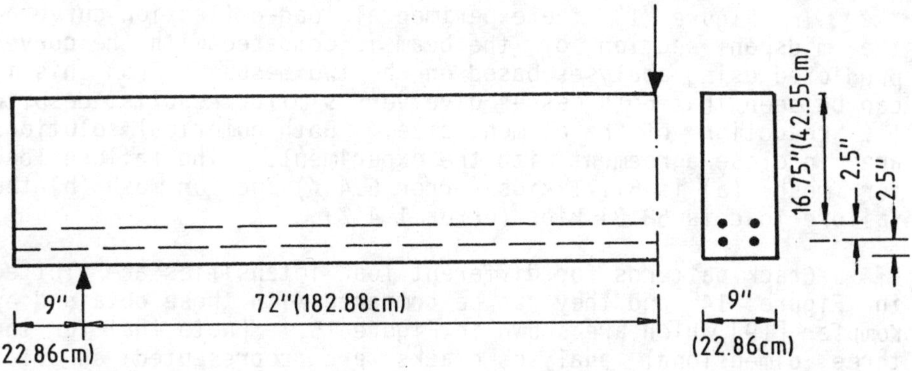

Figure 12 Bresler-Scordelis beam

The beam is subjected to a central concentrated load and the experimentally determined failure load by crushing of the concrete is 58 kips.

The experimental results are considered to be very reliable so the beam has almost become a benchmark for testing of analytical and numerical formulations.

6.2.2 Element meshes and material constants

Taking advantage of symmetry only one half of the beam is idealised and two different meshes are considered:

mesh (a) - which consists of five equal element along the span, and

mesh (b) - which consists of ten equal elements disposed in two layers through the thickness.

Material constants used in the analysis are:

E_c = 3300. kips/in² $\quad$ ε_u = 0.002
f_c' = 3.16 kips/in² $\quad$ E_s = 27800. kips/in²
f_t' = 0.333 kips/in² $\quad$ H' = 0. kips/in²
G_f = 0.0006 kips/in $\quad$ f_y = 60. kips/in²
ν = 0.2

6.2.3 Comparison with test results

In Figure 13 the experimental load-deflection curve at the midspan section of the beam is compared with the curves predicted using analyses based on the two meshes. From this it can be seen that both meshes give very similar results, despite the reduction of the element size. Both numerical solutions are in close agreement with the experiment. The failure load for mesh (a) is 61.71 kips (error 6.4 %) and for mesh (b) the failure load is 58.84 kips (error 1.4 %).

Crack patterns for different load intensities are plotted in Figure 14 and they can be compared with those obtained by Kompfer [19] which are shown in Figure 15. [Note that for the three dimensional analyses cracks are represented as flat squares in the physical crack plane. The distortion of the squares in the figures gives an idea of the inclination of the cracks.] Only two orthogonal cracks are allowed to open at each sampling point. When two non-orthogonal cracks appear at a point in one of the figures, or more than two cracks are present, this is because that particular point is actually two sampling points for two adjacent elements and cracks in both elements are plotted at the same position.

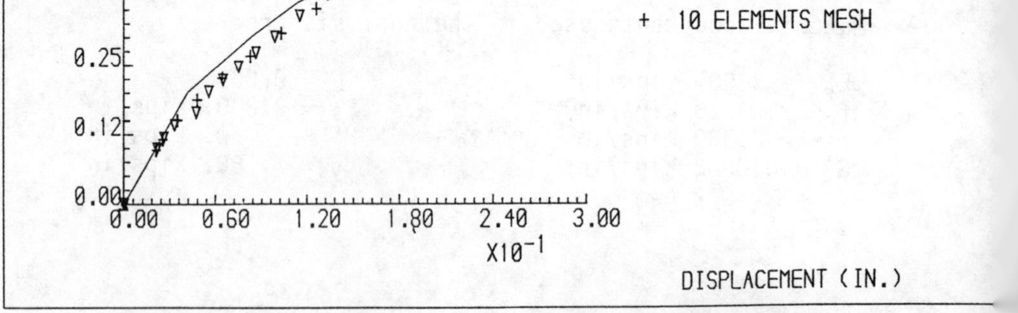

Figure 13 Load-displacement curve for Bresler-Scordelis beam

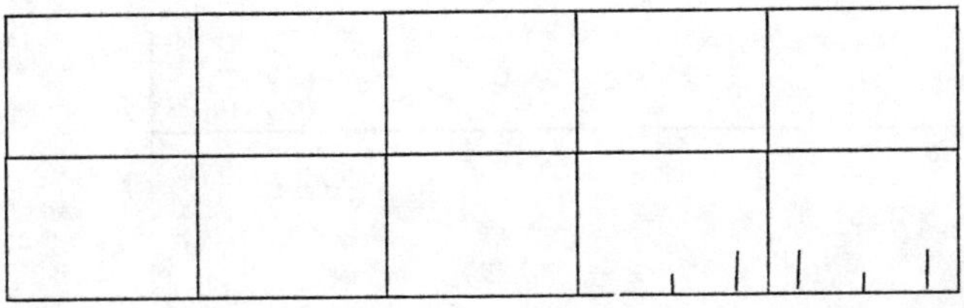

P = 15.0 kips

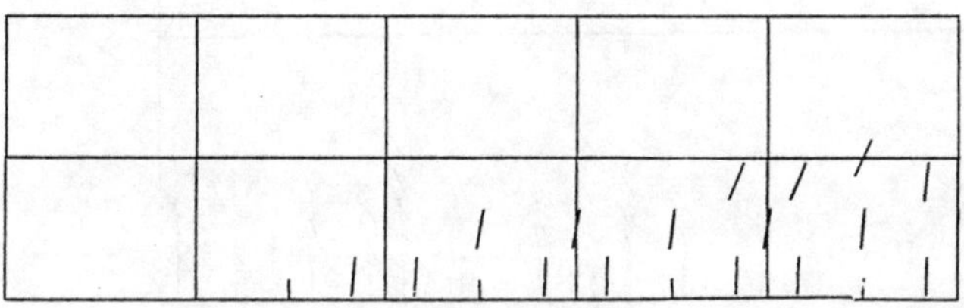

P = 32.5 kips

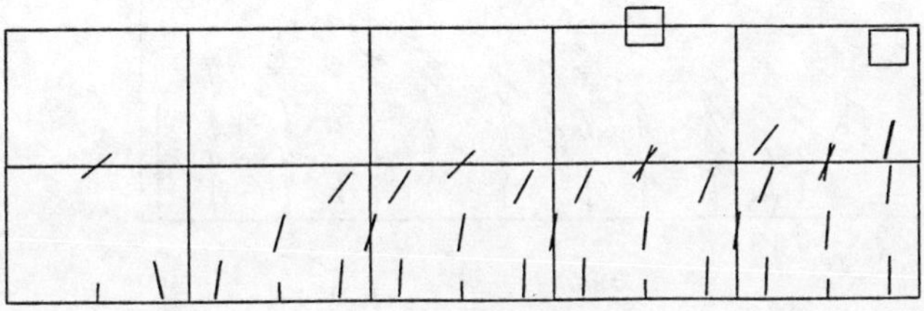

P = 54.0 kips

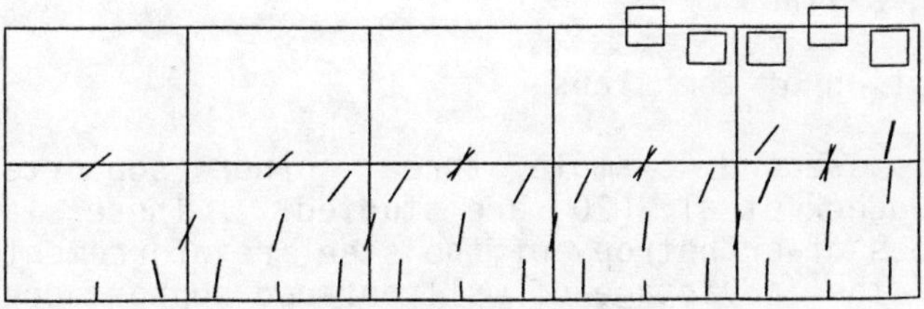

P = 58.8 kips

Figure 14 Crack patterns for Bresler-Scordelis beam

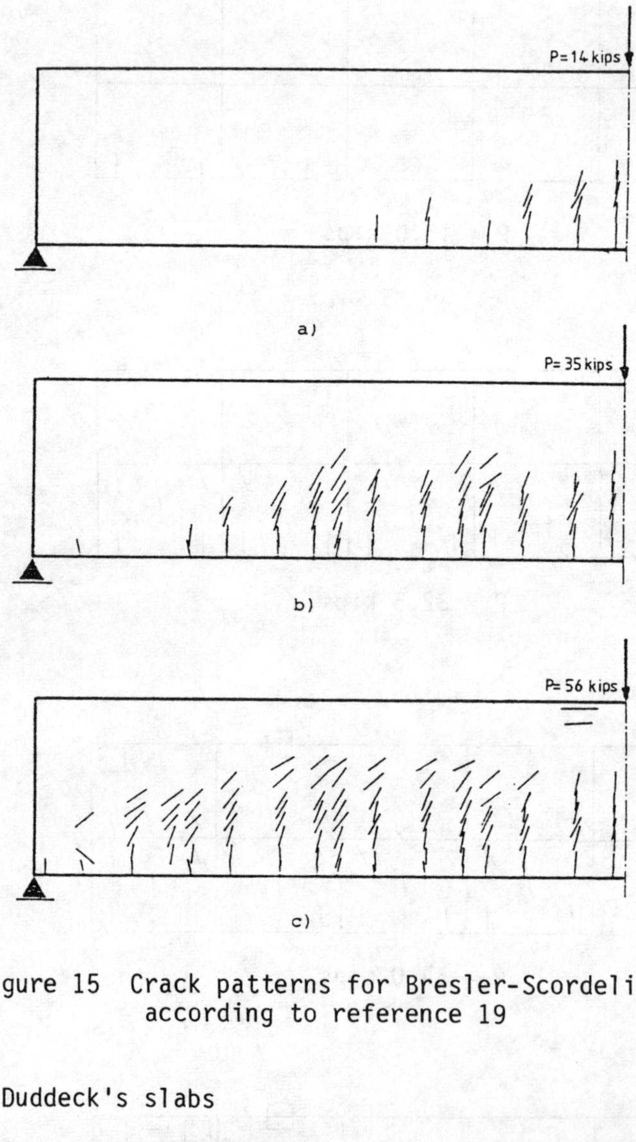

Figure 15 Crack patterns for Bresler-Scordelis beam according to reference 19

6.3 Duddeck's slabs

6.3.1 Description of the slabs

In this second example three corner supported slabs tested by Duddeck et al. [20] are studied. These slabs have various degrees of orthotropy of the steel reinforcement. This example has the advantage of well defined support conditions where only transverse deflections are restrained at the corner supports.

The material properties are rather poorly defined. The values assumed by Figueiras [3] are used here:

$E_c = 16400.\ N/mm^2$	$E_s = 201000.\ N/mm^2$
$f_c' = 43.\ N/mm^2$	$H^s = 0.\ N/mm^2$
$f_t' = 3.\ N/mm^2$	$f_y = 670.\ N/mm^2$
$G_f = 0.2\ N/mm$	
$\nu = 0.$	
$\varepsilon_u = 0.0035$	

The equivalent reinforcement layers for the three slabs are given below in millimetres:

	Top layers		Bottom layers	
	x-dir	y-dir	x-dir	y-dir
S1	0.193	0.193	0.397	0.397
S2	0.252	0.133	0.520	0.273
S3	0.283	0.103	0.582	0.212

All three slabs were tested under the same loading conditions of a concentrated load at the slab centre.

6.3.2 Finite element idealization

In the finite element analysis, by taking advantage of symmetry, it is only necessary to consider one quarter of the slab. A 3x3x1 mesh of 20-noded elements is adopted, as shown in Figure 16. The mesh is refined near the centre of the slab where higher stresses are expected due to the application of the central point load.

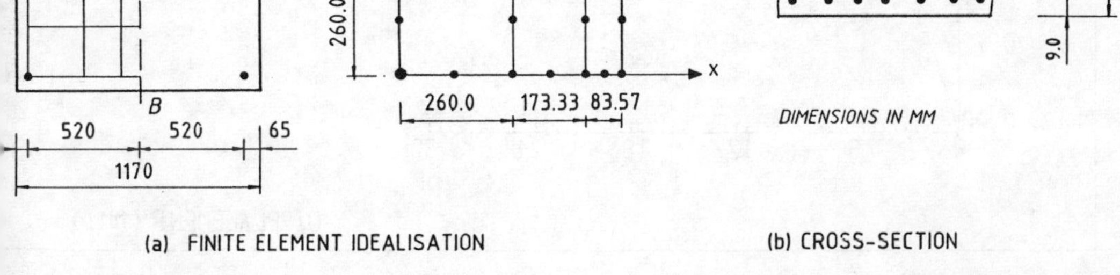

Figure 16 Structure and idealization of Duddeck's slabs

6.3.3 Results

The failure loads (expressed in Kilo-Newtons) obtained in the numerical analysis are now listed:

	Numerical	Experimental	Error
S1	59.61	61.06	-3.32 %
S2	45.05	43.46	+3.66 %
S3	36.29	34.25	+5.95 %

Excellent agreement with the experiments is observed and is also found through the whole loading process as can be seen from the load-central deflection curves of Figure 17a-c.

The crack patterns obtained for the bottom surface at different load levels are shown in Figures 18a-c and they can be compared with results obtained by Abdel Rahman [17] which are shown in Figures 19a-c.

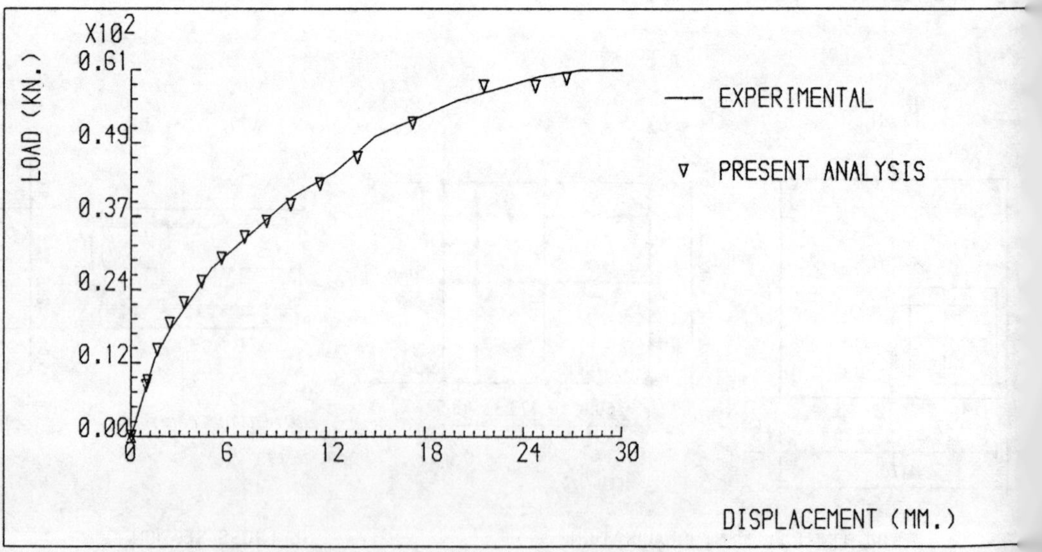

Figure 17(a) Load-displacement curve for Duddeck's slab S1

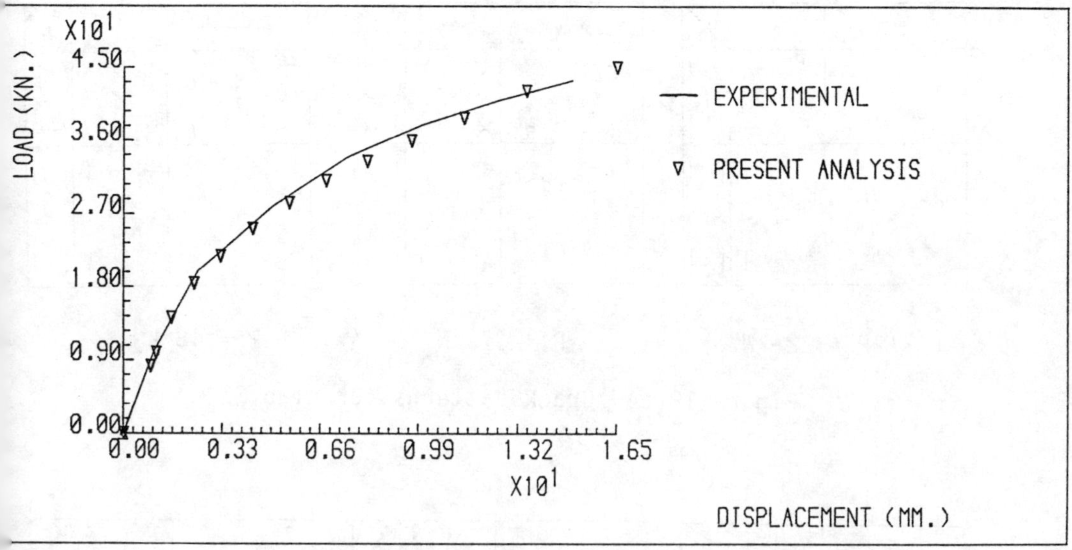

Figure 17(b) Load-displacement curve for Duddeck's slab S2

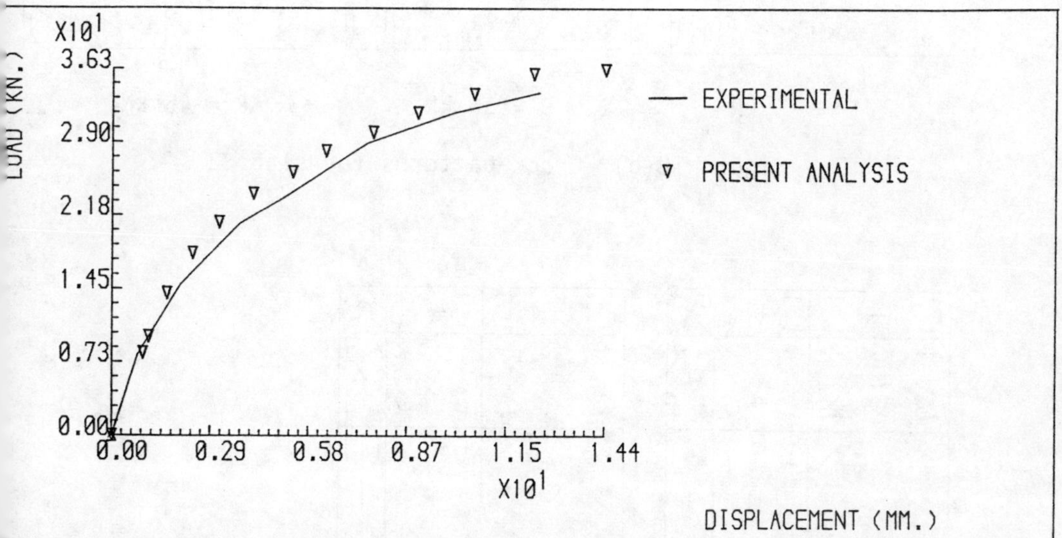

Figure 17(c) Load-displacement curve for Duddeck's slab S3

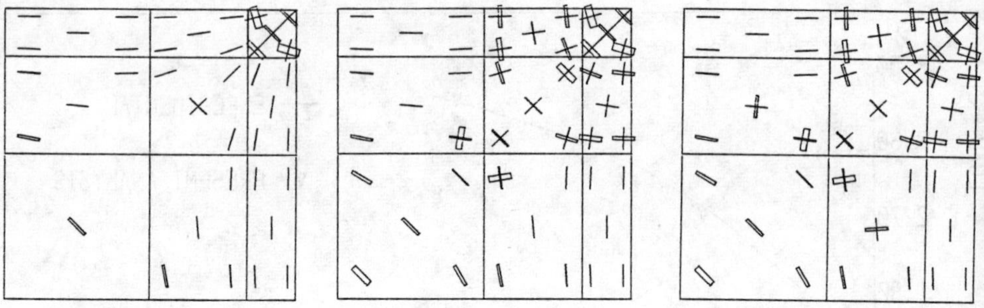

P = 22 kN P = 37 kN P = 48 kN

Figure 18(a) Crack patterns for slab S1

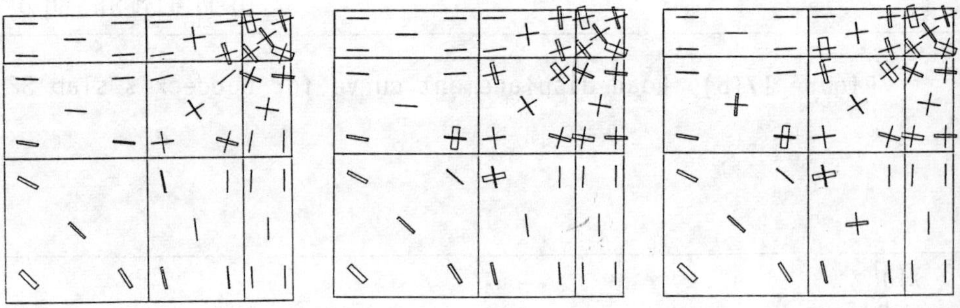

P = 25 kN P = 36 kN P = 45 kN

Figure 18(b) Crack patterns for slab S2

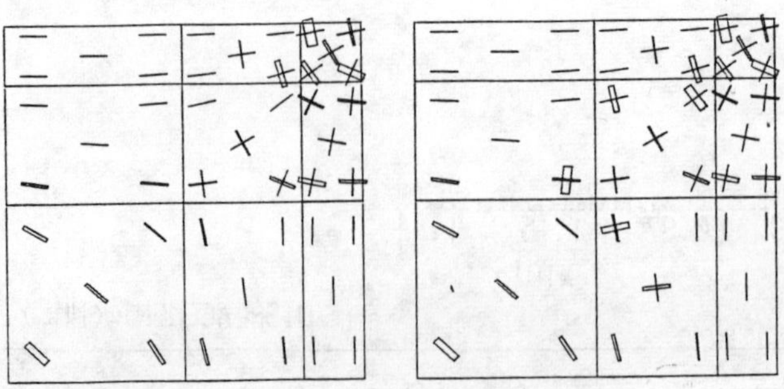

P = 26 kN P = 36 kN

Figure 18(c) Crack patterns for slab S3

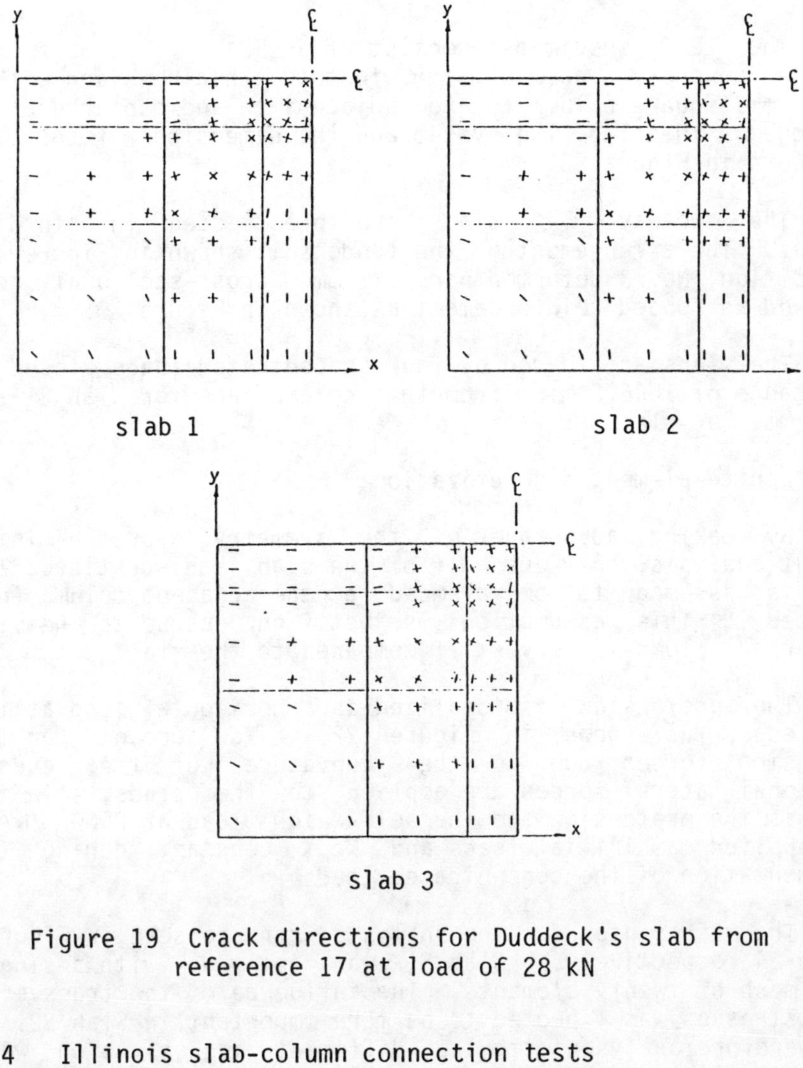

Figure 19 Crack directions for Duddeck's slab from reference 17 at load of 28 kN

6.4 Illinois slab-column connection tests

Finally, two tests from a series carried out by Sunidja [21] at the University of Illinois in Urbana-Champaign are studied. The object of the tests was to study the strength and behaviour of unbonded prestressed concrete plate-edge column connections representative of those used in prestressed flat plate buildings and subjected to static vertical loading.

6.4.1 Description of the slabs

The test specimens consist of a 1524 mm. square prestressed concrete slab, 101.6 mm. in total thickness and with a 304.8 mm. square column located adjacent to and centered along one edge of the slab. The system and the dimensions of the slab are shown in Figure 20.

The two-way flat slabs are prestressed in both directions. The arrangement of the tendons is shown in Figure 21. In addition, No. 3 deformed bars (71 mm. cross-sectional area) are used as bonded reinforcement as shown in Figure 20.

The slabs are loaded by four concentrated loads placed at a distance of 1066.8 mm. from the column face for slab S1 and 609.6 mm. for slab S2.

6.4.2 Finite element discretization

By taking advantage of the symmetry, for the finite element analysis, only one half of the slab is discretized. The slab is assumed to be clamped at the adjacent column (not analyzed). This assumption is justified because the heavily reinforced column is very stiff compared to the slab.

The pretension is idealized as a horizontal load at the slab edges as shown in Figure 22. To account for the pretension forces due to the curvature of the tendons additional lateral forces are applied to the slabs. In the analysis the pretension and the self-weight load of 25000 N/m^3 are applied as initial loads and kept constant during the incrementation of the concentrated loads.

The meshes used for the analysis can be seen in Figures 23 and 24 respectively. Slab S1 was analysed with a single layer mesh of twenty elements. The influence of the transverse shear stresses was expected to be more important in Slab S2. It was therefore analysed with two different meshes which were identical in plan. However, one mesh had one layer of elements through the thickness, whereas the other had two layers.

The material properties used in the analysis are:

E_c	= 29000. N/mm^2	E_s	= 200000. N/mm^2
f_c'	= 50. N/mm^2 for S1	H'	= 0.
	= 43. N/mm^2 for S2	f_y	= 500. N/mm^2
f_t'	= 3.67 N/mm^2 for S1		
	= 3.15 N/mm^2 for S2		
G_f	= 0.1 N/mm		
ν	= 0.15		
ε_u	= 0.003		

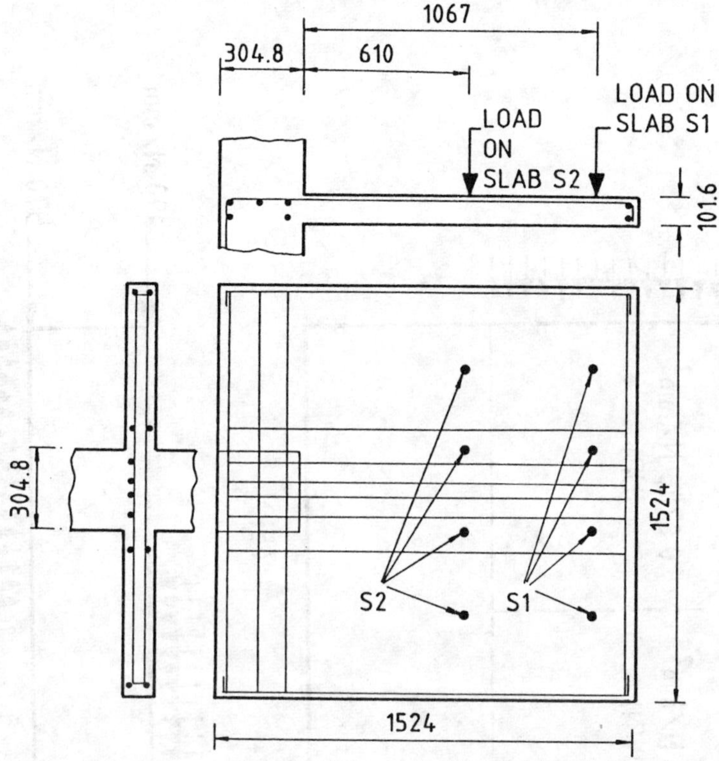

Figure 20 Test specimens with bonded reinforcement

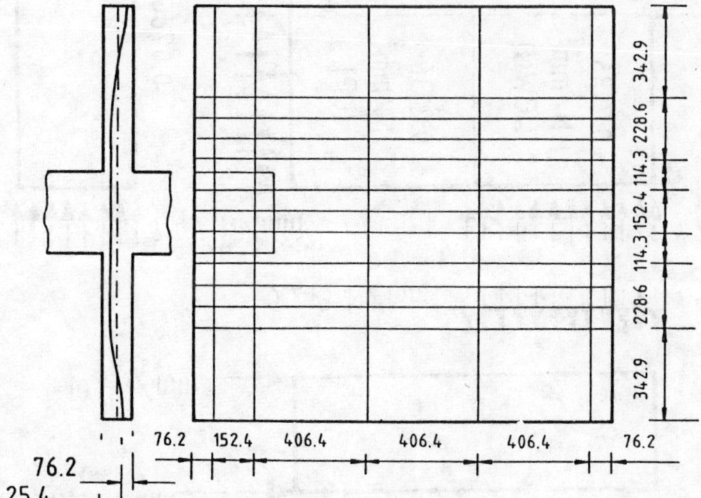

Figure 21 Unbonded tendons

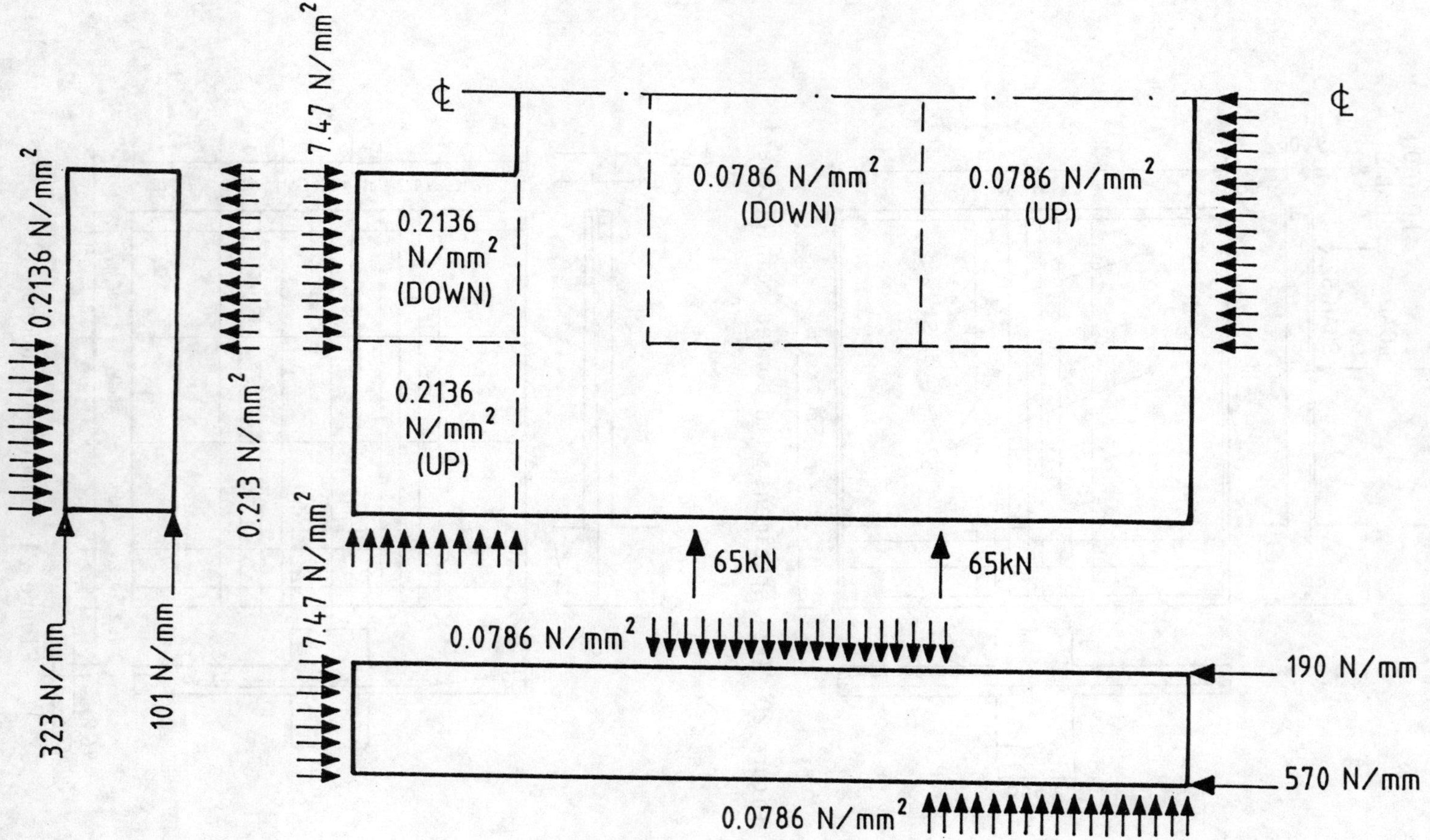

FIGURE 22 LOAD PATTERN FOR PRETENSION

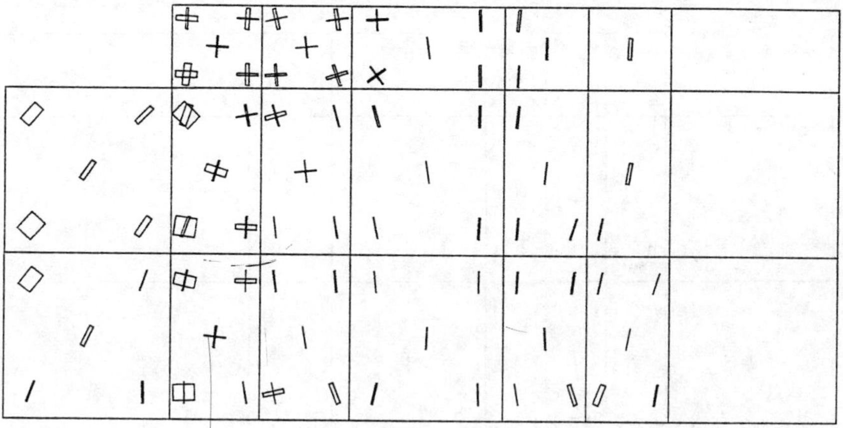

(a) top surface

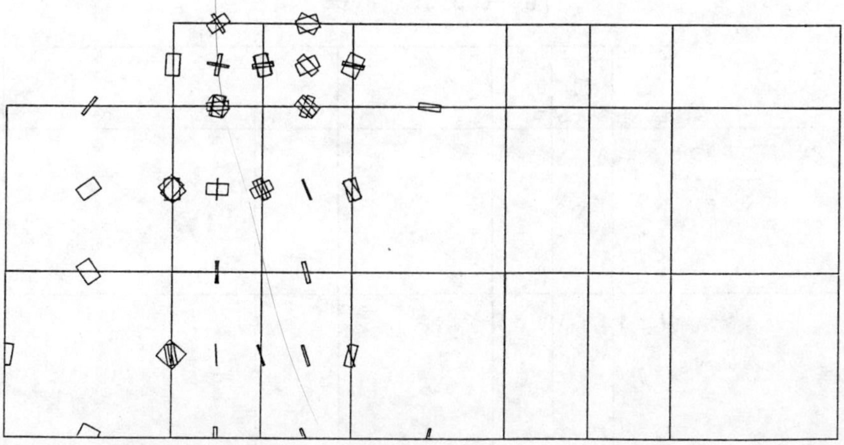

(b) middle surface

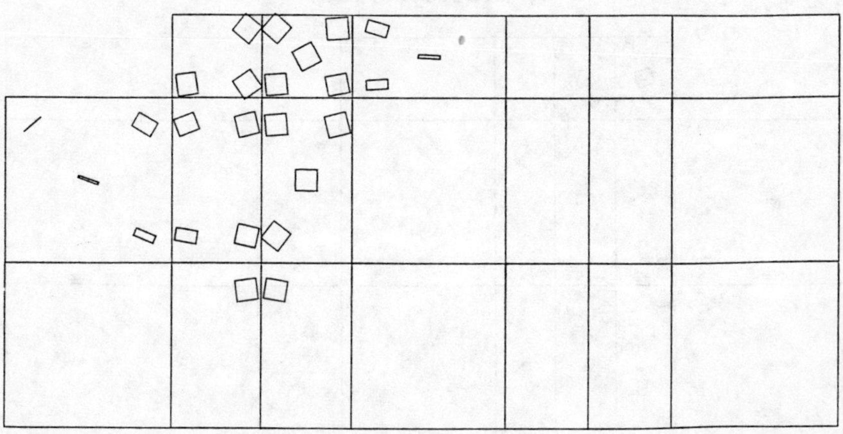

(c) bottom surface

Figure 23 Cracking in slab S1

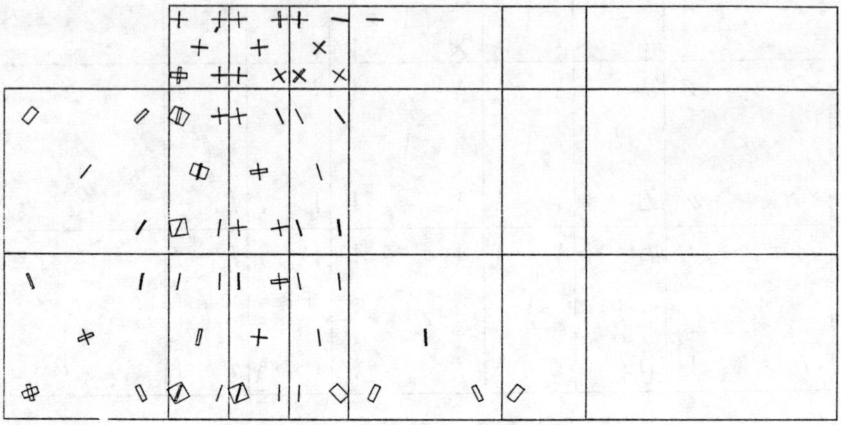

(a) top surface

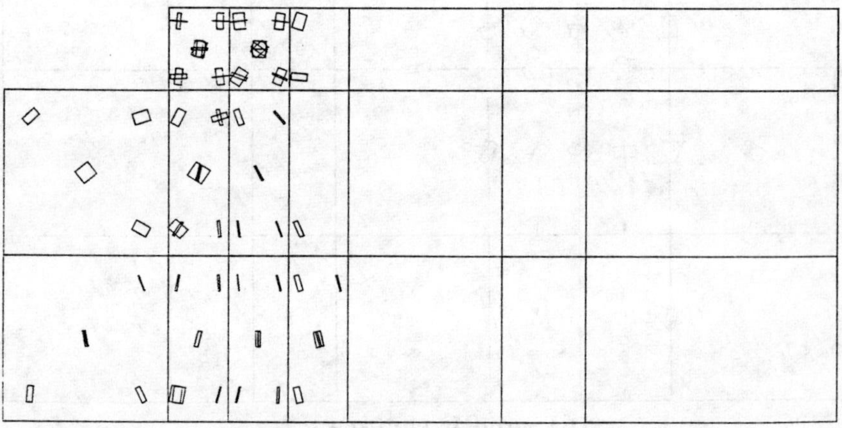

(b) middle surface

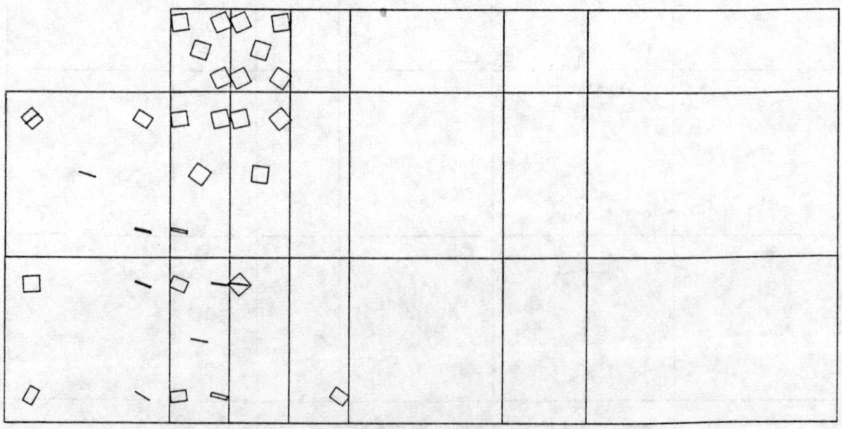

(c) bottom surface

Figure 24 Cracking in slab S2

6.4.3 Comparison with test results

In Figures 25 and 26 the moment-deflection relations obtained from the analysis are compared with the test results. The moment is obtained by multiplying the total applied load by the distance from the load to the face from the column. The initial moment caused by the selfweight of the slab is 288 kN/cm. The deflections shown are the additional value due to the concentrated loads and do not include the deformation caused by either the dead load or by the pretension. The deflection is measured at the east edge of the slab, and in the symmetry axis.

Experimentally, slab S1 failed in a flexural mode, while slab S2 collapsed due to a dominant shear crack. Figure 25 shows that both the initial stiffness and the progressively nonlinear response of slab S1 is accurately predicted by the analysis. The failure load is only slightly higher than the experimental one. Results for slab S2 shown in Figure 26, are not in such close agreement with the experiment. As expected, discretisation through the thickness slightly affects the load-deflection curve and reduces the failure load. Better results can be obtained if a reduced value for the fracture energy is used. A more flexible response and a lower failure load is obtained. This departure of the analysis from the test results can be explained by recalling that the smeared crack approach may not be the optimal approach for the analysis of structures failing due to a dominant crack. Such idealisations tend to spread the discontinuity, thus producing a stiffer response.

Crack patterns at peak load can be seen in Figures 23 (for slab S1) and 24 (for slab S2) for the top, middle and bottom surfaces.

As an additional test for the cracking model a loading-unloding test was performed on slab S1. Experimental data is available and plotted in Figure 27. The numerical analysis was performed using the modified Newton KT1 method when loading and the initial elastic matrix when unloading and reloading. The result from the numerical analysis is plotted in Figure 28. Some remarks on these results will now be presented:

(a) The unloading paths tend to go back to the origin. This is caused by the secant model for the closing of cracks and it agrees with the experiment.

(b) Upon reloading the path returns (within the tolerance used in the analysis) to the point where the unloading had begun.

(c) The actual unloading and reloading paths must depend strongly on the nonlinear algorithm used, that is, on the stiffness matrix used for iterating, because this is the only factor controlling the order in which cracks will be closed and re-opened.

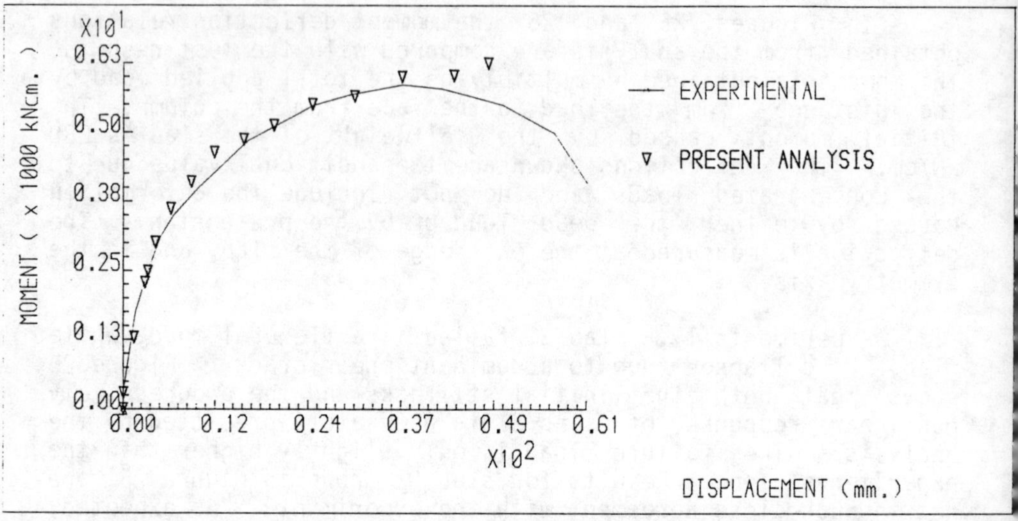

Figure 25 Load-displacement curve for Illinois slab S1 (monotonic loading)

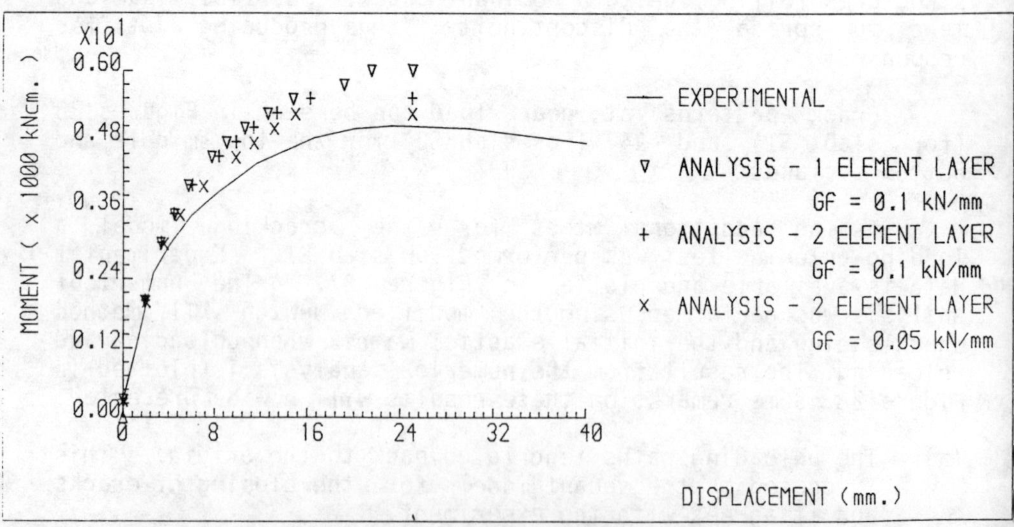

Figure 26 Load-displacement curve for Illinois slab S2 (monotonic loading)

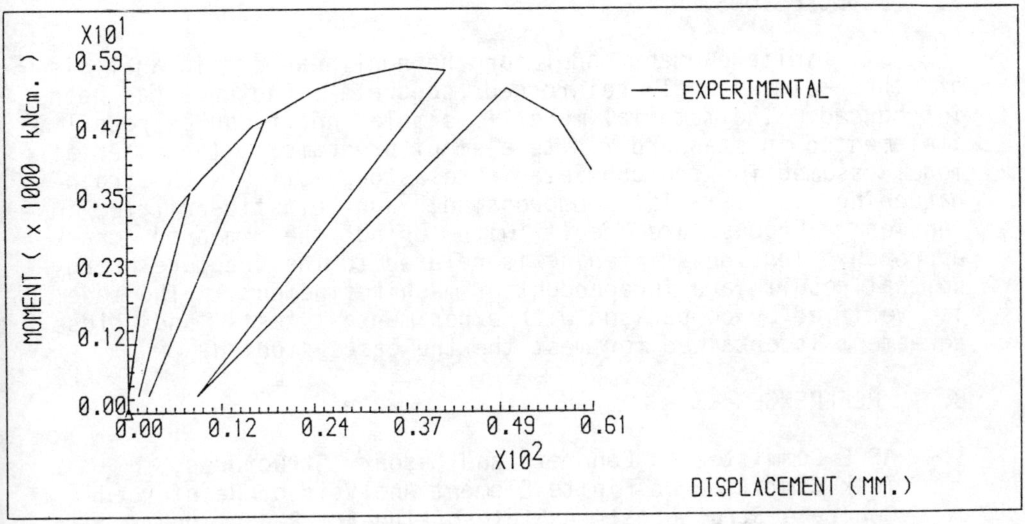

Figure 27 Load-displacement curve for Illinois slab S1

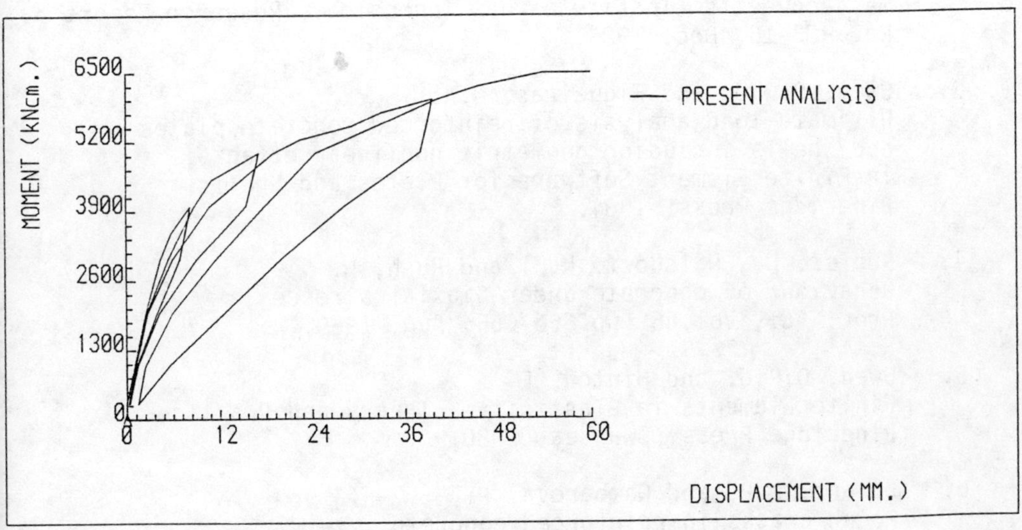

Figure 28 Load-displacement curve for Illinois slab S1 (loading- unloading)

7. CONCLUSIONS

A finite element model for the nonlinear static analysis of three-dimensional reinforced concrete structures has been introduced. The material model is simple and it can be readily implemented in standard finite element programs. The material model assumed for the concrete is elasto-plastic with strain-hardening in triaxial compression, and elastic-brittle in tension. Cracks are dealt with using the smeared crack approach. Tension-stiffening is related to the fracture energy so that results are independent of mesh refinement. The model is verified by comparison with experimental tests and close agreement is obtained for most the the cases studied.

8. REFERENCES

1. ASCE Committee on Concrete and Masonry Structures, Task Committee on Finite Element Analysis of Reinforced Concrete Structures: A State-of-the-Art Report on Finite Element Analysis of Reinforced Concrete Structures, ASCE Spec. Pub. 1981.

2. Shareef, A.S. and Buyukozturk, O. Constitutive modelling of conrete in finite element analysis, Massachusetts Institute of Technology, Research Report No. R83-16, Dec. 1983.

3. Owen, D.R.J. and Figueiras, J.A. Ultimate load analysis of reinforced concrete plates and shells including geometric nonlinear effects, in Finite Element Software for Plates and Shells, Pineridge Press, 1984.

4. Kupfer, H., Hilsdorf, K.H. and Rush, H. Behaviour of concrete under biaxial stresses, Proc. ACI, Vol.66, pp.656-666, Aug. 1969.

5. Owen, D.R.J. and Hinton, E. Finite Elements in Plasticity - Theory and Practice, Pineridge Press, Swansea, 1980.

6. Bazant, Z.P. and Gambarova, P. Rough cracks in reinforced concrete, ASCE Vol.106, No. ST4, pp.819-842, April 1980.

7. Gambarova, P. On aggregate interlock mechanism in reinforced concrete plates with extensive cracking, IABSE Colloquium, Delft, pp.105-134, 1981.

8. Walraven, J.C. Aggregate interlock: A theoretical and experimental analysis, Delft University Press, 1980.

9. Cope, R.J.
Material modelling of real reinforced concrete slabs, in Proc. of Int. Conf. on Computer Aided Analysis and Design of Concrete Structures, held in Split, Sept. 1984, Pineridge Press, Swansea 1984.

10. Jian-Jing, J.
Finite element techniques for static analysis of structures in reinforced concrete,
Dept. of Structural Mechanics, Chalmers University of Technology, Pub 83:2, 1983.

11. Glemberg, R.
Dynamic analysis of concrete structures,
Chalmers University of Technology, Pub. 84:1, 1984.

12. Rots, J.G., Nauta, P., Kusters, G.M.A. and Blaauwendraad, J.
Smeared crack approach and fracture localisation in concrete,
HERON, Vol. 30, No. 1, 1985.

13. Hinton, E. and Owen, D.R.J.
An introduction to finite element computations,
Pineridge Press, Swansea, 1979.

14. Zienkiewicz, O.C.
The finite element method,
McGraw-Hill, Third Edition, London, 1977.

15. Irons, B.M.
Quadrature rules for brick-based finite elements,
Int. J. Num. Meth. Eng., Vol. 3, pp.293-294, 1971.

16. Crisfield, M.A.
A fast incremental/iterative solution procedure that handles snap-through,
Computers and Structures, Vo.13, pp.55-62, 1981.

17. Abdel-Rahman, H.H.
Computational models for the nonlinear analysis of reinforced concrete flexural slab systems,
Ph.D. thesis C/Ph/66/82, University of Wales, May, 1982.

18. Franklin, H.A.
Nonlinear analysis of reinforced concrete frames and panels,
Report No. SESM 70-5, Dept. of Civil Engng., University of California, 1970.

19. Kompfer, T.A.,
Ein finites elementmodell fur die geometrisch und physikalisch nicktlineare Berechnung von Stahlbetonschalen,
Institute fur Baustatik, University of Stuttgart, Report No. 2, 1983.

20. Duddeck, H., Griebenow, G. and Schaper, G.
Material and time-dependent nonlinear behaviour of cracked concrete slabs,
in 'Nonlinear Behaviour of Reinforced Concrete Spatial Structures - Vol.1, Preliminary Report, IASS Symp., Darmstadt, 1978.

21. Sunidja, H., Foutch, D.A. and Gamble, W.L.
Response of prestressed concrete plate-edge column connections,
Civil Eng. Studies, Structural Research Series No. 498, University of Illinois in Urbana-Champaign, 1982.

AUTHOR INDEX

SUBJECT INDEX